Communications
in Computer and Information Science 690

Commenced Publication in 2007
Founding and Former Series Editors:
Alfredo Cuzzocrea, Dominik Ślęzak, and Xiaokang Yang

Editorial Board

More information about this series at http://www.springer.com/series/7899

Ana Fred · Hugo Gamboa (Eds.)

Biomedical Engineering Systems and Technologies

9th International Joint Conference, BIOSTEC 2016
Rome, Italy, February 21–23, 2016
Revised Selected Papers

 Springer

Editors
Ana Fred
Instituto de Telecomunicações/IST
Lisbon
Portugal

Hugo Gamboa
New University of Lisbon
Lisbon
Portugal

ISSN 1865-0929   ISSN 1865-0937   (electronic)
Communications in Computer and Information Science
ISBN 978-3-319-54716-9   ISBN 978-3-319-54717-6   (eBook)
DOI 10.1007/978-3-319-54717-6

Library of Congress Control Number: 2017932794

Printed on acid-free paper

This Springer imprint is published by Springer Nature
The registered company is Springer International Publishing AG
The registered company address is: Gewerbestrasse 11, 6330 Cham, Switzerland

Preface

The present book includes extended and revised versions of a set of selected papers from the 9th International Joint Conference on Biomedical Engineering Systems and Technologies (BIOSTEC 2016), held in Rome, Italy, February 21–23, 2016.

BIOSTEC 2016 received 321 paper submissions from 54 countries, of which 7% are included in this book.

The papers were selected by the event chairs and their selection is based on a number of criteria that include the evaluations and comments provided by the Program Committee members, the session chairs' assessment, and also the program chairs' global view of all papers included in the technical program. The authors of selected papers were then invited to submit a revised and extended version of their papers having at least 30% innovative material.

The purpose of the International Joint Conference on Biomedical Engineering Systems and Technologies is to bring together researchers and practitioners interested in both theoretical advances and applications of information systems, artificial intelligence, signal processing, electronics, and other engineering tools in knowledge areas related to biology, medicine, and health care.

BIOSTEC is composed of five complementary and co-located conferences, each specialized in at least one of the aforementioned main knowledge areas:

- International Conference on Biomedical Electronics and Devices — BIODEVICES
- International Conference on Bioimaging — BIOIMAGING
- International Conference on Bioinformatics Models, Methods and Algorithms — BIOINFORMATICS
- International Conference on Bio-inspired Systems and Signal Processing — BIOSIGNALS
- International Conference on Health Informatics — HEALTHINF

The papers selected to be included in this book contribute to the understanding of relevant trends of current research on biomedical engineering systems and technologies.

In the biomedical electronics and devices area we have included a set of three papers focusing on aspects related to arterial pulse measurement, radiation exposure monitoring, and speech rehabilitation. Although these address diverse domains, they share two common features: The developments reported are aimed at increasing the usability of the devices, as a way of making them more acceptable to patients and allowing use by non-trained individuals. They also demonstrate the importance of integrated novel sensing devices with sophisticated signal processing techniques — which illustrates the value of BIOSTEC in bringing together researchers in complementary fields.

The area of bioimaging is approached also in a set of three papers. Complying with the well-known phrase, "A picture is worth a thousand words," the world of medical imaging has been playing a very crucial role in the modern clinic. It encompasses many fields starting from diagnosis through functional assessment to image-guided therapy. It is almost impossible to imagine any field in medicine today that does not utilize some form of imaging. Attesting to this fact are the three manuscripts chosen for publication in this volume. The first work suggests an image-based tool for quantitative assessment of synaptic densities. The second offers modeling strategies for the design of orthodontic aligners. And the third utilizes dynamic computed tomography (CT) imaging for assessment of ventilation distribution reproducibility. These works are taken from three seemingly unrelated fields, but they all share the same basis, which is the reliance on images and image processing tools. I hope you will find them interesting and stimulating, conveying the message that, "medical imaging has no boundaries."

The papers selected from the Bioinformatics conference reflect the recent direction in bioinformatics research defined by a particular focus on using advanced computational methodologies to maximize the ability to extract useful information from the available raw biological data. It can be argued that bioinformatics research is entering a second phase in its development, with the initial phase primarily focusing on the use of simple algorithmic techniques to analyze the unexpected large amount of biological data. Recently, with the increasing volume, variability, and complexity of the available data, algorithmic approaches with a higher degree of sophistication are needed to extract much-needed knowledge from the 'big' biological data. The selected papers employ advanced statistical approaches, complex graph theoretic models, and various machine-learning techniques to handle the difficult task of analyzing bioinformatics data. The papers also reflect the unique nature of BIOSTEC conferences; they cover various aspects of biomedical informatics in a harmonious and complementary way. It is becoming increasingly difficult to draw hard or clear lines between the different types of advances in technology or informatics that support biomedical research. As a result, there are references and connections in the selected papers to the other four BIOSTEC conferences, particularly the conferences in health informatics and biomedical imaging. Taking full advantage of the available data sources, whether they contain omics data, imaging data, clinical/patient data or ontology data, is critical in advancing biomedical informatics and positively impacting health-care practices. We hope that reading the selected papers will inspire the readers to further explore the exciting discipline of bioinformatics and find ways to contribute to its emerging subdisciplines.

From the conference's track on bio-inspiring systems and signal processing, seven papers were chosen, which illustrate its current trends. Their main application domains address issues in monitoring, well-being, and rehabilitation. Common methodological solutions include various forms of machine-learning algorithms, as well as experimental set-ups based on brain–computer interfaces. Three papers focus on monitoring physiological features: Hörmann et al. study the relation between modern working environment demands and the concomitant increase in cognitive workload; Boutaayamou et al. introduce a new signal-processing algorithm for gait analysis; and Uchida et al. present a real-time remote monitoring approach, aimed at estimating the heart rate of a subject from various types of images of his face. Two papers address a clustering or classification task: Arini and Torkar compare various machine-learning classifiers in their ability to discriminate ventricular premature beats in rabbits; Wang et al. introduce a new approach for the problem of dynamic time warping, and illustrate its improved efficiency in the context of human sleep stage clustering. Bustamante et al. present a robot-based strategy for rehabilitation of lower limb movements, using electromyographic control signals. Their methodology appears to be extendable to other joints, as well as commanding signals. Finally, aiming also at recovering lost physical abilities, Gonzalez et al. describe a new system, based on permanent magnetic articulography, which enables patients having undergone laryngectomy to communicate verbally.

Health informatics encompasses a wide spectrum of health-related topics including among others: health-care IT; decision support systems; clinical data mining and machine learning; physiological and behavioral modeling; software systems in medicine and public health; and mobile, wearable, and assistive technologies in health care. This book includes a selected set of five papers that address the design, implementation, and evaluation of a variety of health informatics systems and frameworks: e-health and m-health systems that encourage well-being and disease control by leveraging social community support (Elloumi et al. for physical activity and exercise, and Smedberg et al. for stress management); an evaluation framework for connected health systems to holistically assess their value and potential impact (Carroll et al.); a continuous, real-time, and non-invasive system for monitoring and calibrating patient post operatory breathing (Seppänen et al.); and a system for formally representing computer-interpretable guidelines, detecting potential interactions in multimorbidity cases, and coping with possibly conflicting pieces of evidence in clinical studies (Zamborlini et al.).

We would like to thank all the authors for their contributions, and express also our gratitude to the reviewers who have helped ensure the quality of this publication.

February 2016   Ana Fred
                                                                      Hugo Gamboa

Organization

Conference Co-chairs

Ana Fred   Instituto de Telecomunicações/IST, Portugal
Hugo Gamboa   LIBPHYS-UNL/FCT, New University of Lisbon,
                            Portugal

Program Co-chairs

BIOSIGNALS

Carla Quintão   FCT-UNL, Portugal

BIOIMAGING

Haim Azhari   Technion – Israel Institute of Technology, Israel

BIODEVICES

James Gilbert   University of Hull, UK

HEALTHINF

Jan Sliwa   Bern University of Applied Sciences (BUAS),
                            Switzerland
Carolina Ruiz   WPI, USA

BIOINFORMATICS

Hesham Ali   University of Nebraska at Omaha, USA

BIOSIGNALS Program Committee

Jean-Marie Aerts   M3-BIORES, Katholieke Universitëit Leuven, Belgium
Sergio Alvarez   Boston College, USA
Sridhar P. Arjunan   RMIT University, Australia
Ofer Barnea   Tel Aviv University, Israel
Richard Bayford   Middlesex University, UK
Eberhard Beck   Brandenburg University of Applied Sciences, Germany
Peter Bentley   UCL, UK
Tolga Can   Middle East Technical University, Turkey
Guy Carrault   University of Rennes 1, France
Maria Claudia F. Castro   Centro Universitário da FEI, Brazil
M. Emre Celebi   Louisiana State University in Shreveport, USA
Sergio Cerutti   Polytechnic University of Milan, Italy
Jan Cornelis   VUB, Belgium

Unal Sakoglu	Texas A&M University at Commerce, USA
Carlo Sansone	University of Naples, Italy
Anita Sant'Anna	Halmstad University, Sweden
Andres Santos	Universidad Politécnica de Madrid, Spain
Roberto Sassi	Università degli studi di Milano, Italy
Gerald Schaefer	Loughborough University, UK
Reinhard Schneider	Fachhochschule Vorarlberg, Austria
Lotfi Senhadji	University of Rennes 1, France
Tapio Seppänen	University of Oulu, Finland
Erchin Serpedin	Texas A&M University, USA
Samuel Silva	Universidade de Aveiro, Portugal
Jordi Solé-Casals	University of Vic, Central University of Catalonia, Spain
Asser Tantawi	IBM, USA
Wallapak Tavanapong	Iowa State University, USA
João Paulo Teixeira	Polytechnic Institute of Bragança, Portugal
Carlos Eduardo Thomaz	Centro Universitário da FEI, Brazil
Ana Maria Tomé	University of Aveiro, Portugal
Carlos M. Travieso	University of Las Palmas de Gran Canaria, Spain
Ahsan Ahmad Ursani	Mehran University of Engineering and Technology, Pakistan
Michal Vavrecka	Czech Technical University, Czech Republic
Giovanni Vecchiato	IRCCS Fondazione Santa Lucia, Italy
Jacques Verly	University of Liege, Belgium
Pedro Gómez Vilda	Universidad Politécnica de Madrid, Spain
Quan Wen	University of Electronic Science and Technology of China, China
Didier Wolf	Research Centre for Automatic Control, CRAN CNRS UMR 7039, France
Pew-Thian Yap	University of North Carolina at Chapel Hill, USA
Chia-Hung Yeh	National Sun Yat-sen University, Taiwan
Nicolas Younan	Mississippi State University, USA
Damjan Zazula	University of Maribor, Slovenia
Rafal Zdunek	Wroclaw University of Technology, Poland

BIOSIGNALS Additional Reviewers

Geng Chen	BRIC UNC, USA
Yan Jin	University of North Carolina at Chapel Hill, USA
Timo Lauteslager	Imperial College London, UK
Gabriele Piantadosi	Università Federico II di Napoli, Italy

BIOIMAGING Program Committee

| Ahmet Ademoglu | Bogazici University, Turkey |
| Sos S. Agaian | University of Texas San Antonio, USA |

Sameer K. Antani	National Library of Medicine, National Institutes of Health, USA
Ergin Atalar	Bilkent University, Turkey
Peter Balazs	University of Szeged, Hungary
Ewert Bengtsson	Centre for Image Analysis, Uppsala University, Sweden
Obara Boguslaw	University of Durham, UK
Abdel-Ouahab Boudraa	Ecole Navale, France
Alberto Bravin	European Synchrotron Radiation Facility, France
Katja Buehler	Vrvis Research Center, Austria
Enrico G. Caiani	Politecnico di Milano, Italy
Begoña Calvo	University of Zaragoza, Spain
Rita Casadio	University of Bologna, Italy
Alessia Cedola	CNR, Institute of Nanotechnology, Italy
M. Emre Celebi	Louisiana State University in Shreveport, USA
Heang-Ping Chan	University of Michigan, USA
Dean Chapman	University of Saskatchewan, Canada
Chung-Ming Chen	National Taiwan University, Taiwan
Paola Coan	LMU Munich, Germany
Giacomo Cuttone	INFN, Laboratori Nazionali del Sud Catania, Italy
Ivo Dinov	University of Michigan, USA
Chikkannan Eswaran	Multimedia University, Malaysia
Dimitrios Fotiadis	University of Ioannina, Greece
Carlos F.G.C. Geraldes	Universidade de Coimbra, Portugal
Mario Rosario Guarracino	National Research Council of Italy, Italy
Tzung-Pei Hong	National University of Kaohsiung, Taiwan
Xiaoyi Jiang	University of Münster, Germany
Cheng Guan Koay	National Intrepid Center of Excellence, USA
Patrice Koehl	University of California, USA
Dimitris D. Koutsouris	National Technical University of Athens, Greece
Adriaan A. Lammertsma	VU University Medical Center Amsterdam, The Netherlands
Roger Lecomte	Université de Sherbrooke, Canada
Xiongbiao Luo	XMU-TMMU, China
Jan Mares	University of Chemistry and Technology, Czech Republic
G.K. Matsopoulos	National Technical University of Athens, Greece
Erik Meijering	Erasmus University Medical Center, The Netherlands
Kunal Mitra	Florida Institute of Technology, USA
Jayanta Mukhopadhyay	Indian Institute of Technology Kharagpur, India
Joanna Isabelle Olszewska	University of Gloucestershire, UK
Evren Özarslan	Bogazici University in Turkey, Turkey
Dirk Padfield	Amazon, USA
Kalman Palagyi	University of Szeged, Hungary

Joao Papa	UNESP, Universidade Estadual Paulista, Brazil
Ales Prochazka	Institute of Chemical Technology, Czech Republic
Miroslav Radojevic	Erasmus MC, Biomedical Imaging Group Rotterdam, Rotterdam, The Netherlands
Alessandra Retico	Istituto Nazionale di Fisica Nucleare, Italy
Kate Ricketts	University College London, UK
Jorge Ripoll	Universidad Carlos III Madrid, Spain
Giovanna Rizzo	Consiglio Nazionale delle Ricerche, Italy
Olivier Salvado	CSIRO, Australia
Carlo Sansone	University of Naples, Italy
K.C. Santosh	The University of South Dakota, USA
Emanuele Schiavi	Universidad Rey Juan Carlos, Spain
Jan Schier	Institute of Information Theory and Automation of the Czech Academy of Sciences, Czech Republic
Gregory C. Sharp	Massachusetts General Hospital, USA
Dar-Bin Shieh	National Cheng Kung University, Taiwan
Natasa Sladoje	Centre for Image Analysis, Uppsala University, Sweden
Bassel Solaiman	Institut Mines Telecom, France
Andriyan Bayu Suksmono	Institute of Technology Bandung, Indonesia
Aladar A. Szalay	StemImmune Inc., USA
Pécot Thierry	Inria, France
Arkadiusz Tomczyk	Lodz University of Technology, Poland
Carlos M. Travieso	University of Las Palmas de Gran Canaria, Spain
Benjamin M.W. Tsui	Johns Hopkins University, USA
Vladimír Ulman	Masaryk University, Czech Republic
Sandra Rua Ventura	School of Allied Health Technologies, Portugal
Quan Wen	University of Electronic Science and Technology of China, China
Stefan Wesarg	Fraunhofer IGD, Germany
Zeyun Yu	University of Wisconsin at Milwaukee, USA
Habib Zaidi	Geneva University Hospital, Switzerland
Yicong Zhou	University of Macau, China

BIOIMAGING Additional Reviewers

Gabriele Piantadosi	Università Federico II di Napoli, Italy

BIODEVICES Program Committee

Elli Angelopoulou	University of Erlangen-Nuremberg, Germany
Steve Beeby	University of Southampton, UK
W. Andrew Berger	University of Scranton, USA
Dinesh Bhatia	North Eastern Hill University, India
Jan Cabri	Norwegian School of Sport Sciences, Norway
Carlo Capelli	Norwegian School of Sport Sciences, Norway
Wenxi Chen	The University of Aizu, Japan

Michael J. Schöning	FH Aachen, Germany
Fernando di Sciascio	Institute of Automatics National University of San Juan, Argentina
Mauro Serpelloni	University of Brescia, Italy
Anita Lloyd Spetz	Linköpings Universitet, Sweden
João Paulo Teixeira	Polytechnic Institute of Bragança, Portugal
John Tudor	University of Southampton, UK
Renato Varoto	University of Campinas, Brazil
Pedro Vieira	Universidade Nova de Lisboa, Portugal
Bruno Wacogne	FEMTO-ST, France
Tim Wark	Commonwealth Scientific and Industrial Research Organisation, Australia
Huikai Xie	University of Florida, USA
Sen Xu	Merck & Co., Inc., USA
Hakan Yavuz	Çukurova Üniversity, Turkey

BIODEVICES Additional Reviewers

Simone Benatti	University of Bologna, Italy
Bojan Milosevic	FBK, Italy

HEALTHINF Program Committee

Flora Amato	Università degli Studi di Napoli Federico II, Italy
Francois Andry	Philips, USA
María Teresa Arredondo	Universidad Politécnica de Madrid, Spain
Wassim Ayadi	LERIA, University of Angers, France and LaTICE, University of Tunis, Tunisia
Philip Azariadis	University of the Aegean, Greece
Adrian Barb	Penn State University, USA
Rémi Bastide	Jean-Francois Champollion University, France
Isabel Segura Bedmar	Universidad Carlos III de Madrid, Spain
Bert-Jan van Beijnum	University of Twente, The Netherlands
Anna Belardinelli	University of Tübingen, Germany
Marta Bienkiewicz	TUM, Technische Universität München, Germany
Alécio Binotto	IBM Research, Brazil
Edward Brown	Memorial University of Newfoundland, Canada
Federico Cabitza	Università degli Studi di Milano-Bicocca, Italy
Eric Campo	LAAS CNRS, France
Nick Cercone	York University, Canada
Sergio Cerutti	Polytechnic University of Milan, Italy
Philip K. Chan	Florida Institute of Technology, USA
James Cimino	University of Alabama at Birmingham, USA
Miguel Coimbra	Instituto de Telecomunicações, Portugal
Emmanuel Conchon	University of Limoges, XLIM, France
Carlos Costa	Universidade de Aveiro, Portugal
Chrysanne DiMarco	University of Waterloo, Canada

Carlos Eduardo Pereira	Federal University of Rio Grande Do Sul, UFRGS, Brazil
Enrico Maria Piras	Fondazione Bruno Kessler, Trento, Italy
Rosario Pugliese	Università di Firenze, Italy
Juha Puustjärvi	University of Helsinki, Finland
Arkalgud Ramaprasad	University of Illinois at Chicago, USA
Zbigniew W. Ras	University of North Carolina at Charlotte, USA
Alvaro Rocha	University of Coimbra, Portugal
Marcos Rodrigues	Sheffield Hallam University, UK
Elisabetta Ronchieri	INFN, Italy
Carolina Ruiz	WPI, USA
George Sakellaropoulos	University of Patras, Greece
Ovidio Salvetti	National Research Council of Italy, CNR, Italy
Akio Sashima	AIST, Japan
Jacob Scharcanski	Universidade Federal do Rio Grande do Sul, Brazil
Boris Schauerte	Karlsruhe Institute of Technology, Germany
Bettina Schnor	Potsdam University, Germany
Luca Dan Serbanati	Politehnica University of Bucharest, Romania
Kulwinder Singh	University of South Florida, USA
Jan Sliwa	Bern University of Applied Sciences, Switzerland
Irena Spasic	University of Cardiff, UK
Jan Stage	Aalborg University, Denmark
Robyn Tamblyn	McGill University, Canada
Ioannis G. Tollis	University of Crete, Greece
Vicente Traver	ITACA, Universidad Politécnica de Valencia, Spain
Gary Ushaw	Newcastle University, UK
Aristides Vagelatos	CTI, Greece
Justin Wan	University of Waterloo, Canada
Rafal Wcislo	AGH, University of Science and Technology in Cracow, Poland
Janusz Wojtusiak	George Mason University, USA
Lixia Yao	University of North Carolina at Charlotte, USA
Clement T. Yu	University of Illinois at Chicago, USA
André Zúquete	IEETA, IT, Universidade de Aveiro, Portugal

HEALTHINF Additional Reviewers

Ghazar Chahbandarian	IRIT, France
Juan Bautista Montalvá Colomer	UPM, Spain
Claudio Eccher	Fondazione Bruno Kessler, Italy
Daniela Fogli	Università degli Studi di Brescia, Italy
Wolfgang Fuhl	University of Tübingen, Germany
Aude Motulsky	McGill University, Canada
Antonio Piccinno	University of Bari, Italy
Daniala Weir	McGill University, Canada

BIOINFORMATICS Program Committee

Mohamed Abouelhoda	Nile University, Egypt
Rini Akmeliawati	International Islamic University Malaysia, Malaysia
Tatsuya Akutsu	Kyoto University, Japan
Jens Allmer	Izmir Institute of Technology, Turkey
Sameer K. Antani	National Library of Medicine, National Institutes of Health, USA
Rolf Backofen	Albert-Ludwigs-Universität, Germany
Dhundy Bastola	University of Nebraska at Omaha, USA
Tim Beissbarth	University of Göttingen, Germany
Shifra Ben-Dor	Weizmann Institute of Science, Israel
Michael Biehl	University of Groningen, The Netherlands
Inanc Birol	BC Cancer Agency, Canada
Ulrich Bodenhofer	Johannes Kepler University Linz, Austria
Carlos Brizuela	Centro de Investigación Científica y de Educación Superior de Ensenada, Baja California, Mexico
Lutgarde M.C. Buydens	Radboud University, The Netherlands
Stefan Canzar	Toyota Technological Institute at Chicago, USA
Claudia Consuelo Rubiano Castellanos	Universidad Nacional de Colombia, Bogota, Colombia
Mia D. Champion	Active Cloudomics LLC, USA
Kwang-Hyun Cho	Korea Advanced Institute of Science and Technology, South Korea
Mark Clement	Brigham Young University, USA
Netta Cohen	University of Leeds, UK
Kathryn Cooper	University of Nebraska at Omaha, USA
Francisco Couto	Universidade de Lisboa, Portugal
Antoine Danchin	Institute of Cardiometabolism and Nutrition, France
Thomas Dandekar	University of Würzburg, Germany
Sérgio Deusdado	Instituto Politecnico de Bragança, Portugal
Richard Edwards	University of Southampton, UK
George Eleftherakis	CITY College, International Faculty of the University of Sheffield, Greece
Gang Fang	NYU Shanghai, China
Fabrizio Ferre	University of Rome Tor Vergata, Italy
António Ferreira	Universidade de Lisboa, Portugal
Liliana Florea	Johns Hopkins University, USA
Gianluigi Folino	Institute for High Performance Computing and Networking, National Research Council, Italy
Bruno Gaëta	University of New South Wales, Australia
Dario Ghersi	University of Nebraska at Omaha, USA
Manolo Gouy	Claude Bernard University Lyon 1, France
Arndt von Haeseler	Center of Integrative Bioinformatics Vienna, MFPL, Austria
Christopher E. Hann	University of Canterbury, New Zealand

Shan He	University of Birmingham, UK
Volkhard Helms	Universität des Saarlandes, Germany
Song-Nian Hu	Chinese Academy of Sciences, China
Daisuke Ikeda	Kyushu University, Japan
Sohei Ito	National Fisheries University, Japan
Bo Jin	MilliporeSigma, Merck KGaA, USA
Giuseppe Jurman	Fondazione Bruno Kessler, Italy
Michael Kaufmann	Witten/Herdecke University, Germany
Ed Keedwell	University of Exeter, UK
Seyoung Kim	Carnegie Mellon University, USA
Jirí Kléma	Czech Technical University in Prague, Czech Republic
Peter Kokol	University of Maribor, Slovenia
Malgorzata Kotulska	Wroclaw University of Technology, Poland
Dimitris D. Koutsouris	National Technical University of Athens, Greece
Ivan Kulakovskiy	Engelhardt Institute of Molecular Biology, Russian Federation
Yinglei Lai	George Washington University, USA
Carlile Lavor	University of Campinas, Brazil
Matej Lexa	Masaryk University, Czech Republic
Li Liao	University of Delaware, USA
Leo Liberti	Ecole Polytechnique, France
Alberto Magi	University of Florence, Italy
Paolo Magni	Università degli Studi di Pavia, Italy
Thérèse E. Malliavin	CNRS/Institut Pasteur, France
Elena Marchiori	Radboud University, The Netherlands
Majid Masso	George Mason University, USA
Petr Matula	Masaryk University, Czech Republic
Ivan Merelli	ITB CNR, Italy
Saad Mneimneh	Hunter College CUNY, USA
Pedro Tiago Monteiro	INESC-ID/IST, Universidade de Lisboa, Portugal
Burkhard Morgenstern	University of Göttingen, Germany
Vincent Moulton	University of East Anglia, UK
Jean-Christophe Nebel	Kingston University, UK
José Luis Oliveira	Universidade de Aveiro, Portugal
Hakan S. Orer	Koc University, Turkey
Florencio Pazos	National Centre for Biotechnology, Spain
Marco Pellegrini	Consiglio Nazionale delle Ricerche, Italy
Matteo Pellegrini	University of California, Los Angeles, USA
Horacio Pérez-Sánchez	Catholic University of Murcia, Spain
Guy Perrière	Université Claude Bernard Lyon 1, France
Olivier Poch	ICube UMR7357 CNRS-Université de Strasbourg, France
Alberto Policriti	Università degli Studi di Udine, Italy
Giuseppe Profiti	University of Bologna, Italy
Junfeng Qu	Clayton State University, USA
Mark Ragan	The University of Queensland, Australia

Gajendra Raghava   Institute of Microbial Technology, India
Christian M. Reidys   Virginia Polytech University, USA
Javier Reina-Tosina   University of Seville, Spain
Laura Roa   University of Seville, Spain
Simona E. Rombo   Università degli Studi di Palermo, Italy
Eric Rouchka   University of Louisville, USA
Carolina Ruiz   WPI, USA
Indrajit Saha   National Institute of Technical Teachers' Training
                            and Research, India
J. Cristian Salgado   University of Chile, Chile
Alessandro Savino   Politecnico di Torino, Italy
Jaime Seguel   University of Puerto Rico at Mayaguez, USA
Erchin Serpedin   Texas A&M University, USA
Joao C. Setubal   Universidade de São Paulo, Brazil
Christine Sinoquet   University of Nantes, France
Neil R. Smalheiser   University of Illinois Chicago, USA
Pavel Smrz   Brno University of Technology, Czech Republic
Peter F. Stadler   Universität Leipzig, IZBI, Germany
Andrew Sung   University of Southern Mississippi, USA
David Svoboda   Masaryk University, Czech Republic
Peter Sykacek   BOKU, University of Natural Resources and Life
                            Sciences, Austria
Gerhard Thallinger   Graz University of Technology, Austria
Jerzy Tiuryn   Warsaw University, Poland
Takashi Tomita   Japan Advanced Institute of Science and Technology,
                            Japan
Silvio C.E. Tosatto   Università di Padova, Italy
Ioannis Tsamardinos   University of Crete, Greece
Alexander Tsouknidas   Aristotle University of Thessaloniki, Greece
Gabriel Valiente   Technical University of Catalonia, Spain
Juris Viksna   University of Latvia, Latvia
Jason T.L. Wang   New Jersey Institute of Technology, USA
Sebastian Will   Universität Leipzig, Germany
Yanbin Yin   Northern Illinois University, USA
Malik Yousef   Zefat Academic College, Israel
Jingkai Yu   Institute of Process Engineering, Chinese Academy
                            of Sciences, China
Nazar Zaki   United Arab Emirates University, United Arab
                            Emirates
Helen Hao Zhang   University of Arizona, USA
Xingming Zhao   Tongji University, China

BIOINFORMATICS Additional Reviewers

Konstantinos Dimopoulos	University of Sheffield International Faculty, CITY College, Greece
Artem Kasianov	VIGG, Russian Federation

Invited Speakers

Hayit Greenspan	Tel Aviv University, Israel
Marcus Cheetham	Universität Zürich, Switzerland
Salvador Pané i Vidal	Swiss Federal Institute of Technology (ETH), Switzerland
Luisa Torsi	University of Bari A. Moro, Italy
Haim Azhari	Technion, Israel Institute of Technology, Israel

Contents

Biomedical Electronics and Devices

Arterial Pulse Waveform Monitoring via a Flexible PET-Based
Microfluidic Sensor 3
  Dan Wang and Zhili Hao

Towards an Intraoral-Based Silent Speech Restoration System
for Post-laryngectomy Voice Replacement 22
  Lam A. Cheah, James M. Gilbert, Jose A. Gonzalez, Jie Bai,
  Stephen R. Ell, Phil D. Green, and Roger K. Moore

Bioimaging

SynapCountJ: A Validated Tool for Analyzing Synaptic Densities
in Neurons 41
  Gadea Mata, Germán Cuesto, Jónathan Heras, Miguel Morales,
  Ana Romero, and Julio Rubio

4DCT-Derived Ventilation Distribution Reproducibility Over Time 56
  Geoffrey G. Zhang, Kujtim Latifi, Vladimir Feygelman,
  Thomas J. Dilling, and Eduardo G. Moros

Modelling Strategies for the Advanced Design of Polymeric
Orthodontic Aligners 67
  Sandro Barone, Alessandro Paoli, Armando Viviano Razionale,
  and Roberto Savignano

Bioinformatics Models, Methods and Algorithms

Implicitly Weighted Robust Classification Applied to Brain
Activity Research 87
  Jan Kalina and Jaroslav Hlinka

Finding Median and Center Strings for a Probability Distribution
on a Set of Strings Under Levenshtein Distance Based
on Integer Linear Programming 108
  Morihiro Hayashida and Hitoshi Koyano

Accelerating the Exploitation of (bio)medical Knowledge
Using Linked Data 122
  Mohammad Shafahi, Hamideh Afsarmanesh, and Hayo Bart

A Stochastic Framework for Neuronal Morphological Comparison:
Application to the Study of *imp* Knockdown Effects in Drosophila
Gamma Neurons.   145
   A. Razetti, X. Descombes, C. Medioni, and F. Besse

Sensitivity Analysis of Granularity Levels in Complex
Biological Networks   167
   Sean West and Hesham Ali

Bio-inspired Systems and Signal Processing

Optimal Lead Selection for Evaluation Ventricular Premature Beats
Using Machine Learning Approach   191
   Pedro David Arini and Drago Torkar

Detailed Estimation of Cognitive Workload with Reference to a Modern
Working Environment   205
   Timm Hörmann, Marc Hesse, Peter Christ, Michael Adams,
   Christian Menßen, and Ulrich Rückert

Continuous Real-Time Measurement Method for Heart Rate Monitoring
Using Face Images   224
   Daisuke Uchida, Tatsuya Mori, Masato Sakata, Takuro Oya,
   Yasuyuki Nakata, Kazuho Maeda, Yoshinori Yaginuma,
   and Akihiro Inomata

Algorithm for Temporal Gait Analysis Using Wireless Foot-Mounted
Accelerometers   236
   Mohamed Boutaayamou, Vincent Denoël, Olivier Brüls,
   Marie Demonceau, Didier Maquet, Bénédicte Forthomme,
   Jean-Louis Croisier, Cédric Schwartz, Jacques G. Verly,
   and Gaëtan Garraux

Online Simulation of Mechatronic Neural Interface Systems:
Two Case-Studies   255
   Samuel Bustamante, Juan C. Yepes, Vera Z. Pérez, Julio C. Correa,
   and Manuel J. Betancur

Modeling and Clustering of Human Sleep Time Series Using Dynamic
Time Warping: Sequential and Distributed Implementations   276
   Chiying Wang, Sergio A. Alvarez, Carolina Ruiz, and Majaz Moonis

Voice Restoration After Laryngectomy Based on Magnetic Sensing of
Articulator Movement and Statistical Articulation-to-Speech Conversion   295
   Jose A. Gonzalez, Lam A. Cheah, James M. Gilbert, Jie Bai,
   Stephen R. Ell, Phil D. Green, and Roger K. Moore

Health Informatics

Connecting Multistakeholder Analysis Across Connected Health Solutions. . .   319
  Noel Carroll, Marie Travers, and Ita Richardson

Continuous Postoperative Respiratory Monitoring with Calibrated
Respiratory Effort Belts: Pilot Study   340
  Tiina M. Seppänen, Olli-Pekka Alho, Merja Vakkala, Seppo Alahuhta,
  and Tapio Seppänen

Generalizing the Detection of Clinical Guideline Interactions Enhanced
with LOD   360
  Veruska Zamborlini, Rinke Hoekstra, Marcos da Silveira, Cedric Pruski,
  Annette ten Teije, and Frank van Harmelen

Online Stress Management for Self- and Group-Reflections
on Stress Patterns   387
  Åsa Smedberg, Hélène Sandmark, and Andrea Manth

Requirements, Design and Pilot Study of a Physical Activity Activation
System Using Virtual Communities   405
  Lamia Elloumi, Margot Meijerink, Bert-Jan van Beijnum,
  and Hermie Hermens

Author Index   427

Biomedical Electronics and Devices

Arterial Pulse Waveform Monitoring via a Flexible PET-Based Microfluidic Sensor

Dan Wang[(✉)] and Zhili Hao

Department of Mechanical and Aerospace Engineering,
Old Dominion University, Norfolk, VA KH238, USA
{dwang009,zlhao}@odu.edu

Abstract. For the purpose of cardiovascular system monitoring, this paper presents a flexible microfluidic sensor for measuring arterial pulse waveforms. Sitting on a flexible substrate, the core of the sensor is a polymer microstructure embedded with an electrolyte-enabled 5×1 resistive transducer array. As a time-varying load, a pulse signal deflects the microstructure and registers as a resistance change by the transducer at the site of the pulse. Radial, carotid and temporal pulse signals are all originally recorded as an absolute resistance signal by the sensor. Moreover, carotid pulse signals are measured on two subjects at-rest and post-exercise. A wavelet-based cascaded adaptive algorithm is written in Matlab to remove baseline drift (or motion artifacts) in a recorded pulse signal. The obtained pulse waveforms are consistent with the related findings in the literature. The sensor features low cost for mass production and easy use by an untrained individual.

Keywords: Cardiovascular diseases (CVD) · Arterial pulse waveform · Motion artifacts · Microfluidics · Wearable sensors

1 Introduction

Cardiovascular diseases (CVD) are the leading cause of death in the world [1]. The cardiovascular system consists of the heart and its intricate conduits (arteries, veins and capillaries) that transverse the whole human body carrying blood. Ejection of blood from the left ventricle during systole initiates an arterial pressure wave that travels toward the periphery. At points of impedance mismatch, wave reflection occurs. As a consequence of differing elastic qualities and wave reflection, the shape of the arterial pulse waveform changes continuously throughout the arterial tree. As such, arterial pulse waveforms are intimately associated with the physiological conditions of the whole cardiovascular system. The early diagnosis and optimized treatment of patients with CVDs requires continuous monitoring of their arterial pulse waveforms.

To date, quite a few arterial tonometric devices have been developed for arterial pulse waveform measurement. For instance, a flexible pulse monitoring system from PPS (Pressure Profile System) [2] and CASPal from HealthStats [3] are commercially available and have been successfully employed for arterial pulse waveform measurement. However, these tonometric devices are unsuitable for wearing with relative comfort and for an untrained individual to use at home.

© Springer International Publishing AG 2017
A. Fred and H. Gamboa (Eds.): BIOSTEC 2016, CCIS 690, pp. 3–21, 2017.
DOI: 10.1007/978-3-319-54717-6_1

Recently, based on microfluidics technology, a polyethylene terephthalate (PET)-based sensor array to monitor arterial pressure waveforms was developed [4]. Although this PET-based sensor array offers quite a few attractive features for daily use by an untrained individual, it entails a complex fabrication process, including bonding three layers together and injecting electrolyte into each individual sensor in the sensor array. Meanwhile, its spatial resolution of 5 mm × 5 mm does not pose any problems with measuring a carotid artery (the mean diameter of the carotid artery is 6.10 ± 0.80 mm in women and 6.52 ± 0.98 mm in men [5]), but it is well above the typical size of a radial artery (the mean diameter of the right radial artery is 2.3 ± 0.4 mm [6]). Thus, the sensor array risks none of individual sensor being aligned at the site of a targeted radial artery.

To address the above-mentioned two issues while preserving the attractive features of the sensor array, this paper presents a PET-based sensor for arterial pulse waveform measurement. Sitting on a flexible substrate, this sensor bears the same design of a Pyrex-based sensor previously developed by our group [7, 8]. The core of the sensor is a single polydimethylsiloxane (PDMS) microstructure embedded with an electrolyte-enabled 5×1 resistive transducer array underneath. The spatial resolution of 1.5 mm in the 6 mm-long transducer array allows an untrained individual to easily align one transducer right at the site of an artery. A simple, low-cost fabrication process is developed for realizing this PET-based sensor, where a new bonding process is employed to strengthen the bonding between the PDMS microstructure and the PET substrate with indium titanium oxide (ITO) electrodes. Three fabricated sensors of the identical design are utilized to collectively record the carotid, radial and temporal arterial pulse signals of four subjects as absolute resistance signals. A signal-processing algorithm is written in Matlab to remove the baseline drift in an originally recorded pulse signal and convert the drift-free signal to the sensor deflection as a function of time, which represents the pulse waveform. The measured results are compared with the related findings in the literature for validating the feasibility of the sensor for monitoring arterial pulse waveforms and the robustness of the sensor and its signal-processing algorithm to motion artifacts.

2 Flexible PET-based Sensor

2.1 Sensor Design

Figure 1 depicts the configuration of the flexible PET-based sensor. The sensor encompasses a rectangular PDMS microstructure embedded with an electrolyte-filled microchannel, and a set of ITO electrode pairs distributed along the microchannel length. The portion of electrolyte across an electrode pair functions as a resistive transducer, whose resistance varies with the bottom deflection of the microstructure at its location and is routed out by the electrode pair. Thus, together with the set of electrode pairs, one body of electrolyte in the microchannel forms a 5×1 transducer array with a length of 6 mm and a spatial resolution of 1.5 mm.

Induced by a pulse signal, distributed load acting on the top of the microstructure translates to the bottom deflection of the microstructure and thus geometrical changes of the microchannel, which register as resistance changes by the transducer array.

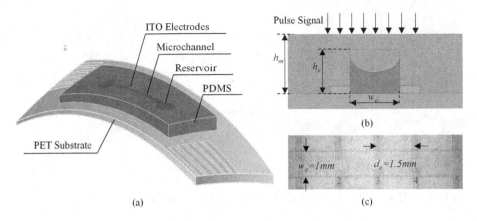

Fig. 1. A flexible PET-based wearable sensor: (a) 3D configuration; (b) side view with key design parameters; (c) the transducer array being labeled with numbers.

Table 1 summarizes the key design parameters of the sensor. The details of the sensor design can be found in the literature [7, 8].

Table 1. Key design parameters of the flexible PET-based sensor.

Parameter	Value	Symbol
Microchannel cross-section	1 mm × 80 μm	$w_e \times h_e$
Microchannel length	30 mm	L_e
Spatial resolution	1.5 mm	d_e
Microstructure thickness	1.2 mm	h_m

2.2 Fabrication Process

Figure 2 illustrates a low-cost, two-mask fabrication process for realizing the PET-based sensor. The process starts with a commercial ITO/PET sheet (a 0.2 mm-thick PET substrate coated with 120 nm–160 nm-thick ITO layer). To pattern ITO electrodes on the PET substrate, a 15 μm-thick dry film (Alpho NIT 215, NichigoMorton Co., Ltd.) is laminated onto the PET substrate. Via the first mask, electrode pattern is transferred to the dry film, which is followed by wet etching of the ITO layer to form ITO electrodes. Afterward, the dry film is removed using ethanol. Via the second mask, a SU8 mold is created on a Pyrex substrate. Then, a mixture of curing agent to PDMS elastomer (Sylgard 184, Dow Corning) with a weight ratio of 1:10 is poured over the SU8 mold. After being cured at room temperature over 24 h, the microstructure is peeled off from the SU8 mold and a hole is punched into each reservoir using a needle.

To strengthen the bonding strength between the PDMS microstructure and the PET substrate with patterned ITO electrodes, a chemical gluing strategy is adopted [9, 10]. First, photoresist is placed onto the microchannel of the microstructure. Then, the patterned ITO electrodes and the microstructure are activated with hydroxyl groups by an oxygen plasma treatment for 1 min, which are followed by immersing the

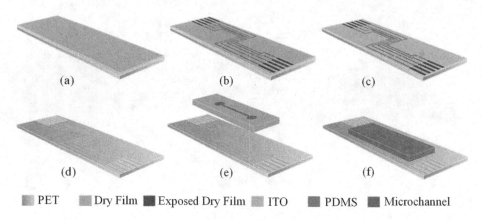

■ PET   ▨ Dry Film ▦ Exposed Dry Film ▧ ITO   ■ PDMS ▨ Microchannel

Fig. 2. Fabrication process for the flexible PET-based sensor: (a) dry film lamination; (b) dry film exposure under UV light; (c) patterned dry film; (d) patterned ITO electrodes; (e) alignment and bonding of patterned ITO electrodes and PDMS microstructure; (f) 3D view of the sensor.

microstructure and ITO electrodes into 1% (v/v) 3-Glycidyloxypropyltrimethoxysilane (GOPTS) and 5% (v/v) 3-Aminopropyltriethoxysilane (APTES) for 20 min, respectively. Afterward, the microstructure is rinsed with acetone, isopropanol and DI water, sequentially. The PET substrate with patterned ITO electrodes is rinsed with ethanol and DI water. Finally, the microstructure and the PET substrate are aligned and bonded under a contact pressure at 100 °C for 5 min, then at 50 °C for 24 h. Either 1-ethyl-3-methylimidazolium tricyanomethanide (EMIM TCM) or 1-ethyl-3-methylimidazolium dicyanamide (EMIDCA) is injected into the microchannel via a reservoir using a syringe. Each reservoir is then sealed with a 1:10 PDMS drop. Conductive epoxy is used to make electrical connection between the contact pads of the sensor and the associated electronics on PCBs.

(a)   (b)

Fig. 3. Pictures of a fabricated PET-based sensor: (a) the transparent sensor without PDMS drops sealing the reservoirs; (b) the reservoir-sealed sensor with electrical connections to wires made with epoxy and further fixed with duck tapes.

Figure 3 shows a couple of pictures of a fabricated PET-based sensor. Three sensors of the identical design are fabricated at different times. While Sensor #1 and Sensor #2 are injected with EMIDCA, Sensor #3 is injected with EMIM TCM.

3   Pulse Measurement

3.1   Subjects

Three fabricated sensors of the same design are used for measuring pulse signals of the following four subjects at-rest: a 28yr-old female (28yr-F), a 16yr-old female (16yr-F), a 29yr-old female (29yr-F) and a 28yr-old male (28yr-M). All four subjects are in good health and are not known to have cardiovascular diseases. As will be seen in Sect. 4, the three sensors differ in their performance, due to fabrication variation and the electrolyte used. The three sensors are used to collectively measure the carotid, radial, and temporal pulse signals of the four subjects at-rest. Each pulse signal is measured on the right side of a subject in a quiet environment, after the subject resting for several minutes and no drinking and eating for over 1 h. Additionally, the 28yr-M and 16yr-M subjects are required to do 5 min-long strenuous exercise and their carotid pulse signals are taken immediately after exercise from the same location as at-rest.

3.2   Pulse Signal Recording

As shown in Fig. 4, the sensor is placed at the site of the radial, carotid and temporal artery, respectively. Care is taken to align the transducer array at the site of an artery. The sensor is held and pressed against an artery with two fingers. As such, the hold-down pressure against an artery is uncontrollable and may vary during a pulse signal measurement. The pulse signal of an artery exerts a time-varying load on the top of the PDMS microstructure and then registers as a resistance change by the transducer at the site of the artery. Since both the location of the transducer array and the location of the two fingers vary relative to an artery among the measurements, the transducer capturing the strongest pulse signal keeps changing too. Later on, only the strongest pulse signal in each measurement is processed for obtaining its waveform.

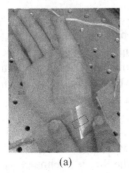
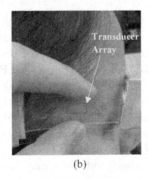
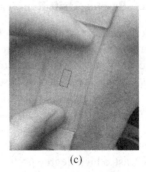

(a)   (b)   (c)

Fig. 4. Pictures of recording a pulse signal with the flexible PET-based sensor: (a) sensor being aligned above the radial artery; (b) sensor being aligned above the temporal artery; (c) sensor being aligned above the carotid artery.

To monitor resistance changes in the transducer array, a function generator is used to apply an AC (Alternating Current) signal (100 kHz, peak-peak amplitude: 220 mv) as the common input for all the transducers [7]. The output of a transducer is connected to its electronics implemented on PCB for both amplifying the AC signal coming out from the transducer and converting the AC signal to a DC (Direct Current) voltage output, which is recorded by a LabVIEW program. The DC voltage output, V_{out}, is related to the resistance, R, of a transducer by [7, 8]:

$$V_{\text{out}} = \frac{v_{pp}^2 R_F^2}{8R^2}$$

(1)

where v_{pp} is the peak-to-peak value of the AC signal, and R_F is the feedback resistance of the electronics used. Therefore, the resistance of a transducer is obtained by:

$$R = \frac{v_{pp} R_F}{2\sqrt{2V_{out}}}$$

(2)

As such, the originally recorded data for a pulse signal is the absolute resistance of a transducer as a function of time. Note that the same design of electronics is used for all the transducers and can be found in our previous work [7]. Each pulse signal is recorded for a 10 s period. The sampling rate is kept at 500 Hz for Sensor #1. Later on, 25 data points per second are utilized for the extracted pulse waveform from an originally recorded pulse signal. The sampling rate is kept at 1 kHz for Sensor #2 and Sensor #3. Note that the measurements conducted using Sensor #1 and Sensor #3 were roughly half a year earlier than those conducted using Sensor #2.

4   Signal-Processing Algorithm

4.1   Baseline Drift Removal

Motion artifacts (i.e., motion of the sensor and the respiration and body motion of a subject during measurement) can introduce baseline drift to a recorded pulse signal. The Discrete Meyer Wavelet Transformation (DMWT) and Cubic Spline Estimation (CSE) have been implemented for removing baseline drift from the recorded data [11]. DMWT is well known for representing localized variations in a signal simultaneously in the time and frequency domains. CSE is commonly used to detect the amplitude envelope of a signal.

Because the baseline drift introduced by motion artifacts has nonlinear and quasi-periodic contents, linear interpolation estimation has been proven ineffective [11]. In contrast, a high-degree polynomial is smooth, but it may cause the Runge phenomenon, which increases the error of the signal [11]. Thus, CSE is widely used and is adopted in this work to remove the baseline drift when the Energy Ratio (ER) of an originally recorded pulse signal reaches a threshold.

In DMWT, a function is defined in frequency domain:

$$\psi(w) = \begin{cases} (2\pi)^{-1/2} \times e^{jw/2} \times \sin\left(\dfrac{\pi}{2} \times v\left(\dfrac{3}{2\pi}|w| - 1\right)\right), & \dfrac{2\pi}{3} \le |w| \le \dfrac{4\pi}{3} \\[3mm] (2\pi)^{-1/2} \times e^{jw/2} \times \cos\left(\dfrac{\pi}{2} \times v\left(\dfrac{3}{4\pi}|w| - 1\right)\right), & \dfrac{4\pi}{3} \le |w| \le \dfrac{8\pi}{3} \\[3mm] 0, & |w| \notin \left[\dfrac{2\pi}{3}, \dfrac{8\pi}{3}\right] \end{cases} \tag{3}$$

where $v(x)$ is an auxiliary function expressed as [11]:

$$v(x) = x^4(35 - 84x + 70x^2 - 20x^3), x \in [0, 1] \tag{4}$$

Figure 5 depicts the flow chart of the signal-processing algorithm for removing the baseline drift from an originally recorded pulse signal, while Fig. 6 illustrates the corresponding intermediate results of the signal processing. First, we utilize CSE to obtain the baseline envelope of the originally recorded absolute resistance signal. Figure 6(a) shows the recorded absolute resistance signal and the baseline of the signal as a function of time, which is estimated via CSE.

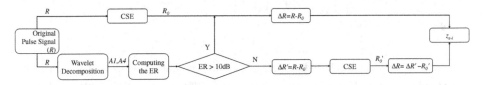

Fig. 5. Flow chart of the signal-processing algorithm for removing baseline drift and obtaining pulse waveform of an originally recorded pulse signal [11].

Motion artifacts in a recorded pulse signal are low frequency components. The main frequency of baseline drift is less than 1 Hz. The cutoff frequency of the fourth-level scale function, A4, is 1.56 Hz. The frequency content of the pulse waveform is less than 40 Hz. Thus (A1–A4) and A4 are used to approximate a recorded pulse signal and its baseline drift, respectively. DMWT is applied to the absolute resistance signal to obtain its first-level approximation, $A1$, and its fourth-level approximation, $A4$, as shown in Fig. 6(b). The two approximations are utilized to compute the ER of a recorded pulse signal as below:

$$ER = 20 \log_{10} \frac{\|A1 - A4 - mean(A1 - A4)\|}{\|A4 - mean(A4)\|} \tag{5}$$

where ‖ ‖ represents the order-two norm, and $mean\|A1–A4\|$ represents the average of A1–A4.

ER is used to quantify the extent of the baseline drift. It is found that the baseline drift in a pulse signal is relatively low, when its ER is higher than 10 dB [11]. Therefore, 10 dB is selected as the ER threshold for removing the baseline drift from a recorded resistance signal.

If ER is higher than 10 dB for a recorded pulse signal, its baseline drift is low and is removed by subtracting the estimated baseline in Fig. 6(a) from the recorded resistance

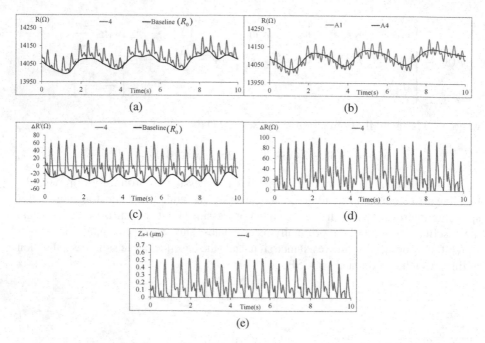

Fig. 6. The intermediate results of the signal-processing algorithm using DMWT and CSE for a recorded pulse signal of the 16yr-F subject post-exercise (4th transducer, Sensor #1): (a) absolute resistance, R, with estimated baseline (or estimated initial resistance, R_0); (b) the first-level and fourth-level approximation content of pulse decomposition; (c) the resistance change, $\Delta R'$, with estimated baseline (or second estimated initial resistance, R'_0); (d) the resistance change, ΔR, free of baseline drift; (e) drift-free sensor deflection representing the pulse waveform.

signal. Thus, the estimated baseline is equivalent to the initial resistance, R_0, of a transducer, which is clearly a function of time, due to motion artifacts. It is found that all the pulse signals measured at-rest have their ER higher than 10 dB.

If ER is lower than 10 dB, the baseline drift is high and is first removed by subtracting $A4$ in Fig. 6(b) from the absolute resistance signal in Fig. 6(a). Then, as shown in Fig. 6(c), the resistance change, $\Delta R'$, is obtained, but it still contains a little baseline drift. CSE is used again to obtain the baseline drift of the resistance change, as shown in Fig. 6(c). Afterward, the resistance change in Fig. 6(c) is subtracted by this new baseline drift or second estimated initial resistance, R'_0, to obtain the resistance change, ΔR, in Fig. 6(d), which is believed to be drift-free. The two pulse signals measured post-exercise are found to have an ER lower than 10 dB and thus are treated with the estimates of the baseline drift twice, as described here. Then, the initial resistance of a recorded pulse signal post-exercise is the sum of $A4$ and the baseline drift estimated from the resistance change, $(A4 + R'_0)$. Finally, the resistance change is Fig. 6(d) is converted to the sensor deflection, as shown in Fig. 6(e). This sensor deflection is free of baseline drift and captures the waveform of an originally recorded pulse signal, as detailed in the following subsection.

4.2   Conversion of Absolute Resistance to Sensor Deflection

As mentioned earlier, a pulse signal deflects the microstructure and then leads to a geometrical change in a transducer, which is recorded as a resistance change by the transducer. As such, the bottom deflection at a transducer represents the pulse waveform, instead of the resistance change. The following analysis describes the procedure for obtaining the sensor deflection from the resistance change.

Owing to fabrication variation in transducer height, h_e, (the smallest design parameter), the original resistance (defined as the resistance of a transducer when it is free of load) may vary among the transducers. The original resistance of the ith transducer is roughly calculated as [12]:

$$R_{0-i} = \frac{\rho \cdot w_e}{d_e/2 \cdot h_{e-i}} \tag{6}$$

where ρ is the electrical conductivity of the electrolyte used, and d_e, w_e and h_{e-i} are the length, width and height of the ith transducer, respectively. We further define the resistance of a transducer after being pressed against an artery as its initial resistance, R'_{0-i}. As such, the resistance change is calculated relative to the initial resistance, instead of the original resistance:

$$\Delta R = R_i - R'_{0-i}$$
$$= \frac{\rho \cdot w_e}{d_e/2 \cdot h'_{e-i}} \cdot \left[\left(1 - \frac{z_{s-i}}{h'_{e-i}} \right)^{-1} - 1 \right] \tag{7}$$
$$= \frac{\rho \cdot w_e}{d_e/2 \cdot (h'_{e-i})^2} \cdot z_{s-i} = \frac{(R'_{0-i})^2 \cdot d_e/2}{\rho \cdot w_e} \cdot z_{s-i}$$

where z_{s-i} is the deflection for the ith transducer, and h'_{e-i} is the initial height of the transducer.

Since the hold-down pressure against an artery is not controllable, the initial resistance of the sensor varies among all the measurements and keeps changing with time in each measurement. According to Eq. (7), the sensor deflection at the ith transducer can be obtained:

$$z_{s-i} = \frac{\Delta R}{(R'_{0-i})^2} \cdot \frac{\rho \cdot w_e}{d_e/2} \tag{8}$$

The above equation explains the reason that the signal-processing algorithm is critical for obtaining how the initial resistance varies over time in each pulse measurement, due to motion artifacts.

5   Measured Results

Our goal for the sensor in this work is to acquire the pulse waveforms of different arteries of a subject so as to evaluate the physiological condition of his cardiovascular system, instead of blood pressures. Note that the measured pulse signal amplitude varies with the hold-down pressure and thus cannot reflect blood pressure without calibration. To demonstrate the feasibility of the sensor for pulse waveform measurement, carotid, radial and temporal pulse waveforms are measured by the sensor on the four subjects at-rest. Moreover, carotid pulse waveforms are measured on two subjects both at-rest and post-exercise. Since we recently started to explore the sensor for pulse waveform measurement, the data collected here are not comprehensive. Nevertheless, the measured pulse waveforms are compared with the related findings in the literature for validating the feasibility of the sensor for pulse waveform monitoring.

5.1 Robustness to Motion Artifacts

Owing to the respiration of a subject and the handshaking during a measurement, motion artifacts in a pulse measurement are unavoidable. Sometimes, it is desirable to assess arterial pulse waveform difference between at-rest and post-exercise. As such, the sensor needs to be immune to motion artifacts. Figure 7 shows the absolute resistance signals of the five transducers from the 16yr-F subject post-exercise measured by Sensor #1.

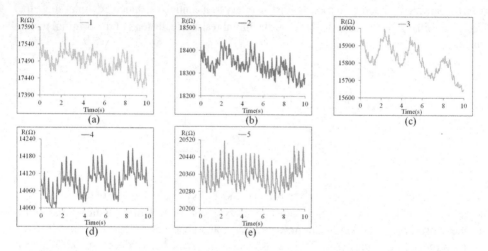

Fig. 7. Carotid pulse signals of the 16yr-F subject post-exercise measured using Sensor #1: (a)–(e) originally recorded absolute resistance signals of the five transducers.

The 4th and 5th transducers capture not only a clear patterned pulse signal, but also a heavily-breathing pattern, indicating that these two transducers at the site of the carotid artery. The breathing pattern introduces extremely large baseline drift to the recorded pulse signal. However, as will be seen in the next subsection, the sensor is capable of obtaining the undistorted pulse waveform under such severe motion artifacts. The rest three transducers obtain random signals with the heavily-breathing pattern, indicating that these transducers are away from the carotid artery. Owing to fabrication variation, the original resistance varies among the five transducers. As such, this measurement indicates that the transducers do not interfere with each other in a pulse measurement. Thus, the sensor provides a large alignment margin for an untrained individual to use.

5.2 Carotid Pulse Waveforms at-Rest and post-Exercise

Sensor #1 is used to conduct all the pulse measurements on the carotid artery of the 16yr-F and 28yr-M subjects. Figure 8 illustrates the carotid pulse signals of the 16yr-F subject at-rest and post-exercise. The 4th transducer captures the strongest pulse signal of the 16yr-F subject at-rest and post-exercise. While Figs. 8(a) and (c) illustrate the originally recorded absolute resistance signals and the baseline drifts at-rest and post-exercise, Figs. 8(b) and (d) show the corresponding drift-free pulse waveforms at-rest and post-exercise, respectively.

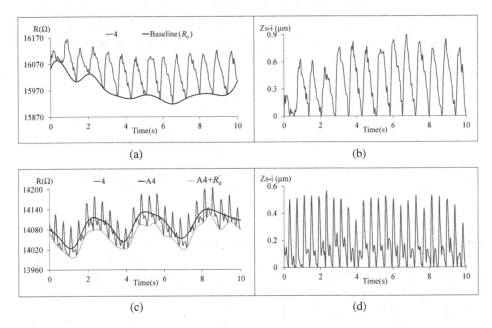

(a)   (b)

(c)   (d)

Fig. 8. Carotid pulse signal of the 16yr-F subject at-rest and post-exercise measured using Sensor #1 (4th transducer): (a) absolute resistance and estimated initial resistance at-rest; (b) sensor deflection at-rest; (c) absolute resistance, A4, and estimated initial resistance post-exercise; (d) sensor deflection post-exercise.

Since the initial resistance at-rest is higher than the initial resistance post-exercise, a higher hold-down pressure is used at-rest than post-exercise. The sensor deflection amplitude at-rest is higher than the one post-exercise, indicating that a strong pulse signal from post-exercise does not directly translate to a large sensor deflection, without a high hold-down pressure. While the baseline drift at-rest is random, the baseline drift post-exercise contains the heavily-breathing pattern, indicating that the fingers holding the sensor moved very slightly during the measurement.

Figure 9 illustrates the carotid pulse signals of the 28yr-M subject at-rest and post-exercise. The 4th transducer and the 5th transducer capture the strongest pulse signal of the subject at-rest and post-exercise, respectively. While Figs. 9(a) and (c) illustrate the originally recorded absolute resistance signals at-rest and post-exercise, Figs. 9(b) and (d) show the corresponding drift-free pulse waveforms at-rest and post-exercise, respectively. As shown in Fig. 9(c), the initial resistance keeps decreasing in this measurement, indicating that the hold-down pressure goes down significantly in this measurement. Therefore, the heavily-breathing pattern is not as obvious as the one in Fig. 8(c). Also, the baseline drift at-rest for this subject remains random.

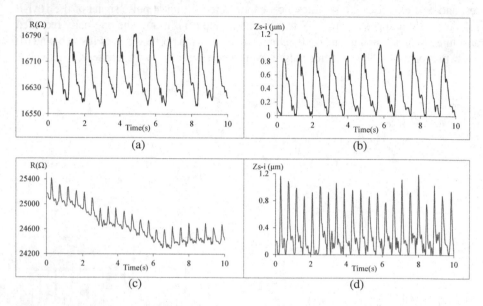

(a)   (b)

(c)   (d)

Fig. 9. Carotid pulse signals of the 28yr-M subject at-rest and post-exercise measured using Sensor #1: (a) absolute resistance and (b) sensor deflection of the 4th transducer at-rest; (c) absolute resistance and sensor deflection of the 5th transducer post-exercise.

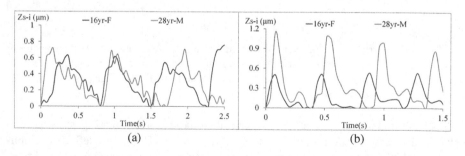

(a)   (b)

Fig. 10. Comparison of carotid pulse waveforms of the 16yr-F and 28yr-M subjects between at-rest and post-exercise: (a) sensor deflection at-rest; (b) sensor deflection post-exercise.

Figure 10 compares the carotid pulse waveforms of the two subjects between at-rest and post-exercise. The 16yr-F has a faster pulse rate than the 28yr-M both at-rest and post-exercise. A significant difference in the pulse waveform between at-rest and post-exercise is that the diastolic notch in the pulse waveforms of the two subjects post-exercise is much lower than its counterpart at-rest. This observation is consistent with the finding in the literature, as shown in Fig. 11. The same sensor is used for all the carotid pulse measurements. Therefore, difference in the pulse waveform between at-rest and post-exercise is believed to arise mostly from the corresponding cardiovascular system conditions, although the hold-down pressure may affect the pulse waveform slightly.

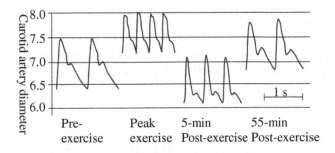

Fig. 11. Carotid artery distension wave and blood pressure recordings before, during and after dynamic exercise. Carotid artery distension wave recordings and corresponding blood pressure values in one representative subject during the control period, at peak-exercise, and at 5 and 55 min post-exercise. Time scale = 1 s. [13].

5.3 Radial Pulse Waveforms

The radial pulse signals are measured on the 28yr-F, 28yr-M and 29yr-F subjects at-rest using Sensor #2. There are illustrated in Figs. 12, 13 and 14. The upstroke swing is steep in all the radial pulse waveforms. The diastolic notch in the pulse waveform of the 28yr-M is much deeper than those of the 28yr-F and 29yr-F subjects. The inflection point in the 29yr-F subject is more conspicuous and farther away from the systolic peak than the one in the 28yr-F subject. In contrast, the pulse waveform of the 29yr-M subject shows a very faint inflection point. Taken together, the sensor is capable of resolving the salient difference in the radial pulse waveform among the three subjects.

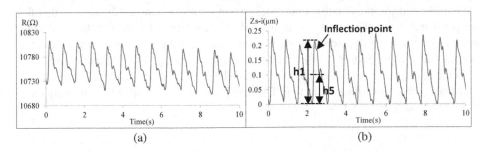

Fig. 12. Radial pulse waveforms of the 28yr-F subject at-rest measured using Sensor #2: (a) absolute resistance; (b) sensor deflection of the 1th transducer.

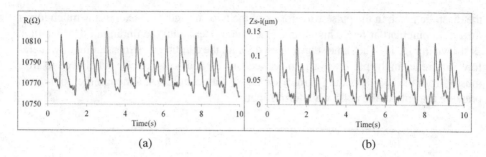

(a)   (b)

Fig. 13. Radial pulse waveforms of the 28yr-M subject at-rest measured using Sensor #2: (a) absolute resistance; (b) sensor deflection of the 1th transducer.

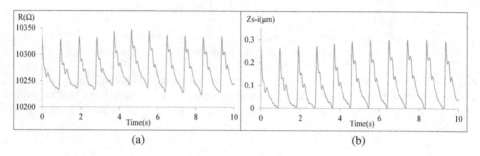

(a)   (b)

Fig. 14. Radial pulse waveforms of the 29yr-F subject at-rest measured using Sensor #2: (a) absolute resistance; (b) sensor deflection of the 2nd transducer.

5.4   Temporal Pulse Waveforms

Figures 15, 16 and 17 illustrate the temporal pulse signals of the 28yr-F, 28yr-M, and 29yr-F subjects at-rest, respectively, which are measured using Sensor #3 and Sensor #2. As compared with those radial pulse waveforms, the upstroke swing is a little bit inclined in these temporal pulse waveforms and the diastolic peak is closer to the systolic peak, simply because the temporal artery is closer to the heart. Comparison between the temporal pulse amplitudes and the radial pulse amplitudes measured using Sensor #2 indicates that the temporal pulse signal is weaker than the radial pulse signal for all the three subjects. Evidently, the temporal pulse amplitude measured using Sensor #3 is much stronger than the one measured using Sensor #2, with the latter being quite noisy. This indicates that Sensor #3 is much more sensitive than Sensor #2. As can be seen in Eq. (8), a low initial resistance translates to a higher sensor deflection. The initial resistance of Sensor #3 is roughly three times lower than the one of Sensor #2.

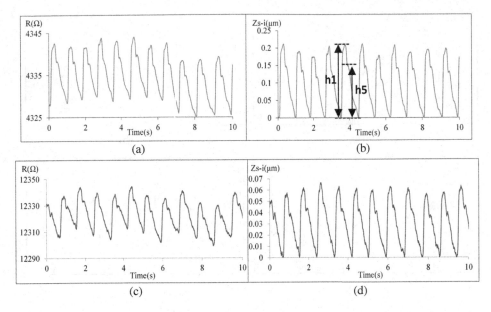

Fig. 15. Temporal pulse waveforms from the 28yr-F subject at-rest: (a) absolute resistance; (b) sensor deflection of the 5th transducer at-rest measured using sensor #3; (c) absolute resistance; (d) sensor deflection of the 3rd transducer measured using sensor #2.

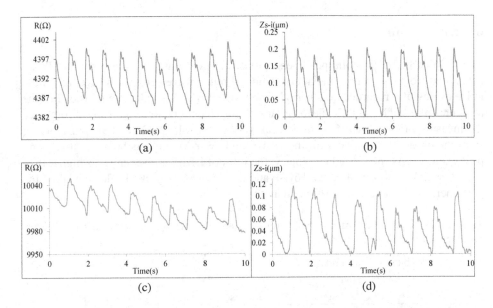

Fig. 16. Temporal pulse waveforms from the 28yr-M subject adult at-rest: (a) absolute resistance; (b) sensor deflection of the 5th transducer at-rest measured using sensor #3; (c) absolute resistance; (d) sensor deflection of the 2nd transducer measured using sensor #2.

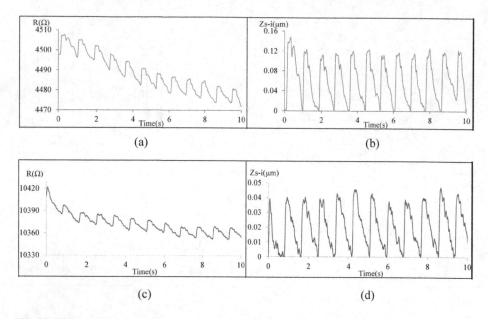

Fig. 17. Temporal pulse waveforms from the 29yr-F subject at-rest: (a) absolute resistance; (b) sensor deflection of the 5th transducer at-rest measured using sensor #3; (c) absolute resistance; (d) sensor deflection of the 1st transducer measured using sensor #2.

6   Discussion

The ratio of h1/h5 (systolic peak versus diastolic peak) and the pulse period, T, are calculated from the carotid, radial and temporal pulse waveforms as the average of the pulse cycles of a recorded 10 s period. The ratio of h1/h5 is calculated, instead of the absolute values of h1 and h5, due to the lack of a device for calibration. Table 2 compares the measured ratio of h1/h5 (systolic peak versus diastolic peak) and the pulse period from the carotid pulse signals of the 16yr-F and 28yr-M subjects between at-rest and post-exercise. As mentioned previously, the ratio of h1/h5 post-exercise is higher than at-rest, which is observed on the 16yr-F and the 28yr-M subjects. Understandably, the pulse period post-exercise is shorter than the one at-rest.

Table 2. Pulse peroid and ratio of h1/h5 from the carotid pulse signals of the 16yr-F and 28yr-M subjects at-rest and post-exercise.

Variables	Carotid artery			
	16yr-F		28yr-M	
	At-rest	Post-exercise	At-rest	Post-exercise
h1/h5	1.5506	3.2021	1.4149	3.1956
T(s)	0.7200	0.4000	0.8800	0.4400

Table 3 compares the pulse periods and the ratio of h1/h5 from the radial and temporal pulse signals of the three subjects measured using the two sensors. The ratio of h1/h5 of the radial pulse waveforms measured using Sensor #2 is higher than the ratio of the temporal pulse waveforms measured using Sensor #2. This is consistent with the finding in the literature [14]: as the pulse transmits from central (temporal) to peripheral (radial), the ratio of h1/h5 goes up. Meanwhile, the ratio of h1/h5 of the temporal pulse waveforms measured using the two sensors are comparable. This indicates that the sensitivity of the sensor does not affect the measured pulse waveform. However, the pulse period of the same subject varies noticeably among different measurements.

Table 3. Pulse periods and ratio of h1/h5 from the radial and temporal pulse signals of the 28yr-F, 28yr-M and 29yr-F subjects at-rest.

Sensor	Variables	Radial artery			Temporal artery		
		28yr-F	29yr-F	28yr-M	28yr-F	29yr-F	28yr-M
#3	T(s)	–	–	–	0.8962	0.8505	0.9394
	h1/h5	–	–	–	1.4052	1.1837	1.4099
#2	T(s)	0.7804	0.8923	0.7808	0.8705	0.8623	1.0381
	h1/h5	2.1900	1.9337	1.8198	1.4358	1.4241	1.3535

Baseline drift affects the interaction between the sensor and the artery, and thus affects the initial resistance. When the initial resistance is high, the sensor deflects the artery more and acquires a stronger pulse signal. In contrast, when the initial resistance is low, the sensor deflects the artery less and obtains a weaker pulse signal. Evidently, the initial resistance keeps drifting even in one pulse cycle. As such, the interaction between the sensor and the artery varies during the cycle. However, such drift in one cycle is relative small and thus is believed to not severely modify the measured pulse waveform.

We envision that this flexible PET-based sensor can be attached on a wristband and can then be worn on the wrist loosely, together with integrated circuit for signal routing and wireless transmission. Whenever an individual needs to measure his pulse waveform of an artery, he can locate his artery of interest using his fingers and then align the sensor at the site of the artery and press the sensor against the artery with his fingers. Owing to the small size and flexibility of the sensor, it is expected to be comfortable to wear a wristband with the sensor.

7   Conclusion

This paper presents a PET-based wearable sensor for arterial pulse waveform measurement for untrained individuals to conduct the arterial pulse waveform measurement at home. The sensor contains a PDMS microstructure embedded with a 5×1 resistive transducer array, spanning 6 mm and having a spatial resolution of 1.5 mm. Built on a PET substrate, the sensor is fabricated using a low-cost, two-mask fabrication process. Three sensors of the identical design are fabricated and used to demonstrate its feasibility for arterial pulse waveform measurement. Carotid, radial and temporal pulse signals of four subjects are collectively measured using the three sensors. The originally recorded pulse signal is the absolute resistance signal of a transducer. Motion artifacts during a measurement introduce baseline drift to the recorded signal and are further removed by a combination of DMWT and CSE implemented in Matlab. Ultimately, a drift-free pulse signal is represented by the sensor deflection as a function of time. All the measured pulse waveforms of the radial and temporal arteries of three subjects at-rest are consistent with their counterparts in the literature, demonstrating the feasibility of the sensor for monitoring the arterial pulse waveform. Meanwhile, the measured carotid pulse waveforms of two subjects at-rest and post-exercise are in good agreement with the related findings in the literature, demonstrating the robustness of the sensor for pulse measurement with severe motion artifacts.

Acknowledgments. This work is financially supported by the NSF CMMI division under Grant No. 1265785.

References

1. Lin, W.H., Zhang, H., Zhang, Y.T.: Investigation on cardiovascular risk prediction using physiological parameters. Comput. Math. Methods Med. **2013**, 272691 (2013)
2. Hu, C.S., Chung, Y.F., Yeh, C.C., Luo, C.H.: Temporal and spatial properties of arterial pulsation measurement using pressure sensor array. Evid. Complement. Altern. Med. eCAM **2012**, 745127 (2012)
3. Saugel, B., Fassio, F., Hapfelmeier, A., Meidert, A.S., Schmid, R.M., Huber, W.: The T-Line TL-200 system for continuous non-invasive blood pressure measurement in medical intensive care unit patients. Intensive Care Med. **38**, 1471–1477 (2012)
4. Digiglio, P., Li, R., Wang, W., Pan, T.: Microflotronic arterial tonometry for continuous wearable non-invasive hemodynamic monitoring. Ann. Biomed. Eng. **42**, 2278–2288 (2014)
5. Krejza, J., Arkuszewski, M., Kasner, S.E., Weigele, J., Ustymowicz, A., Hurst, R.W., Cucchiara, B.L., Messe, S.R.: Carotid artery diameter in men and women and the relation to body and neck size. Stroke J. Cereb. Circ. **37**, 1103–1105 (2006)
6. Ashraf, T., Panhwar, Z., Habib, S., Memon, M.A., Shamsi, F., Arif, J.: Size of radial and ulnar artery in local population. JPMA-J. Pak. Med. Assoc. **60**, 817 (2010)
7. Cheng, P., Gu, W., Shen, J., Ghosh, A., Beskok, A., Hao, Z.: Performance study of a PDMS-based microfluidic device for the detection of continuous distributed static and dynamic loads. J. Micromech. Microeng. **23**, 085007 (2013)
8. Gu, W., Cheng, P., Ghosh, A., Liao, Y., Liao, B., Beskok, A., Hao, Z.: Detection of distributed static and dynamic loads with electrolyte-enabled distributed transducers in a polymer-based microfluidic device. J. Micromech. Microeng. **23**, 035015 (2013)
9. Tang, L., Lee, N.Y.: A facile route for irreversible bonding of plastic-PDMS hybrid microdevices at room temperature. Lab Chip **10**, 1274–1280 (2010)
10. Tsuwaki, M., Kasahara, T., Edura, T., Matsunami, S., Oshima, J., Shoji, S., Adachi, C., Mizuno, J.: Fabrication and characterization of large-area flexible microfluidic organic light-emitting diode with liquid organic semiconductor. Sens. Actuators A Phys. **216**, 231–236 (2014)
11. Xu, L., Zhang, D., Wang, K., Li, N., Wang, X.: Baseline wander correction in pulse waveforms using wavelet-based cascaded adaptive filter. Comput. Biol. Med. **37**, 716–731 (2007)

12. Yang, Y., Guo, S., Hao, Z.: A two-dimensional (2D) Distributed-deflection sensor for tissue palpation with correction mechanism for its performance variation. IEEE Sens. J. **16**, 4219–4229 (2016)
13. Studinger, P., Lenard, Z., Kovats, Z., Kocsis, L., Kollai, M.: Static and dynamic changes in carotid artery diameter in humans during and after strenuous exercise. J. Physiol. **550**, 575–583 (2003)
14. McEniery, C.M., Cockcroft, J.R., Roman, M.J., Franklin, S.S., Wilkinson, I.B.: Central blood pressure: current evidence and clinical importance. Eur. Heart J. **35**, 1719–1725 (2014)

Towards an Intraoral-Based Silent Speech Restoration System for Post-laryngectomy Voice Replacement

Lam A. Cheah[1](✉), James M. Gilbert[1], Jose A. Gonzalez[2], Jie Bai[1],
Stephen R. Ell[3], Phil D. Green[2], and Roger K. Moore[2]

[1] School of Engineering, University of Hull, Kingston upon Hull, UK
{l.cheah,j.m.gilbert,j.bai}@hull.ac.uk
[2] Department of Computer Science, University of Sheffield, Sheffield, UK
{j.gonzalez,p.green,r.k.moore}@sheffield.ac.uk
[3] Hull and East Yorkshire Hospitals Trust, Castle Hill Hospital, Cottingham, UK
srell@doctors.org.uk

Abstract. Silent Speech Interfaces (SSIs) are alternative assistive speech technologies that are capable of restoring speech communication for those individuals who have lost their voice due to laryngectomy or diseases affecting the vocal cords. However, many of these SSIs are still deemed as impractical due to a high degree of intrusiveness and discomfort, hence limiting their transition to outside of the laboratory environment. We aim to address the hardware challenges faced in developing a practical SSI for post-laryngectomy speech rehabilitation. A new Permanent Magnet Articulography (PMA) system is presented which fits within the palatal cavity of the user's mouth, giving unobtrusive appearance and high portability. The prototype is comprised of a miniaturized circuit constructed using commercial off-the-shelf (COTS) components and is implemented in the form of a dental retainer, which is mounted under roof of the user's mouth and firmly clasps onto the upper teeth. Preliminary evaluation via speech recognition experiments demonstrates that the intraoral prototype achieves reasonable word recognition accuracy and is comparable to the external PMA version. Moreover, the intraoral design is expected to improve on its stability and robustness, with a much improved appearance since it can be completely hidden inside the user's mouth.

Keywords: Silent speech interface · Assistive technology · Wireless intraoral device · Permanent Magnet Articulography · Magnetoresistive sensors

1 Introduction

Speech is perhaps the most convenient and natural form of human communication. Patients who have had a laryngectomy (e.g. surgical removal of larynx as part of treatment for cancer or other diseases affected the vocal cords) lose their voices and often struggle with their daily communication. Hence, they may experience severe impact on their lives which can lead to social isolation, loss of identity and depression [1, 2]. However, there are currently only a limited number of post-laryngectomy voice restoration methods available for these individuals: esophageal speech, the electrolarynx and speech valves. Unfortunately, these methods are often limited by their usability and/or the abnormal voice produced, which may

© Springer International Publishing AG 2017
A. Fred and H. Gamboa (Eds.): BIOSTEC 2016, CCIS 690, pp. 22–38, 2017.
DOI: 10.1007/978-3-319-54717-6_2

be hard to understand for listeners [1, 3, 4]. On the other hand, typing-based augmented and alternative communication (AAC) devices are limited by slow manual text input [5]. Although some improvements have been achieved in term of the voice quality of the electrolarynx and esophageal speech [6, 7], emerging assistive technologies (ATs) such as silent speech interfaces (SSIs) have shown promising potential in recent years as an alternate solution.

SSIs are devices that enable speech communication to take place in the absence of audible acoustic signals [8]. To date, a number of SSIs have been proposed in an attempt to extract non-acoustic information generated during speech production and reproduce audible speech using different sensing modalities, such as measuring electrical activities of the brain [9–11] or the articulator muscles [12–14], or by capturing movements of the speech articulators themselves [3, 5, 8, 15–19]. A comprehensive summary on different SSIs technologies were presented in [8]. Because of their unique feature, SSIs can also be deployed in acoustically challenging environment or where privacy/confidentially is desirable, and not limited to its use as a communication aid for speech impaired individuals.

Despite the attractive attributes of SSIs, there are still challenges in the form of hardware (e.g. portability, lightweight, unobtrusiveness and wearability) and processing software (e.g. efficiency, robustness and intelligibility speech generation). Preliminary discussions on the influential factors affecting the SSIs' implementation were presented in [8], based upon criteria such as ability to operate in silence and noisy environments, usability by laryngectomees, issue of invasiveness market reediness and cost.

In the present work we employ the Permanent Magnet Articulography (PMA), which is a type SSI that is based on sensing the changes in the magnetic field generated by a set of permanent magnet markers attached onto the vocal apparatus (i.e. lips and tongue) during speech articulation by using an array of magnetic sensors located around the mouth [1, 3]. Although PMA shares some similarities with Electromagnetic Articulography (EMA) [5, 17], it does not explicitly provide the Cartesian position/orientation of the markers, but rather a summation of the magnetic fields from magnets that are associated with a particular articulatory gesture. The focus here is to build upon our previous work of [20], to further improve and alleviate the shortcomings from a hardware perspective. The proposed prototype has several distinctive features, such as being miniature in size, highly portable, discreet and unobtrusive since it is hidden from sight within the user's mouth.

The rest of the chapter is organized as follows. Section 2 overviews the PMA technique and its development to date. Next, Sect. 3 outlines the design challenges of the intraoral version of the PMA device. Then, Sect. 4 describes the architecture of the intraoral PMA prototype. Section 5 describes the experimental methods used to assess performance, followed by the results of that evaluation in Sect. 6. The final section concludes this chapter and provides an outlook for future work.

2   System Overview

PMA is a sensing technique for capturing the magnetic field resulting from movement of a set of permanent magnets attached onto the lips and tongue during speech articulation. The variations of the magnetic field can then be used to determine the speech which the user wishes to produce by first performing automatic speech recognition (ASR) on the PMA data and then synthesising the recognised text using a text-to-speech (TTS) synthesizer [3, 18–20].

A number of PMA prototypes have been investigated in recent years. Earlier prototypes [3, 18, 19] provided acceptable speech recognition performance, but were not particularly satisfactory in terms of their appearances, comfort and ergonomic factors for the users. To address these challenges, a PMA prototype in the form of a wearable headset (designed based on a customized pair of spectacles or a headband) comprising of miniaturized sensing modules and wireless capability was developed [20]. The second generation prototype was re-designed based on a user-centered approach utilizing feedback from questionnaires completed by potential users and through discussion with stakeholders including clinicians, potentials users and their families. The appearance and comfort of the prototype was much improved and it demonstrated comparable recognition performances to its predecessors.

As illustrated in Fig. 1, the second generation PMA system consists of a set of six cylindrical Neodynium Iron Boron (NdFeB) permanent magnets, four on the lips (ø1 mm × 5 mm), one at the tongue tip (ø2 mm × 4 mm) and one on the tongue blade (ø5 mm × 1 mm). These magnets are currently attached using Histoacryl surgical tissue adhesive (Braun, Melsungen, Germany) during experimental trials, but will be surgically implanted for long term usage. The remainder of the PMA system is composed of a set of four tri-axial Anisotropic Magnetoresistive (AMR) magnetic sensors mounted on the wearable headset, a set of microcontrollers, rechargeable battery and a processing unit (e.g. computer/tablet PC). Detailed information on these hardware modules and their operations is presented in [20].

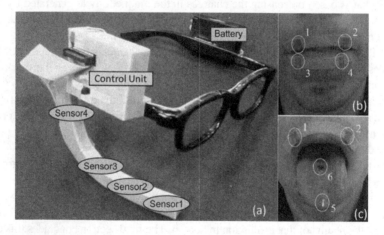

Fig. 1. (a) A wearable PMA prototype designed in a form of spectacles. (b) & (c) Placement of six magnets on lips (pellets 1–4), tongue tip (pellet 5) and tongue blade (pellet 6).

3   Design Challenges

Although the second generation prototype has many desirable hardware features, it is not without drawbacks. Firstly, the performance of the external headset cannot be maintained in certain real-life conditions (i.e. exaggerated movement or sports activity) due to issues

Location	Volume (cm³)	Circuitry	Volume (cm³)
Front palatal left (FPL)	0.422	Sensor board	0.27
Front palatal right (FPR)	0.422	-	-
Palatal roof (PR)	1.436	Sensor array board	1.105
Palatal side wall left (PSWL)	1.845	Coin battery	1.28
Palatal side wall right (PSWR)	1.845	Bluetooth module	1.02
Total	5.97		3.675

Fig. 2. Space within the palatal cavity.

with instability. If there is a considerable movement of the headset on the user's head, the PMA system may need re-calibration/re-training to avoid degradation in performance. In addition, wearing the headset over long periods may not be comfortable, despite the fact that the device was designed to be lightweight and ergonomically friendly. Lastly, and potentially most importantly the external version of the PMA device may still be cosmetically unacceptable to some users. Previous studies indicated that the appearance is one of the most important factors that affect the acceptability of any AT by their potential end users [21–23].

In order to overcome these limitations, an intraoral version of the PMA prototype, which fits under the palate inside the user's mouth in a form of a dental retainer, was proposed. Being tightly clamped onto the upper teeth means that the device would be more stable than the previous wearable headset. Due to the fact that the device is completely hidden from sight during normal use, it is cosmetically inconspicuous. In addition, since the sensors are much closer to the articulators than the external headset, the size of the implants can be significantly reduced. Similar intraoral-based designs have been previously implemented for other non-speech related ATs with various degree of success [24–26].

4 System Description

4.1 Space Budget

The latest intraoral-based PMA system is made up of: three tri-axial magnetic sensors, a wireless communication module, a microprocessor to synchronize data capture and communications and a suitable power source capable of providing an appropriate operating lifetime. This must be accommodated within the oral cavity, without excessively interfering with the natural tongue articulation during speech. A recent study [27] suggested that the palatal cavity is suitable to house the intraoral circuitry because of its relatively flat surfaces and proximity to the articulators. As illustrated in Fig. 2, a 3D palatal model was created and divided into five possible locations to accommodate the intraoral circuitry: front palatal left (FPL), front palatal right (FPR), palatal roof (PR), palatal side wall left (PSWL) and palatal side wall right (PSWR). The estimated space available in the palatal cavity on our test subject is 5.97 cm³ as shown in Fig. 2 (assuming a uniform 3 mm thickness), whereas the estimated volume of the intraoral circuitry, as described subsequently, is approximately 3.68 cm³.

4.2 Intraoral Circuitry

A crucial design element for the intraoral circuitry is to drastically reducing the size of the electronics and rechargeable battery of the external version of PMA prototype, so that all necessary circuitry can be fitted inside the mouth. The major components of the PMA prototype are shown in Fig. 3. These are implemented using a low-power ATmega328P microcontroller, three tri-axial HMC5883L magnetic sensors (AMR), a rechargeable Li-Ion coin battery (capacity of 40 mAh, 3.7 V and 20 mm diameter × 3.2 mm thickness), and a wireless transceiver (Bluetooth 2.0 module). The remainder of the system shown in Fig. 4 consists of a processing unit (e.g. computer/tablet PC) and a set six permanent magnets (NdFeB) attached onto lips and tongue in the same locations as illustrated in Fig. 1. The elements of the intraoral sensing system (which have a total volume of 3.68 cm^3) are arranged as shown in Fig. 4(a). These may be encapsulated and placed in the oral cavity as shown in Fig. 4(d).

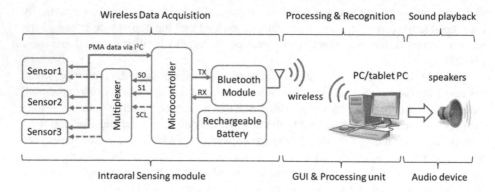

Fig. 3. Simplified operation block diagram.

Although there were many design changes for the intraoral design, the positions of the magnets remained unchanged from the earlier prototype. However, because of the proximity of the sensors, significantly smaller magnets (see Fig. 4c) can be used (note that the magnetic field strength decreases with cube of the distance away from the magnets): four on lips (ø1 mm × 4 mm), one on the tongue tip (ø1 mm × 1 mm) and one on the tongue blade (ø1 mm × 1 mm).

4.3 Circuit Operation

The operational block diagram of the intraoral version of the PMA system is presented in Fig. 3. A command is sent wirelessly from the processing unit to the intraoral sensing module via Bluetooth to trigger data acquisition. All three tri-axial magnetic sensors then measure the three components of magnetic field and digitize it with 12-bit resolution. The microcontroller acquires these measurements (9 PMA channels sampled at 80 Hz) through managing a multiplexer using three control signals (S0, S1 and SCL). The multiplexer acts as a switching device to route the serial clock (SCL) to the desired magnetic sensor through the I^{2}C interface. The acquired samples are then transmitted back to the processing unit

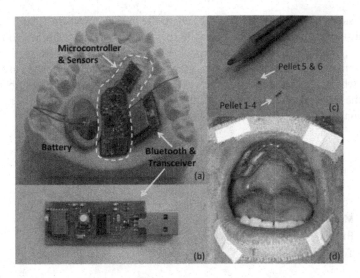

Fig. 4. (a) & (b) Circuitry of the intraoral version of the PMA system. (c) Placement of magnets on lips (pellets 1–4), tongue tip (pellet 5) and tongue blade (pellet 6). (d) View of the device when worn by the user.

wirelessly via the Bluetooth transceiver and custom designed Bluetooth dongle (in Fig. 3(b)) for further processing. Unlike the external version of the PMA prototype, the intraoral device is restricted to only operate wirelessly from inside the mouth. Hence, wired connectivity is impossible, as the sensing modules are to be sealed and packaged inside a dental retainer. In terms of software, a bespoke MATLAB-based graphical user interface (GUI) developed in [20] was adapted, where all speech processing and recognition algorithms were embedded.

4.4 Power Budget

As the circuitry is to be sealed into a dental retainer, the only way the intraoral device can acquire power is from a battery. With limited space available, only a small battery can be accommodated (in the current design, the battery takes 27% of the total volume of the circuitry). The battery can be recharged through a charging point located on the under-side of the dental retainer. In addition, any measures to extend the battery life will be of interest. Power hungry components such as the microcontroller, the magnetic sensors and the Bluetooth module may be set to *standby mode* or *sleep mode* to reduce the current consumption when they are inactive. As shown in Table 1, *sleep mode* gives a saving of 93% over *standby mode* or a saving of 97% over *active mode*.

Figure 5 shows a summary of the discharging cycle of the battery with the circuit in *active* and *sleep* modes. Neither of these operating regimes is fully representative of the expected use since they correspond to continuous speech and no speech respectively. If the system is to operate continuously (in *active mode*), the battery will last approximately one hour before being depleted below the minimum operating voltage (cut-off voltage) required by the Bluetooth module of 2.1 V. In contrast, if the system was inactive at all times (in *sleep mode*) the battery would last about 32 h. Based on the measurements in Table 1 and Fig. 5, a more realistic regime would be to allow 30 min of speech with a further 16 h in *sleep mode*. Hence, the estimated usage time is considered to be sufficient for a typical day before charging is required. This assumes that the circuit is active only while utterance is underway, which implies that a user interface is required to allow speech to be initiated. Note that the intraoral circuit can be 'woken up' by Bluetooth command sent from the processing unit, so a variety of user interfaces could be devised.

Table 1. Current consumption in difference operational modes.

Current consumption	Active mode (mA)	Stanby mode (mA)	Sleep mode (mA)
Sensors	5.1	0.006	0.006
Microcontroller	5.4	4.4	0.7
Bluetooth	19.0	7.22	0.007
Total	29.5	11.626	0.776

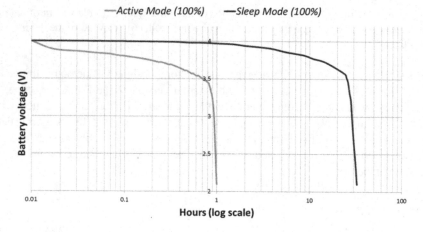

Fig. 5. Battery discharging over time under *active mode* and *sleep mode*.

4.5 System Implementation

The intraoral circuitry described above must be encapsulated to protect it from damage and short circuits due to saliva and to ensure it is held in place within the palate. The retainer must be customized according to the individual's oral anatomy. This may be achieved by forming it on a dental impression of the user's oral cavity (seen in the background of Fig. 4a). The intraoral PMA prototype was implemented in the form of dental retainers utilizing both soft and semi-rigid materials, as illustrated in Fig. 6. We will refer to these as *Type I* and *Type II*, respectively.

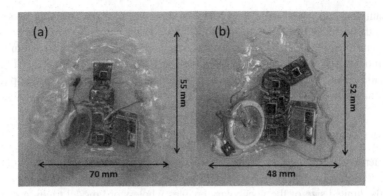

Fig. 6. PMA circuitry embedded inside a (a) soft bite raiser like dental retainer (*Type I*) and (b) semi-rigid dental retainer (*Type II*).

Type I (soft) retainer is similar to a soft bite raising appliance and is made of polypropylene or polyvinylchloride (PVC) material. On the other hand, *Type II* (semi-rigid) retainer is made from Essix C+ plastic. To allow stable fitting in the palate, the *Type I* retainer is fitted over the entire arch of the upper teeth. In contrast, the *Type II* retainer utilizes a set of curved edges to clasp tightly onto the upper teeth.

In generally, both intraoral and external PMA devices are speaker dependent systems, because their designs need to be individually tailored, based on the user's head or oral anatomy for optimal performance. In the case of the external device, this involves moving sensor arm so that it is close to the user's cheek and lips while in the case of the intraoral device, it must be encapsulated and formed on an impression of the user's palate.

5   Methods

5.1   Experimental Design

The data used for evaluating the new intraoral prototype were collected from a male native English speaker who is proficient in the usage of the external PMA device. Magnets were temporary attached on the subject using Histoacryl surgical tissue adhesive (Braun, Melsungen, Germany).

Recordings of PMA and audio data for training and evaluation were performed via using a customized Matlab GUI. The software provides a visual prompt of randomized utterances to the subject at interval of 5 s during the training session. The subject's head was not restrained during the recording sessions, but the subject was requested to avoid any large head movements. This was necessary to ensure that interference induced by movement relative to earth's magnetic field was at its minimum, so that it did not corrupt or distort the desired signal. This is because the current prototype is not yet equipped with a background cancellation/removal mechanism.

The recordings were conducted in an acoustically isolated room for optimal sound quality. The audio data were recorded using a shock-mounted AKG C1000S condenser microphone via a dedicated stereo USB-sound card (Lexicon Lambda) to a PC, with a

16 kHz sampling rate. Meanwhile, the PMA data were captured at a sampling frequency of 80 Hz via the intraoral PMA device and transmitted to the same PC wirelessly via Bluetooth, as illustrated in Fig. 3. Since both data streams (PMA and audio) are acquired from separate modality, synchronization between the two data streams is necessary. Therefore, an automatic timing re-alignment mechanism was implemented utilizing start-stop markers generated in additional to both data streams.

5.2 Data Corpus and Recording

Our long term goal is to explore the feasibility of using the intraoral device for continuous speech reconstruction. For preliminary testing, the TIDigits database [28] was selected because the limited size of the vocabulary enables whole-word model training from relatively sparse data and because of the simplicity of the language involved. The corpus consists of sequences of connected English digits with up to seven digits per utterance. The vocabulary is made up of eleven individual digits, i.e. from 'one' to 'nine', plus 'zero' and 'oh' (both representing digit 0).

The experimental data were collected from two independent sessions, with each session consisted of four datasets containing 77 sentences each. A total of 308 utterances containing 1012 individual digits were recorded during each session. To prevent subject fatigue, short breaks in between each recording session were allowed.

5.3 HMM Training and Recognition

Prior to the training and recognition processes, the acquired PMA data were segmented and checked using the audio data. Inappropriate endpoints were manually corrected if necessary. In addition, any mislabeled utterances were corrected using the acquired audio data.

The PMA data was then subjected to offset removal via median subtraction over 2 s windows with 50% overlap and followed by data normalization. Next, the delta parameters were computed for all PMA channels and added to its original time series data, resulting in a feature vector of size 18. The delta-delta parameters were not included as part of the feature vector as they did not produced significant improvement in performance [18, 19]. The recognition performance based on the audio data was also evaluated for comparison purposes. In this case, 13 Mel-frequency cepstral coefficients (MFCCs) were extracted from the audio signals using 25 ms analysis windows with 10 ms overlap. Next, the delta and delta-delta parameters were computed and appended to the static parameters, resulting in a feature vector of dimension 39. An overview on the PMA and audio parameters used is presented in Table 2.

The extracted PMA and audio features were used for training two independent speech recognizers using the HTK toolkit [29]. In both cases, the acoustic model in the recognizer uses whole-word Hidden Markov Models (HMMs) [30] for each of the eleven digits. Each HMM has 21 states and 5 Gaussians per state. The selected parameters were not optimized, but were known for their performances based on previous work [18, 19]. The HMM training and recognition was carried out in four validation cycles. In each cycle, three out of four sets within a session were used for training and the remaining one for testing. The recognition results were averaged over four cycles and across two independent sessions.

Table 2. Vector sizes of the parameters used in PMA and audio.

Parameters	Original	1st delta	2nd delta	Vector size
Sensor	×			9
SensorD	×	×		18
Audio	×	×	×	39

6 Results and Discussion

6.1 Evaluation of the Intraoral Devices

As seen in Fig. 7, it is obvious that *SensorD* performs significantly better than using *Sensor* data alone across both *Type I* and *Type II* intraoral devices. Similar trends where *SensorD* is superior over *Sensor* were also reported in [7]. In addition, the *Type II* intraoral design outperformed its counterpart (i.e. *Type I*) on both word and sequence recognition. Although the hardware on both devices were similar, the *Type II* device had its front sensor placed to the side, whereas it was positioned at the center for the *Type I* design. This is to eliminate or at least minimize possible saturation at the front sensor due to contact with magnet attached onto the tongue tip. Since the *Type II* intraoral version provided superior performance, this version will be the focus for the rest of this chapter.

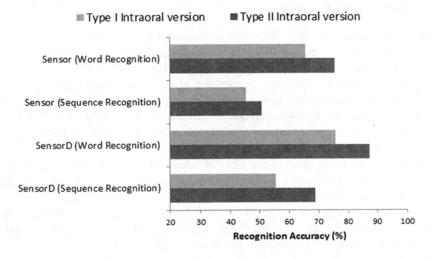

Fig. 7. Comparison of word and sequence accuracies of connected digits between *Type I* and *Type II* intraoral version.

Figure 8 illustrates that an increased number of training sessions yields better performance on both word and sequence recognitions, through the reduction of in word error rate (WER). It also appears that even for word recognition, the inclusion of further training data sets could reduce the WER further. The training sessions were not extended because of the speaker fatigue and increased the likelihood of the magnets becoming detached.

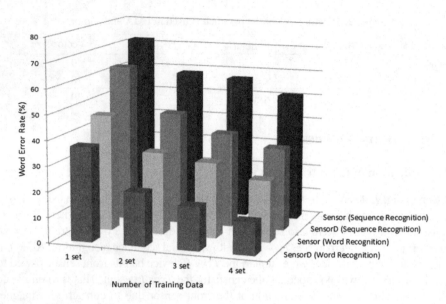

Fig. 8. Decrease in word error rate (WER) with the increase in training sessions.

6.2   Recognition Performance

Both word and sequence recognition results for the intraoral and external versions of the PMA device are presented in Figs. 9 and 10. In addition, the performances of the PMA devices were compared with audio-based recognition. The darker bars indicate the performance achieved using only static PMA data (vector size of 9), whereas the lighter bars are the results achieved using both static and dynamic features (vector size of 18). In addition, the grey-colored bars are the speech-recognition performance achieved using audio data (vector size of 39). We will refer to these three conditions as *Sensor*, *SensorD* and *Audio* features, respectively (see Table 2).

The results reflect the mean of the data collected across the two sessions, but were initially analyzed independently session-by-session. In order to avoid the inconsistency of magnets placement during individual training sessions, data were not merged across different sessions. This however could be solved, as magnets are to be surgically implanted for long term usage. Alternatively, session-independent approaches such as those presented for other SSIs methods could be investigated [14, 31].

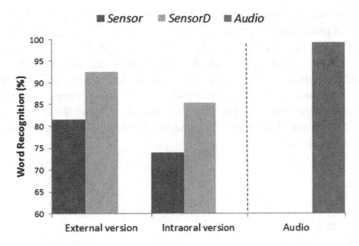

Fig. 9. Comparison of word accuracy in the connected digits.

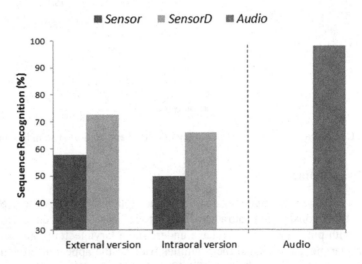

Fig. 10. Comparison of sequence accuracy in the connected digits.

As shown in both Figs. 9 and 10, it is quite obvious that *SensorD* produced better recognition performance on both occasions than using *Sensor* alone Similar trends were also reported [18, 20]. As expected, for this simple task, recognition using *Audio* performed very well (i.e. 99%). Preliminary evaluations indicate a close comparable recognition performance for the intraoral device and the previous external version, as illustrated in Figs. 9 and 10. There are a number of possible explanations for this degradation: (1) the presence of the intraoral prototype affects articulation and, in particular limits the tongue movements. This may lead to inconsistent articulation, (2) the subject was new to the intraoral version, but had prior experiences on the external PMA version, (3) possible drawbacks of operating at a lower sampling rate (up to 80 Hz) due to the design constraint on the intraoral device. Although recognition performance decreases with the used of lower sampling rate, both external and intraoral version showed similar recognition trends (illustrated in Fig. 11), and (4) the magnets are able to come much closer to the sensors in the intraoral device than in the external device, resulting in a more significant non-linear effect (since the field strength decreases with cube of the distance). This means that small unintentional articulator movements (e.g. swallowing, licking the lips and etc.) can generate very large signals in some instances which could have corrupted the data. Further work is required to understand the significance of each of these possible causes.

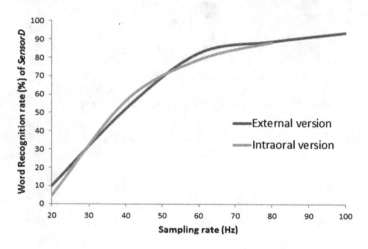

Fig. 11. Decrease in recognition performance with the reduction of sampling rate.

6.3 Hardware Comparison

As discussed in Sect. 3, one major obstacle to the acceptability of an AT (e.g. SSI) is its appearance if it is considered unattractive. Similar views were also concluded through discussions with potential users who have undergone a laryngectomy and an opinion survey of 50 laryngectomees and their families/friends: the appearance of the device was considered to be of a very high priority [20]. To enhance its appeal to users, influential factors such as appearance need to be accounted for during device development. The challenge here is to satisfy the design objective and continue improving the PMA device's appearance but without compromising its speech reconstruction performance. The latest intraoral prototype employs the same functional principles as the previous design reported in [20, 32], but implemented in a different form. A summary of the hardware features of the new intraoral PMA system compared to its predecessor is presented in Table 3.

Despite the improved appearance of the second generation PMA system in the form of a wearable headset, it might not yet to be appealing to all potential users. To address this shortcoming, the latest intraoral circuitry was implemented in the form of a dental retainer. To achieve this, the circuit was re-designed to use fewer and smaller

Table 3. Summary of the PMA devices' specifications and comparison [*Note that although the external sensing system has 12 channels, only 9 are used for speech recognition and 3 are used for cancellation of background magnetic fields].

Specifications		Intraoral Device	External Device
Appearance		Dental retainer	Wearable headset
Operating voltage		2.1 V	5 V
Magnets	Tongue Blade	⌀1 mm × 1 mm	⌀5 mm × 1 mm
	Tongue Tip	⌀1 mm × 1 mm	⌀2 mm × 4 mm
	Lips	⌀1 mm × 4 mm	⌀1 mm × 5 mm
Magnetic Sensing	Dimension	$12 \times 12 \times 3$ mm^3	$12 \times 12 \times 3$ mm^3
	Sensitivity	230 LSb/gauss	440 LSb/gauss
	Sampling rate	80 Hz	100 Hz
	Channels	9	12*
Data Transmission	Type	Bluetooth 2.0	Bluetooth 2.0/USB
	Frequency	2.4 GHz	2.4 GHz
	Data rate	57.6 kbps	500 kbps
Power	Supply	Rechargeable battery	Rechargeable battery/USB
	Battery	Li-Ion 40 mAh	Li-Ion 1080 mAh
	Current consumption	30.5 mA	93.5 (wireless)/67.1 (wired) mA
	Lifetime	1 h	10 h
Prototype	Dimension	$70 \times 55 \times 25$ mm^3	$160 \times 160 \times 150$ mm^3
	Weight	15 g	160 g
	Material	Polypropylene/Essix C+ plastic	VeroBlue/ VeroWhitePlus resin

components. In addition, the power consumption of the circuit was carefully managed to allow it to operate from a small battery suitable for inclusion within the dental retainer. Hence, this led to a much smaller and lighter (i.e. one tenth of previous weight) prototype as compared to its predecessor. In addition, the intraoral prototype is highly portable, it operates and can be controlled wirelessly via Bluetooth using a computer/tablet PC. Also, a higher signal-to-noise ratio (SNR) was obtained with smaller magnetic markers, due to their proximity to the magnetic sensors. The tongue magnets used with the intraoral sensor system had 16 to 25 times smaller volume than those used for the external headset, potentially making them less invasive when implanted.

A significant drawback with the intraoral device is the limited battery size and capacity (i.e. 40 mAh). In contrast, the external version is less restricted in term of size and weight of the battery. Hence, this significantly reduces the operational time of the intraoral device per charging. A number of steps have been introduced to reduce its power consumptions: a lower operating voltage is selected and power-efficient components, lower data sampling and transmission rates were chosen. In addition, software was developed to switch from an *active mode* to *sleep mode* when not in use. Using these measures, it is estimated that the battery life cycle could be extended from one hour to about 16.5 h including 30 min of speech.

7   Conclusion

In this chapter we have described a new intraoral PMA prototype using commercial off-the-shelf (COTS) components embedded inside a dental retainer constructed using the subject's dental impression. Preliminary evaluation of the intraoral prototype indicated a near comparable recognition performance to previous external sensor systems.

Although the intraoral version showed minor degradation in performance, there are several advantages over its predecessor and with a number of avenues for further investigation to improve its performance. It is also considered to be more stable and robust against unintentional movement as it is implemented in a form of a dental retainer, which securely sits in the palatal cavity and is clasping firmly on the upper teeth. Secondly, significantly smaller magnets may be used for the intraoral version (because of their proximity to the magnetic sensors) while also giving a higher SNR. In addition, the dental retainer can be completely hidden inside the user's mouth and out of sight. Hence, this would eliminate the concern of being a sign of disability. However, a downside of the intraoral design would be the possibility of limiting the natural movement of the tongue, because the device occupies part of the user's oral cavity. Further work is required to assess whether users become accustomed to the presence of the device and are able to achieve more consistent articulation.

With these encouraging results obtained, extensive work is needed to: (1) further reduce the size of future intraoral prototypes, (2) improve the circuitry power efficiency, (3) incorporate inductive charging for the battery, and (4) introduce a background cancellation mechanism for movement-induced interference. Though there are still limitations, the present work demonstrates a major step towards creating a viable SSI that would appeal to speech impaired users.

Lastly, an alternative speech generation through direct conversion of PMA data into audible speech without an intermediate recognition step was investigated and preliminary results were encouraging [33]. For further information on the PMA-based SSI and its speech restoration technique, please visit www.hull.ac.uk/speech/disarm.

Acknowledgements. The authors would like to thank Helen Dehkordy from Hull and East Yorkshire Hospitals NHS Trust for prototyping the dental retainers. The work is an independent research funded by the National Institute for Health Research (NIHR)'s Invention for Innovation Programme (Grant Reference Number II-LB-0814-20007). The views stated are those of the authors and not necessary reflecting the thoughts of the sponsor.

References

1. Fagan, M.J., Ell, S.R., Gilbert, J.M., Sarrazin, E., Chapman, P.M.: Development of a (silent) speech recognition system for patients following laryngectomy. Med. Eng. Phys. **30**(4), 419–425 (2008)

2. Braz, D.S.A., Ribas, M.M., Dedivitis, R.A., Nishimoto, I.N., Barros, A.P.B.: Quality of life and depression in patients undergoing total and partial laryngectomy. Clinics **60**(2), 135–142 (2005)
3. Gilbert, J.M., Rybchenko, S.I., Hofe, R., Ell, S.R., Fagan, M.J., Moore, R.K., Green, P.D.: Isolated word recognition of silent speech using magnetic implants and sensors. Med. Eng. Phys. **32**(10), 1189–1197 (2010)
4. Liu, H., Ng, M.: Electrolarynx in voice rehabilitation. Auris Nasus Larynx **30**(3), 327–332 (2007)
5. Wang, J., Samal, A., Green, J.R., Rudzicz, F.: Sentence recognition from articulatory movements for silent speech interfaces. In: Proceedings of 37th ICASSP, Kyoto, Japan, pp. 4985–4988 (2012)
6. Toda, T., Nakagiri, M., Shikano, K.: Statistical voice conversion techniques for body-conducted unvoiced speech enhancement. IEEE Trans. Audio Speech Lang. Process. **20**(9), 2505–2517 (2012)
7. Doi, H., Nakamura, K., Toda, T., Saruwatari, H., Shikano, K.: Esophageal speech enhancement based on statistical voice conversion with Gaussian mixture model. IEICE Trans. Inf. Syst. **93**(9), 2472–2482 (2010)
8. Denby, B., Schultz, T., Honda, K., Hueber, T., Gilbert, J.M., Brumberg, J.S.: Silent speech interfaces. Speech Commun. **52**(4), 270–287 (2010)
9. Brumberg, J.S., Wright, E.J., Andreasen, D.S., Guenther, F.H., Kennedy, P.R.: Classification of intended phoneme production from chronic intracortical microelectrode recordings in speech-motor cortex. Frontiers Neurosci. **65**(5), 1–12 (2011)
10. Brumberg, J.S., Nieto-Castanon, A., Kennedy, P.R., Guenther, F.H.: Brain-computer interfaces for speech communication. Speech Commun. **52**(4), 367–379 (2010)
11. Porbadnigk, A., Wester, M., Calliess, J., Schultz, T.: EEG-based speech recognition – impact of temporal effects. In: Proceedings of 2nd Biosignals, Porto, Portugal, pp. 376–381 (2009)
12. Jou, S.C.S., Schultz, T., Walliczek, M., Kraft, F., Waibel, A.: Towards continuous speech recognition using surface electromyography. In: Proceedings of 9th Interspeech, Pittsburgh, USA, pp. 573–576 (2006)
13. Wand, M., Janke, M., Schultz, T.: Tackling speaking mode varieties in EMG-based speech recognition. IEEE Trans. Biomed. Eng. **61**(10), 2515–2526 (2014)
14. Wand, M., and Schultz, T.: Session-independent EMG-based speech recognition. In: Proceedings of 4th Biosignals, Rome, Italy, pp. 295–300 (2011)
15. Petajan, E.D.: An architecture for automatic lipreading to enhance speech recognition. In: Proceedings of the International Conference on Computer Vision and Pattern Recognition, California, USA, pp. 40–47 (1985)
16. Hueber, T., Benaroya, E.-L., Chollet, G., Denby, B., Dreyfus, G., Stone, M.: Development of a silent speech interface driven by ultrasound and optical images of the tongue and lips. Speech Commun. **52**(4), 288–300 (2010)
17. Toda, T., Black, A.W., Tokuda, K.: Statistical mapping between articulatory movements and acoustic spectrum using a Gaussian mixture model. Speech Commun. **50**(3), 215–227 (2008)
18. Hofe, R., Ell, S.R., Fagan, M.J., Gilbert, J.M., Green, P.D., Moore, R.K., Rybchenko, S.I.: Small-vocabulary speech recognition using silent speech interface based on magnetic sensing. Speech Commun. **55**(1), 22–32 (2013)
19. Hofe, R., Bai, J., Cheah, L.A., Ell, S.R., Gilbert, J.M., Moore, R.K., Green, P.D.: Performance of the MVOCA silent speech interface across multiple speakers. In: Proceedings of 14th Interspeech, Lyon, France, pp. 1140–1143 (2013)

20. Cheah, L.A., Bai, J., Gonzalez, J.A., Ell, S.R., Gilbert, J.M., Moore, R.K., Green, P.D.: A user-centric design of permanent magnetic articulography based assistive speech technology. In: Proceedings of 8th Biosignals, Lisbon, Portugal, pp. 109–116 (2015)
21. Hirsch, T., Forlizzi, J., Goetz, J., Stoback, J., Kurtx, C.: The ELDer project: social and emotional factors in the design of eldercare technologies. In: Proceedings on the 2000 conference of Universal Usability, Arlington, USA, pp. 72–79 (2000)
22. Martin, J.L., Murphy, E., Crowe, J.A., Norris, B.J.: Capturing user requirements in medical devices development: the role of ergonomics. Physiol. Meas. **27**(8), 49–62 (2006)
23. Bright, A.K., Conventry, L.: Assistive technology for older adults: psychological and socio-emotional design requirements. In: Proceedings of 6th International Conference on PErvaesive Technologies Related to Assistive Environments, Rhodes, Greece, pp. 1–4 (2013)
24. Tang, H., Beebe, D.J.: An oral interface for blind navigation. IEEE Trans. Neural Syst. Rehabil. Eng. **14**(1), 116–123 (2006)
25. Lontis, E.R., Lund, M.E., Christensen, H.V., Gaihede, M., Caltenco, H.A., Andreasen-Strujik, L.N.: Clinical evaluation of wireless inductive tongue computer interface for control of computers and assistive devices. In: Proceedings of 32nd IEEE EMBC, Beunos Aires, Argentina, pp. 3365–3368 (2010)
26. Park, H., Kiani, M., Lee, H.M., Kim, J., Block, J., Gosselin, B., Ghovanloo, M.: A wireless magnetoresistive sensing system for an intraoral tongue-computer interface. IEEE Trans. Biomed. Circuits Syst. **6**(6), 571–585 (2012)
27. Bai, J., Cheah, L.A., Ell, S.R., Gilbert, J.M.: Design of an intraoral device based on permanent magnetic articulography. In: Proceedings of Macau Conference on Engineering, Technology and Applied Science, Macau, China, pp. 1–12 (2015)
28. Leonard, R.G.: A database for speaker-independent digit recognition. In: Proceedings of 9th ICASSP, San Diego, USA, pp. 328–331 (1984)
29. Young, S., Everman, G., Gales, M., Hain, T., Kershaw, D., Liu, X., Moore, G., Odell, J., Ollason, D., Povery, D., Valtchev, V., Woodland, P.: The HTK Book (for HTK Version 3.4.1). Cambridge University Press, Cambridge (2009)
30. Rabiner, L.R.: A tutorial on Hidden Markov Models and selected applications in speech recognition. Proc. IEEE **77**, 257–286 (1989)
31. Maier-Hein, L., Metze, F., Schultz, T., Waibel, A.: Session independent non-audible speech recognition using surface electromyography. In: Proceedings of Automatic Speech Recognition and Understanding Workshop, Cancun, Mexico, pp. 331–336 (2005)
32. Gonzalez, J.A., Cheah, L.A., Bai, J., Ell, S.R., Gilbert, J.M., Moore, R.K., Green, P.D.: Analysis of phonetic similarity in a silent speech interface based on permanent magnetic articulography. In: Proceedings of 15th Interspeech, Singapore, pp. 1018–1022 (2014)
33. Gonzalez, J.A., Cheah, L.A., Gilbert, J.M., Bai, J., Ell, S.R., Green, P.D., Moore, R.K.: Direct speech generation for a silent speech interface based on permanent magnet articulography. In: Proceedings of 9th Biosignals, Lisbon, Portugal, pp. 109–116 (2016)

Bioimaging

SynapCountJ: A Validated Tool for Analyzing Synaptic Densities in Neurons

Gadea Mata[1(✉)], Germán Cuesto[3], Jónathan Heras[1], Miguel Morales[2],
Ana Romero[1], and Julio Rubio[1]

[1] Departamento de Matemáticas y Computación,
Universidad de La Rioja, Logroño, Spain
{gadea.mata,jonathan.heras,ana.romero,julio.rubio}@unirioja.es
[2] Institut de Neurociéncies, Universitat Autònoma de Barcelona, Barcelona, Spain
miguelmorales@spineup.es
[3] Facultad de Ciencias de la Salud, Centro de Investigaciones
Biomédicas de Canarias (CIBICAN) Instituto de Tecnologías Biomédicas (ITB),
San Cristóbal de La Laguna, Spain
germancuesto@gmail.com

Abstract. The quantification of synapses is instrumental to measure
the evolution of synaptic densities of neurons under the effect of some
physiological conditions, neuronal diseases or even drug treatments. However, the manual quantification of synapses is a tedious, error-prone,
time-consuming and subjective task; therefore, reliable tools that might
automate this process are desirable. In this paper, we present Synap-
CountJ, an ImageJ plugin, that can measure synaptic density of individual neurons obtained by immunofluorescence techniques, and also can be
applied for batch processing of neurons that have been obtained in the
same experiment or using the same setting. The procedure to quantify
synapses implemented in SynapCountJ is based on the colocalization of
three images of the same neuron (the neuron marked with two antibody
markers and the structure of the neuron) and is inspired by methods
coming from Computational Algebraic Topology. SynapCountJ provides
a procedure to semi-automatically quantify the number of synapses of
neuron cultures; as a result, the time required for such an analysis is
greatly reduced. The computations performed by SynapCountJ have
been validated by comparing the results with those of a formally verified algorithm (implemented in a different system).

1 Introduction

Synapses are the points of connection between neurons, and they are dynamic
structures subject to a continuous process of formation and elimination. Pathological conditions, such as the Alzheimer disease, have been related to synapse loss associated with memory impairments. Hence, the possibility of changing the number of

---

This work was supported by the Ministerio de Economía y Competitividad projects
[MTM2013-41775-P, MTM2014-54151-P, BFU2010-17537]. G. Mata was also supported by a PhD grant awarded by the University of La Rioja [FPI-UR-13].

A. Fred and H. Gamboa (Eds.): BIOSTEC 2016, CCIS 690, pp. 41–55, 2017.
DOI: 10.1007/978-3-319-54717-6_3

synapses may be an important asset to treat neurological diseases [34]. To this aim, it is necessary to determine the evolution of synaptic densities of neurons under the effect of some physiological conditions, neuronal diseases or even drug treatments.

The procedure to quantify synaptic density of a neuron is usually based on the colocalization between the signals generated by two antibodies [6]. Namely, neuron cultures are permeabilized and treated with two different primary markers (for instance, bassoon and synapsin). These antibodies recognize specifically two presynaptic structures. Then, it is necessary a secondary antibody couple attached to different fluorochromes (for instance red and green; note, that several other combinations of color are possible) making these two synaptic proteins visible under the fluorescence microscope. The two markers are photographed in two gray-scale images; that, in turn, are overlapped using respectively the red and green channels. In the resultant image, the yellow points (colocalization of the code channels) are the candidates to be the synapses.

The final step in the above procedure is the selection of the yellow points that are localized either on the dendrites of the neuron or adjacent to them. Tools like MetaMorph [10] or ImageJ [32]—a Java platform for image processing that can be easily extended by means of plugins—can be used to manually count the number of synapses; however, such a manual quantification is a tedious, time-consuming, error-prone, and subjective task; hence, reliable tools that might automate this process are desirable. In this paper, we present *SynapCountJ*, an ImageJ plugin, that semi-automatically quantifies synapses and synaptic densities in neuron cultures. The program is based on Algebraic Topology techniques and has been validated by comparing some intermediate results with those of Kenzo [12], a (partially) formally verified program.

2   Methodology

SynapCountJ supports two execution modes: individual treatment of a neuron and batch processing—the workflow of both modes is provided in Fig. 1.

2.1   Individual Treatment of a Neuron

The input of SynapCountJ in this execution mode are two images of a neuron marked with two antibodies (an image per antibody), see Fig. 2. SynapCountJ is able to read tiff (a standard format for biological images) and lif files (obtained from Leica confocal microscopes)—the latter requires the Bio-Formats plugin [19]. The following steps are applied to quantify the number of synapses in the given images.

In the first step, from one of the two images, the region of interest (i.e. the dendrites where the quantification of synapses will be performed) is specified using NeuronJ [22]—an ImageJ plugin for tracing elongated image structures. In this way, the background of the image is removed. The result is a file containing the traces of each dendrite of the image.

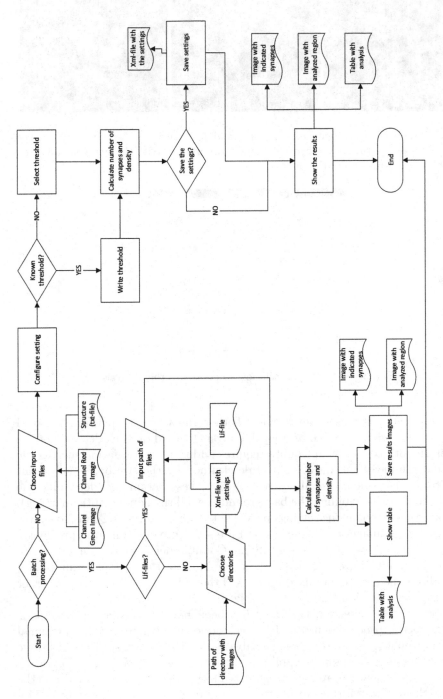

Fig. 1. Workflow of SynapCountJ.

Fig. 2. Neuron with two antibody markers and its structure. *Left.* Neuron marked with the bassoon antibody marker. *Center.* Neuron marked with the synapsin antibody marker. *Right.* Structure of the neuron.

Fig. 3. SynapCountJ window to configure the analysis

Subsequently, the user can decide whether she wants to perform a global analysis of the whole neuron, or a local analysis focused on each dendrite of the neuron. In both cases, SynapCountJ requires additional information such as the scale and the mean thickness (that is determined by the size of the subjacent dendrite) of the region to analyze (see Fig. 3)—these parameters determine the area of the dendrite avoiding the background (i.e. all the non-synaptic marking).

Taking into account the settings provided by the user, SynapCountJ overlaps the two original images of the neuron and the structure of the neuron previously defined. From the resultant image, SynapCountJ identifies the almost white points (the result of green, red, and blue combination) as synaptic candidates, and it allows the user to modify the values of the red and green channels in order to modify the detection threshold (see Fig. 4).

Once that the detection threshold has been fixed, the counting process is started. Such a process is inspired by techniques coming from Computational Algebraic Topology. In spite of being an abstract mathematical subject, Algebraic Topology has been successfully applied in digital image analysis [14, 20, 33]. In our particular case, the white areas are segmented from the overlapped image, and the colors of the resultant image are inverted—obtaining as a result a black-and-white image where the synapses are the black areas. From such an image, the problem

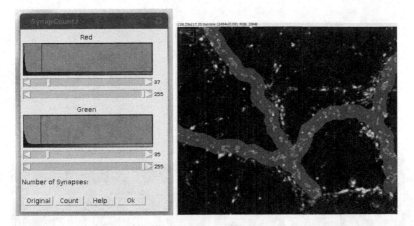

Fig. 4. SynapCountJ window to modify the threshold of the red and green channels. *Left.* Window to fix the threshold of the image. *Right.* Fragment of the neuron image with the synapses indicated as the red areas on the structure of the neuron marked in blue. Moving the scrollbars of left window, the marked areas of the image are changed. (Color figure online)

of quantifying the number of synapses is reduced to compute the homology group in dimension 0 of the image; this corresponds to the computation of the number of connected components of the image. Our algorithm to count synapses can be summarized as presented in Algorithm 1.

Algorithm 1. Counting Synapses in SynapCountJ.

Input: Two images of a neuron marked with two antibodies
Output: Number of synapses of the neuron

1 Create an image with the structure of the neuron using NeuronJ;
2 Overlap the two original images of the neuron and the structure of the neuron;
3 Fix the detection threshold;
4 Segment the white areas of the overlapped image using the fixed threshold;
5 Invert the colours of the segmented image;
6 Count the number of connected components of the image.

Finally, SynapCountJ returns a table with the obtained data (length of dendrites both in pixels and micras, number of synapses, and density of synapses per 100 micron) and two images showing, respectively, the analyzed region and the marked synapses (see Fig. 5).

2.2 Batch Processing

Images obtained from the same biological experiment usually have similar settings; hence, their processing in SynapCountJ will use the same configuration

	Label	Length in pixels	Length in micras	Synapses	Density	Red	Green
1	Tracing N1:	1833.1058	91.6553	71	77.4642	116	164
2	Tracing N2:	867.7840	43.3892	35	80.6652	116	164
3	Tracing N3:	983.5322	49.1766	53	107.7748	116	164
4	Tracing N4:	599.8320	29.9916	41	136.7049	116	164
5	Tracing N5:	437.7388	21.8869	25	114.2234	116	164
6	Tracing N6:	468.8438	23.4422	26	110.9111	116	164
7	Tracing N7:	447.6296	22.3815	31	138.5074	116	164
8	Tracing N8:	574.3691	28.7185	38	132.3191	116	164
9	Tracing N9:	1776.2572	88.8129	69	77.6915	116	164
10	Tracing N10:	1224.7374	61.2369	45	73.4851	116	164
11	Tracing N11:	355.7054	17.7853	26	146.1884	116	164
12	Tracing N12:	905.3750	45.2688	45	99.4063	116	164
13	Total Neuron	10474.9103	523.7455	479	91.4566	116	164

Fig. 5. Results provided by SynapCountJ. *Top.* Table with the results obtained by SynapCountJ. *Bottom Left.* Image with the analyzed region of the neuron. *Bottom Right.* Image with the counted synapses indicated by means of blue crosses.

parameters. In order to deal with this situation, SynapCountJ can be applied for batch processing of several images using a configuration file. It is necessary to study at least one image from experiment to get the optimal settings. The parameters are saved and used to process the set of images from the same experiment.

For batch processing, SynapCountJ reads tiff files organized in folders or a lif file (the kind of files produced by Leica confocal microscopes), and using the configuration file processes the different images. As a result, a table with the information related to each neuron from the batch (the table includes an analysis for both the whole neuron and from each of its dendrites) is obtained. In addition, in the same directory where the lif-file or tiff-files are stored, the plugin saves all the resultant images for each image from experiment (one of them shows the marked synapses and the other one, the region which has been studied).

3   Experimental Results

The original aim of SynapCountJ was the automatic analysis of synaptic density on neurons treated with SB 415286—an organic inhibitor of GSK3, a kinase which inhibition was proposed as a therapy in AD treatment [9]—such a treatment, as it was previously demonstrated, promotes synaptogenesis and spinogenesis in primary cultures of rodent hippocampal neurons and in Drosophyla neurons [7,13]. In this setting, a comparative study has been performed in order to evaluate the results that can be obtained with SynapCountJ.

Primary hippocampal cultures were obtained from P0 rat pups (Sprague-Dawley, strain, Harlan Laboratories Models SL, France). Animals were anesthetized by hypothermia in paper-lined towel over crushed-ice surface during 2–4 min and euthanized by decapitation. Animals were handled and maintained in accordance with the Council Directive guidelines 2010/63EU of the European Parliament. in [23]. Briefly, glass coverslips (12 mm in diameter) were coated with poly-L-lysine and laminin, 100 and 4 μg/ml respectively. Neurons at a 10×104 neurons/cm^2 density were seeded and grown in Neurobasal (Invitrogen, USA) culture medium supplemented with glutamine 0.5 mM, 50 mg/ml penicillin, 50 units/ml streptomycin, 4% FBS and 4% B27 (Invitrogen, CA, USA), as described before in [6]. At days 4, 7 and 14 in culture a 20% of culture medium was replace by fresh medium. Cytosine-D-arabinofuranoside (4 μM) was added to prevent overgrowth of glial cells (day 4).

Synaptic density on hippocampal cultures was identified as previously described in [6]. In short, cultures were rinsed in phosphate buffer saline (PBS) and fixed for 30 min in 4% paraformaldehyde-PBS. Coverslips were incubated overnight in blocking solution with the following antibodies: anti-Bassoon monoclonal mouse antibody (ref. VAM-PS003, Stress Gen, USA) and rabbit polyclonal sera against Synapsin (ref. 2312, Cell Signaling, USA). Samples were incubated with a fluorescence-conjugated secondary antibody in PBS for 30 min. After that, coverslips were washed three times in PBS and mounted using Mowiol (all secondary antibodies from Molecular Probes-Invitrogen, USA). Stack images (pixel size 90 nm with 0.5 μm Z step) were obtained with a Leica SP5 Confocal microscope (40x lens, 1. 3 NA). Percentage of synaptic change is the average of different cultures under the same experimental conditions. As a control, we used sister untreated cultures growing in the same 24 well multi plate.

A total of 13 individual images from three independent cultures has been analyzed. In Fig. 6 we can observe that using a manual method to identify and count synapses, we obtain a mean of 24.12 synapses in control cultures and 16.74 in treated cultures. The results obtained with SynapCountJ are similar, there is a mean of 26.03 synapses in control cultures and 16.50 in the ones which have been treated.

Not with standing the differences in the quantification, in both procedures we obtain almost the same inhibition percentage, a 30.51% manually and 36.61% automatically. This shows the suitability of SynapCountJ to count synapses, meaning a considerably reduction of the time employed in the manual process. Namely, the manual analysis of an image takes approximately 5 min; of a batch, 1 h; and, of a complete study, 4 h. Using SynapCountJ, the time to analyze an image is 30 s; a batch, 2 min; and, a complete study, 6 min.

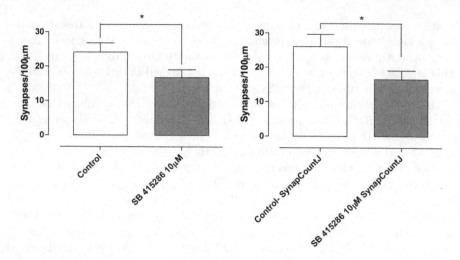

Fig. 6. Quantification of synapses. *Left.* Manual quantification of synapses. *Right.* Quantification of synapses using *SynapCountJ*

4   Scientific Validations of the Computations

Accuracy and reliability are two desirable properties of every software tool, especially in the case of biomedical software. An approach to increase the trust in scientific software is the use of mechanised theorem proving technology to verify the correctness of the programs [2,15]. However, such a formal verification is a challenging task [5]. In our work, we are interested in increasing the reliability of our software; however, due to the difficulty of directly verifying the correctness of our programs, we have followed an indirect approach.

A key component of our algorithm to count synapses is the computation of connected components of a black-and-white image (see Algorithm 1). Such a computation can be performed using two different approaches:

- a direct approach, where the pixels of the image are directly processed; and,
- an indirect approach, where the notion of simplicial complex associated with an image, and techniques from Algebraic Topology (namely, homology groups) are employed to compute the connected components of the image.

The former is efficient and can be easily employed in ImageJ—in fact, it is the one implemented in SynapCountJ—however, its formal verification is a challenging problem. The latter is slower than the former, is difficult to incorporate it into ImageJ; but, it can rely on a previously developed software, the Kenzo system [12], and therefore, it does not require any further development. The formal verification of the Kenzo system is even harder than the verification of the direct approach, but, fortunately, such a task was, at least partially, tackled in the ForMath project [1]—an European project devoted to the development of libraries of formalised mathematics concerning algebra, linear algebra, real number computation, and Algebraic Topology.

In this context, where we have a fast but unverified algorithm, and a slow but verified algorithm, the following strategy can be employed to increase the reliability of the fast version thanks to the verified version. The strategy consists in performing an intensive automated testing checking whether the results obtained with both versions are the same; if that is the case, the reliability of the fast algorithm is increased. In our particular case, we have employed such a strategy to increase the reliability of the computation of connected components of black-and-white images using the fast version implemented in SynapCountJ (the direct approach) thanks to the verified Kenzo system (the indirect approach).

In the rest of this section, we thoroughly explain the two different approaches to compute connected components of a black-and-white image.

4.1   The Direct Approach

The direct approach to compute connected components of a black-and-white image processes directly the pixels of the image by means of an algorithm included in ImageJ which is called *FindMaxima*. This algorithm can be applied to black-and-white, grayscale or color images and determines the local maxima of the image, provided with segmented regions containing all the pixels of the image whose value differs from the corresponding local maxima in less than a chosen threshold. In the case of black-and-white images, the result corresponds to the different connected components.

The algorithm is divided into two steps:

1. First of all, the local maxima of the image are determined, and they are ordered in a decreasing way.
2. Secondly, a filling algorithm is applied for each local maximum to determine its connected region. If a maximum produces a region which was already filled by a previous maximum, the actual local maximum is discarded.

The first step is done by means of a method called *getSortedMaxPoints*. Here, all the pixels in the image are studied comparing them with their adjacent pixels. A pixel is chosen as local maxima if its value is higher than all their adjacent pixels. A threshold is also considered to discard those pixels with value lower than it. The result is an array with the local maxima (with their coordinates) ordered in a decreasing way.

Once the ordered list of local maxima has been obtained, the second step of the algorithm FindMaxima is done by means of a method called *anaylizeAndMark-Maxima*. In this method, a filling algorithm is applied to each local maximum going over the list in a decreasing way. To determine the region associated to a local maximum, an iterative process is applied considering the 8 adjacent pixels to the maximum, selecting those whose difference with the local maximum is lower than a chosen parameter and studying then the adjacent pixels to those selected in the previous step. If a selected pixel is higher than the local maxima then it is stored as the maximum of the region and the previous one is discarded. If it is equal, the new pixel is also stored in order to be able to compute the mean of all the local maxima in the region as we will explain later. The process finishes when all possible adjacent pixels to the previously selected ones have been studied.

Let us observe that when applying the filling algorithm to a local maximum, we could find other maxima (included in the same connected component as the considered one). In that case, the process stops and the second maximum is discarded. Moreover, in case of having several maxima with the same value in a region, the final maximum is computed as the pixel with the same intensity as the local maximum which is closest to the baricenter of all of them.

For a more complete study of the FindMaxima algorithm in ImageJ see [25].

4.2   The Indirect Approach

The indirect approach to compute connected components of a black-and-white image employs the Kenzo system. Kenzo [12] is a Common Lisp system devoted to Algebraic Topology that was developed by Francis Sergeraert. Kenzo has obtained some results not confirmed nor refuted by theoretical or computational means [35], and also has been used to refute some computations obtained by theoretical means [27,28]. Then, the question of Kenzo reliability arose in a natural way, and several works have been focussed on studying the correctness of Kenzo key fragments and algorithms [3,11,18].

The final aim of Kenzo was not the analysis of digital images, but it was extended with a module that tackles such a problem [16]. In particular, such a module computes homological properties, that measure connected components and holes of

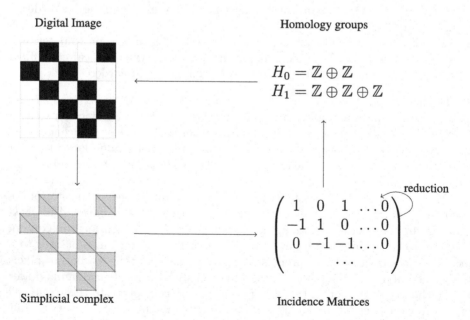

Fig. 7. Workflow to compute homology groups from digital images. The homology groups indicate that the image has two connected components and three holes.

black-and-white images. This Kenzo module for digital images has been employed to validate the results obtained in SynapCountJ using the direct approach.

The Kenzo module for digital images works as follows (see Fig. 7). Given a black-and-white image, a triangulation procedure is employed to obtain a simplicial complex (a generalisation of the notion of graph to higher dimensions)— there are several methods to construct a simplicial complex from a digital image, see [4]. From the simplicial complex, its *boundary (or incidence) matrices* are constructed. Since the size of the boundary matrices coming from biomedical images is too big to be handled directly by Kenzo, a reduction strategy is employed to work with smaller matrices, but preserving their homological properties [29]. From the reduced boundary matrices, homology groups in dimensions 0 and 1 are computed using a diagonalisation process [24]. The homology groups are either null or a direct sum of $\mathbb{Z}$ components, and they should be interpreted as follows: the number of $\mathbb{Z}$ components of the homology groups of dimension 0 and 1 measures respectively the number of connected components and the number of holes of the image. Hence, computing the homology groups associated with a digital image, we can obtain the number of connected components of the image.

The aforementioned workflow to compute homology groups from digital images was fully verified in [17, 26].

5 Discussion

Up to the best of our knowledge, 4 tools have been developed to quantify synapses and measure synaptic density: Green and Red Puncta [36], Puncta Analyzer [37], SynD [31] and SynPAnal [8]—a summary of the general features of these tools can be seen in Table 1. The rest of this section is devoted to compare SynapCountJ with these tools—such a comparison is summarized in Table 2.

Table 1. General features of the analyzed software

Software	Language	Underlying technology	Types of images	Technique for detection
Green and Red Puncta	Java	ImageJ	tiff	Colocalization
Puncta Analyzer	Java	ImageJ2	tiff	Colocalization
SynapCountJ	Java	ImageJ	tiff and lif	Colocalization
SynD	Matlab	Matlab	tiff and lsm	Brightness
SynPAnal	Java		tiff	Brightness

There are two approaches to locate synapses in an RGB image either based on colocalization or brightness. In the former, synapses are identified as the colocalization of bright points in the red and green channels—this is the approach followed by Green and Red Puncta, Puncta Analyzer and SynapCountJ—in the latter, synapses are the bright points of a region of an image—the approach employed

Table 2. Features to quantify synapses and synaptic density of the analyzed software

Software	Detection of dendrites	Threshold	Batch processing	Dendrites length	Density	Export	Save
Green and Red Puncta	Not used	✓					
Puncta Analyzer	Manual ROI	✓				✓	
SynapCountJ	Manual	✓	✓	✓	✓	✓	✓
SynD	Automatic	✓	✓	✓		✓	✓
SynPAnal	Manual	✓		✓		✓	✓

in SynD and SynPAnal. In both approaches, it is necessary a threshold that can be manually adjusted to increase (or decrease) the number of detected synapses; such a functionality is supported by all the tools.

In the quantification of synapses from RGB images, it is instrumental to determine the region of interest (i.e. the dendrites of the neurons where the synapses are located); otherwise, the analysis will not be precise due to noise coming from irrelevant regions or the background of the image—this happens in the Green and Red Puncta tool since it considers the whole image for the analysis. Puncta Analyzer allows the user to fix a rectangle containing the dendrites of the neuron, but this is not completely precise since some regions of the rectangle might contain points considered as synapses that do not belong to the structure of the neuron. SynD is the only software that automatically detects the dendrites of a neuron; however, it can only be applied to neurons with a cell-fill marker, and does not support the analysis from specific regions, such as soma or distal dendrites. SynapCountJ and SynPAnal provide the functionality to manually draw the dendrites of the image; allowing the user to designate the specific areas where quantification is restricted.

The main output produced by all the available tools is the number of synapses of a given image; additionally, SynapCountJ, SynD and SynPAnal provides the length of the dendrites; and, SynapCountJ is the only tool that outputs the synaptic density per micron. All the tools but Green and Red Punctua can export the results to an external file for storage and further processing.

Finally, as we have explained in Subsect. 2.2, images obtained from the same biological experiment usually have similar settings; hence, batch processing might be useful. This functionality is featured by SynapCountJ and SynD, and requires a previous step of saving the configuration of an individual analysis. SynPAnal does not support batch processing, but the configuration of an individual analysis can be saved to be later applied in other individual analysis.

As a summary, SynapCountJ is more complete than the rest of available programs. It can use different types of synaptic markers and can process batch images. Furthermore, a differential feature of SynapCountJ is that it is based on a topological algorithm (namely, computing the number of connected components in a combinatorial structure), allowing us to validate the correctness of our approach by means of formal methods in software engineering.

6 Conclusions and Further Work

SynapCountJ is an ImageJ plugin that provides a semi-automatic procedure to quantify synapses and measure synaptic density from immunofluorescence images obtained from neuron cultures. This plugin has been tested not only with neurons in development, but also with the neuromuscular union of Drosophila; therefore, it can be applied to the study of images that contain two synaptic markers and a determined structure. The results obtained with SynapCountJ are consistent with the results obtained manually; and SynapCountJ dramatically reduces the time required for the quantification of synapses. Moreover, the realiability of Synap-CountJ has been increased by validating some of its computations using the formally verified module for digital images of Kenzo.

As further work, it remains the tasks of improving the usability of the plugin and including post-processing tools to manually edit the obtained results. Additionally, and since the final aim of our project is the complete automation of the whole process, it is necessary a procedure to automatically detect the neuron morphology, and also to automatically fix the threshold for the segmentation of neurons. For such an automation, machine learning techniques like the ones presented in [21] might be employed.

7 Availability and Software Requirements

SynapCountJ is an ImageJ plugin that can be downloaded, together with its documentation, from http://imagejdocu.tudor.lu/doku.php?id=plugin:utilities:synapsescountj:start. SynapCountJ is open source and available for use under the GNU General Public License. This plugin runs within both ImageJ and Fiji [30] and has been tested on Windows, Macintosh and Linux machines.

References

1. Formath: formalisation of mathematics (2010–2013). http://wiki.portal.chalmers.se/cse/pmwiki.php/ForMath/ForMath
2. Amorim, A., et al.: A verified information-flow architecture. In: 41st ACM SIGPLAN-SIGACT Symposium on Principles of Programming Languages (POPL 2014) (2014)
3. Aransay, J., Ballarin, C., Rubio, J.: A mechanized proof of the Basic Perturbation Lemma. J. Autom. Reasoning 40(4), 271–292 (2008)
4. Ayala, R., Domínguez, E., Francés, A., Quintero, A.: Homotopy in digital spaces. Discrete Appl. Math. 125, 3–24 (2003)
5. Benton, N.: Machine Obstructed Proof: how many months can it take to verify 30 assembly instructions? (2006)

6. Cuesto, G., Enriquez-Barreto, L., Caramés, C., et al.: Phosphoinositide-3-kinase activation controls synaptogenesis and spinogenesis in hippocampal neurons. J. Neurosci. **31**(8), 2721–2733 (2011)
7. Cuesto, G., Jordán-Álvarez, S., Enriquez-Barreto, L., et al.: GSK3β inhibition promotes synaptogenesis in Drosophila and mammalian neurons. Plos One **10**(3), e0118475 (2015). doi:10.1371/journal.pone.0118475
8. Danielson, E., Lee, S.H.: SynPAnal: software for rapid quantification of the density and intensity of protein puncta from fluorescence microscopy images of neurons. PLoS ONE **9**(12), e115298 (2014). doi:10.1371/journal.pone.0115298
9. DaRocha-Souto, B., Scotton, T.C., Coma, M., et al.: Brain oligomeric β-amyloid but not total amyloid plaque burden correlates with neuronal loss and astrocyte inflammatory response in amyloid precursor protein/tau transgenic mice. J. Neuropathol. Exp. Neurol. **70**(5), 360–376 (2003)
10. Devices, M.: Metamorph research imaging (2015). http://www.moleculardevices.com/systems/metamorph-research-imaging
11. Domínguez, C., Rubio, J.: Effective homology of bicomplexes, formalized in Coq. Theor. Comput. Sci. **412**, 962–970 (2011)
12. Dousson, X., Rubio, J., Sergeraert, F., Siret, Y.: The Kenzo program. Institut Fourier, Grenoble (1998). https://www-fourier.ujf-grenoble.fr/~sergerar/Kenzo/
13. Franco, B., Bogdanik, L., Bobinnec, Y., et al.: Shaggy, the homolog of glycogen synthase kinase 3, controls neuromuscular junction growth in Drosophila. J. Neurosci. **24**(29), 6573–6577 (2004)
14. González-Díaz, R., Real, P.: On the Cohomology of 3D digital images. Discrete Appl. Math. **147**(2–3), 245–263 (2005)
15. Hales, T.: The Flyspeck Project fact sheet (2005). Project description available at http://code.google.com/p/flyspeck/
16. Heras, J., Pascual, V., Rubio, J.: A certified module to study digital images with the Kenzo system. In: Moreno-Díaz, R., Pichler, F., Quesada-Arencibia, A. (eds.) EUROCAST 2011. LNCS, vol. 6927, pp. 113–120. Springer, Heidelberg (2012)
17. Heras, J., Dénès, M., Mata, G., Mörtberg, A., Poza, M., Siles, V.: Towards a certified computation of homology groups for digital images. In: Ferri, M., Frosini, P., Landi, C., Cerri, A., Fabio, B. (eds.) CTIC 2012. LNCS, vol. 7309, pp. 49–57. Springer, Heidelberg (2012)
18. Lambán, L., Martín-Mateos, F.J., Rubio, J., Ruiz-Reina, J.L.: Verifying the bridge between simplicial topology and algebra: the Eilenberg-Zilber algorithm. Logic J. IGpPL **22**(1), 39–65 (2013)
19. Linkert, M., Rueden, C.T., Allan, C., et al.: Metadata matters: access to image data in the real world. J. Cell Biol. **189**(5), 777–782 (2010)
20. Mata, G., et al.: Zigzag persistent homology for processing neuronal images. Pattern Recogn. Lett. **62**(1), 55–60 (2015)
21. Mata, G., et al.: Automatic detection of neurons in high-content microscope images using machine learning approaches. In: Proceedings of the 13th IEEE International Symposium on Biomedical Imaging (ISBI 2016). IEEE Xplore (2016)
22. Meijering, E., Jacob, M., Sarria, J.C.F., et al.: Design and validation of a tool for neurite tracing and analysis in fluorescence microscopy images. Cytometry Part A **58**(2), 167–176 (2004)
23. Morales, M., Colicos, M.A., Goda, Y.: Actin-dependent regulation of neurotransmitter release at central synapses. Neuron **27**(3), 539–550 (2000)
24. Munkres, J.R.: Elements of Algebraic Topology. Addison-Wesley, Reading (1984)
25. de Greñu de Pedro, J.D.: Análisis Matemático de rutinas de procesamiento de imágenes digitales en Fiji/ImageJ. Technical report, Universidad de La Rioja (2014)

26. Poza, M., Domínguez, C., Heras, J., Rubio, J.: A certified reduction strategy for homological image processing. ACM Trans. Comput. Logic 15(3), 23 (2014)
27. Romero, A., Heras, J., Rubio, J., Sergeraert, F.: Defining and computing persistent Z-homology in the general case. CoRR abs/1403.7086 (2014)
28. Romero, A., Rubio, J.: Homotopy groups of suspended classifying spaces: an experimental approach. Math. Comput. 82, 2237–2244 (2013)
29. Romero, A., Sergeraert, F.: Discrete Vector Fields and Fundamental Algebraic Topology (2010). http://arxiv.org/abs/1005.5685v1
30. Schindelin, J., Argand-Carreras, I., Frise, E., et al.: Fiji: an open-source platform for biological-image analysis. Nat. Methods 9(7), 676–682 (2012)
31. Schmitz, S.K., Hjorth, J.J.J., Joemail, R.M.S., et al.: Automated analysis of neuronal morphology, synapse number and synaptic recruitment. J. Neurosci. Methods 195(2), 185–193 (2011)
32. Schneider, C., Rasband, W., Eliceiri, K.: NIH Image to ImageJ. Nat. Methods 9, 671–675 (2012)
33. Ségonne, F., Grimson, E., Fischl, B.: Topological correction of subcortical segmentation. In: Ellis, R.E., Peters, T.M. (eds.) MICCAI 2003. LNCS, vol. 2879, pp. 695–702. Springer, Heidelberg (2003)
34. Selkoe, D.J.: Alzheimer's diseases is a synaptic failure. Science 298(5594), 789–791 (2002)
35. Sergeraert, F.: Effective homology, a survey. Technical report, Institut Fourier (1992). http://www-fourier.ujf-grenoble.fr/sergerar/Papers/Survey.pdf
36. Shiwarski, D.J., Dagda, R.D., Chu, C.T.: Green and red puncta colocalization (2014). http://imagejdocu.tudor.lu/doku.php?id=plugin:analysis:colocalization_analysis_macro_for_red_and_green_puncta:start
37. Wark, B.: Puncta analyzer v2.0 (2013). https://github.com/physion/puncta-analyzer

4DCT-Derived Ventilation Distribution Reproducibility Over Time

Geoffrey G. Zhang$^{(\boxtimes)}$, Kujtim Latifi, Vladimir Feygelman,
Thomas J. Dilling, and Eduardo G. Moros

Radiation Oncology, Moffitt Cancer Center, Tampa, FL, USA
Geoffrey.zhang@moffitt.org

Abstract. Deriving lung ventilation distribution from 4-dimensional CT (4DCT) using deformable image registration (DIR) is a recent technical development. In this study, we evaluated the serial reproducibility of ventilation data derived from two separate 4DCT data sets, collected at different time points. A total of 33 lung cancer patients were retrospectively analysed. All patients had two stereotactic body radiotherapy treatment courses for lung cancer. Seven patients were excluded due to artifacts in the 4DCT data sets. The ventilation distributions for each patient were calculated using the two sets of planning 4DCT data. The deformation matrices between the expiration and inspiration phases generated by DIR were used to produce ventilation distributions using the ΔV method. Ventilation was analysed in the lung regions that received less than 1 Gy and in the contralateral lung respectively. For the 26 cases, the median Spearman correlation coefficient value was 0.31 (range 0.18 to 0.52, p value < 0.01 for all cases) in the regions of <1 Gy and 0.32 (range 0.19 to 0.51) in the contralateral lungs. The median Dice similarity coefficient value between the upper 30% ventilation regions of the two sets was 0.53 (range 0.40 to 0.64) in the regions of <1 Gy and 0.53 (range 0.44 to 0.70) in the contralateral lungs. We conclude that the two ventilation data sets in each case correlated and the reproducibility over time was fair.

Keywords: Ventilation · Deformable image registration · 4DCT · Reproducibility · Lung cancer

1 Introduction

Perfusion and ventilation can be used to characterize lung function. Clinically, ventilation imaging is mostly performed using SPECT [1] or PET [2]. Deriving lung ventilation distribution from 4-dimensional CT (4DCT) using deformable image registration (DIR) is a recent technical development [3–5]. One of the advantages of the new technique is its higher resolution compared to the nuclear medicine scans [6]. Several studies have shown that this technique agrees reasonably well with other established techniques such as SPECT, Xenon-enhanced dynamic CT and PET [7–10]. Xenon-enhanced dynamic CT can achieve similar resolution. However, because of the dynamic scanning, only a part of the lung can be covered in a ventilation scan, and the technique itself is rather complicated compared to 4DCT [7, 10]. In principle, using the 4DCT/DIR technique, one could determine regions of

© Springer International Publishing AG 2017
A. Fred and H. Gamboa (Eds.): BIOSTEC 2016, CCIS 690, pp. 56–66, 2017.
DOI: 10.1007/978-3-319-54717-6_4

high lung ventilation in thoracic cancer patients and use these regions as avoidance structures in radiotherapy treatment planning, without the need for an additional imaging procedure, since 4DCT is already widely used in thoracic cancer radiotherapy treatment planning [11]. Therefore, 4DCT ventilation represents a relatively simple and low cost way to further personalized radiation therapy treatment planning. Because of these advantages, this new ventilation calculation method using 4DCT has been clinically applied in lung disease detection [12], radiotherapy treatment planning studies [11, 13] and assessment of radiotherapy response [14].

The reproducibility of ventilation of 4DCT technique has been studied by different groups [15–17]. However, short time intervals and small study sizes were the shortcomings of those reports. In this study, we evaluated on a larger cohort of patients the serial reproducibility of ventilation data derived from two separate 4DCT data sets in the same patient, collected at different time points separated by months.

2   Materials and Methods

2.1   Patient Data

A total of 33 lung cancer patients were retrospectively analysed following a protocol approved by our institutional review board. All patients had two lung cancer stereotactic body radiotherapy treatment courses (different isocenters) with a median interval between them of 9.6 months (range: 0.7–39 months). Separate treatment planning 4DCT datasets were acquired for each treatment course. Seven patients were excluded due to obvious mushroom artifacts in the 4DCT datasets, thus 26 valid cases were included in the data analysis.

2.2   Deformable Image Registration

Based on a previous DIR selection study, [18] the Diffeomorphic Morphons (DM) method [19] was used to register the expiration and inspiration phases on each 4DCT dataset. The same DIR method was applied to the expiration phases of the two 4DCT data sets, to map the second ventilation data set to the first one for the comparison.

2.3   Ventilation

The ventilation distributions in the lungs for each patient were calculated using the two 4DCT datasets, with the gross tumor volumes excluded. Blood vessels in the lungs were also excluded by applying an intensity threshold in the ventilation calculation. The resulting deformation matrices were used to generate ventilation distributions using the ΔV method, which is a direct geometrical calculation of the volume change [3, 20]. In the expiration phase of a 4DCT dataset, each voxel is a cuboid defined by 8 vertices. In the inspiration phase, this cuboid is changed into a 12-face polyhedron which is still comprised of the corresponding 8 vertices that can be determined by the DIR calculation. Any hexahedron or a 12-face polyhedron can be divided into 6 tetrahedrons. The volume of a tetrahedron can be calculated using

$$V = (b - a) \cdot [(c - a) \times (d - a)]/6, \tag{1}$$

where a, b, c, d are the vertices' coordinates of the tetrahedron expressed as vectors.

Ventilation is defined as

$$P = \Delta V / V_{ex}, \tag{2}$$

where ΔV is the volume change between expiration and inspiration and V_{ex} is the initial volume at expiration [21].

In each case, the second set of ventilation distribution was mapped onto the first one using DM DIR between the expiration phases of the two 4DCT sets, and the two ventilation distributions were normalized [22] and compared. Since radiation dose can alter regional lung ventilation [23], two approaches were used to avoid high dose regions and thus to prevent introduction of systemic error in the ventilation calculations. In the first one, the high dose regions from the 1st treatment plan were excluded. We chose to analyse lung regions that received less than 1 Gy of radiation dose to minimize the effect of radiation changes on ventilation estimates since it has been established that large dose of radiation reduce ventilation. In the second approach, only the contralateral lung of the 1st treatment was included in the data analysis, assuming that the contralateral lung received acceptable dose. When the contralateral lung was chosen for data analysis, two more cases were excluded, since in one case only a single lung was functional, and in the other one both lungs were initially treated.

As DIR may introduce errors due to artifacts and noise in CT images, subsequent errors may be introduced into ventilation distributions [24]. Smoothing the ventilation distributions may reduce such errors. The calculated ventilation distributions were thus smoothed with a $9 \times 9 \times 9$ mm^3 average filter prior to analysis.

2.4 Correlation and Reproducibility

The voxel-wise Spearman correlation coefficient (SCC) between the ventilation data sets was calculated. The absolute ventilation data (without normalization) were used in the SCC analysis. The Dice similarity coefficient (DSC) [25] was also calculated between the upper 30%, 50% and 70% ventilation regions of the two sets:

$$DSC(A, B) = \frac{2 \times |A \cap B|}{|A| + |B|}, \tag{3}$$

where A and B are the two involved volumes.

Concordance correlation coefficient (CCC) [26] was calculated for the SCC values between the two low dose region analysis approaches, <1 Gy region and contralateral lung, to estimate the correlation between the two approaches.

3   Results

Figure 1 shows a representative dose distribution.

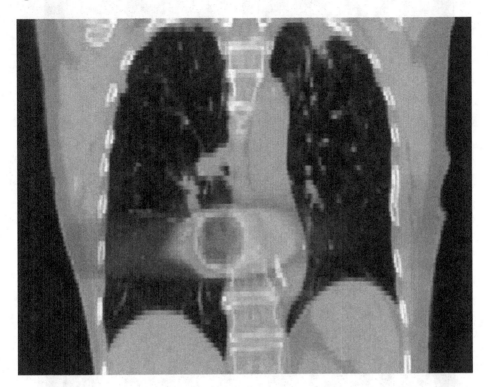

Fig. 1.   Dose distribution for a representative case.

Figure 2 shows the ventilation distributions for the corresponding case in Fig. 1.

For 26 cases, the median SCC value for lung regions with dose less than 1 Gy was 0.31 (range 0.18 to 0.52; original in Fig. 3(A)). The median DSC value for the ventilation regions of upper 70% was 0.75 (range 0.71 to 0.81; original in Fig. 3(B)).

After the ventilation distributions were smoothed with a $9 \times 9 \times 9$ mm^3 average filter, the SCC and DSC improved, with median values of 0.44 (smooth in Fig. 3(A)) and 0.77 (smooth in Fig. 3(B)), respectively.

For the valid 24 cases, the median SCC value for the contralateral lung was 0.32 (range 0.19 to 0.51), p < 0.01 for all cases; original in Fig. 4(A)). The median DSC value for the ventilation regions of upper 70% was 0.76 (range 0.73 to 0.80; original in Fig. 4(B)). After smoothing, the median values of SCC and DSC were improved to 0.45 and 0.78, respectively (range 0.27 to 0.72 and 0.74 to 0.85, respectively, smooth in Fig. 4).

Table 1 lists the median, minimum and maximum DSC values between the 1st and 2nd ventilation data sets for ventilation regions of upper 30%, 50%, 70%, between 30% and 70% for the original and smoothed data sets. The improvement of the smoothed data

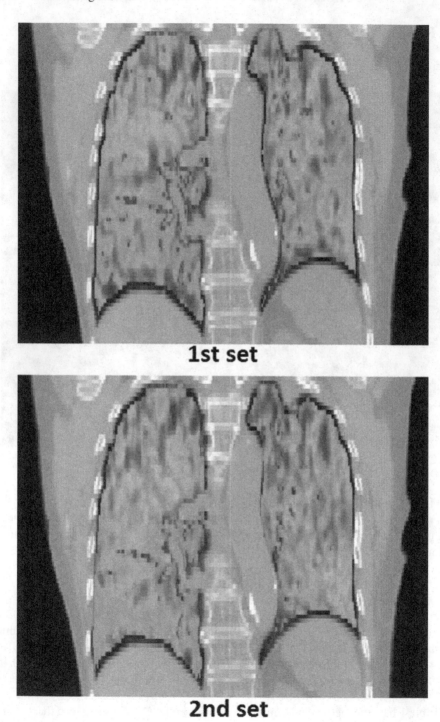

Fig. 2. Ventilation distributions for a representative case.

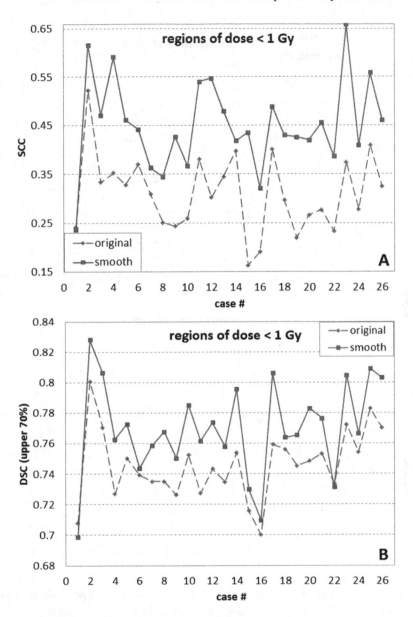

Fig. 3. (A) SCC and (B) DSC in the regions of <1 Gy in lungs for all cases. Blue points are the original data points, while the red ones are after smoothing (p < 0.001). (Color figure online)

compared to the original data was statistically significant based on the paired Student's t-test (p < 0.001 for all the corresponding data pairs). The contralateral lung data show better DSC values for all ventilation regions compared to the <1 Gy data. However, the Student's t-test showed that the differences were not statistically significant (p values were mostly greater than 0.05, ranged 0.007 to 0.885).

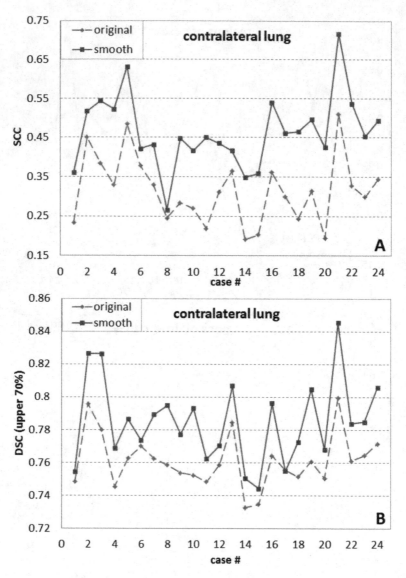

Fig. 4. (A) SCC and (B) DSC in the contralateral lungs for all cases. Blue points are the original data points, while the red ones are after smoothing (p < 0.001). (Color figure online)

The SCC analysis showed the similar trend: higher in the contralateral lung data (median 0.45, range 0.27–0.72) than that in the <1 Gy data (median 0.44, range 0.24–0.66) in the smoothed data comparison. The paired Student's t-test performed on the SCC data sets gave the same results: differences were statistically significant between original and smoothed data sets (p < 0.001) but not significant between contralateral lung and <1 Gy regions (p = 0.400 for the smoothed data, 0.714 for the original data). Figure 5 shows the SCC value comparison between the smoothed datasets for <1 Gy region and contralateral lung (p = 0.714). The two data sets visually demonstrate similar trends. Similar similarity can also be found for the SCC comparison of the original datasets. Weak correlation was found for the SCC values between the two analysis approaches, <1 Gy region and contralateral lung, with the CCC value being around 0.7.

Table 1. DSC values between 1st and 2nd ventilation data sets for various ventilation regions.

	Upper 70%				Upper 50%				Upper 30%				30%–70%			
	<1 Gy		Contral lung		<1 Gy		Contral lung		<1 Gy		Contral lung		<1 Gy		Contral lung	
	Orig*	Smth*	Orig	Smth	Orig	Smth	Orig	Smth	Orig	Smth	Orig	Smth	Orig	Smth	Orig	Smth
Median	0.75	0.77	0.76	0.78	0.60	0.65	0.61	0.66	0.45	0.53	0.46	0.53	0.44	0.47	0.45	0.47
Min	0.70	0.70	0.73	0.74	0.54	0.55	0.57	0.60	0.38	0.40	0.40	0.44	0.40	0.38	0.42	0.40
Max	0.80	0.83	0.80	0.85	0.68	0.74	0.68	0.76	0.55	0.64	0.58	0.70	0.50	0.55	0.52	0.59

*Orig = Original, comparison is between the original data sets; *Smth = Smooth, comparison is between the smoothed data sets.

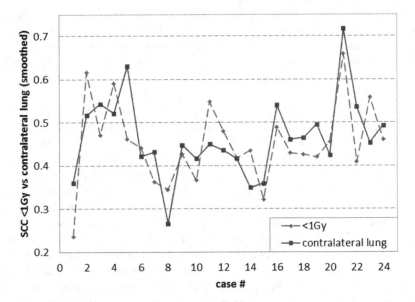

Fig. 5. SCC comparison between the two analysis approaches. The difference between these two datasets was not statistically significant (p = 0.714).

4 Discussion

Mushroom artifacts often appear in 4DCT data due to irregular diaphragmatic motion. This imaging artifact introduces errors in DIR and subsequent errors in the derived ventilation distributions. Because of these errors, the SCC value could be very low, close to 0. This was the reason why 7 cases were excluded from the analysis. Besides the obvious mushroom motion artifacts, there must be other motion artifacts which are not noticeable but may still introduce DIR errors, but at a lower scale.

The DIR errors due to noise in CT images are the other major concern in ventilation calculation using DIR on 4DCT [24]. Smoothing can reduce the effect of such errors as demonstrated in Figs. 3 and 4. Improving the quality of 4DCT should improve the accuracy of ventilation calculation using this technique, and the reproducibility may be higher as a consequence.

In this study, post-treatment ventilation was mapped to pre-treatment ventilation using DIR. This DIR application may introduce additional errors in the final results.

The contralateral lung data always gave higher similarity compared to data of regions <1 Gy dose as demonstrated in Table 1. Because of the ventilation deterioration in high dose regions, the relative ventilation distribution in the whole lungs mostly would change and thus the similarity of the various sub-regions would be deteriorated. For the contralateral lung, the deterioration due to high dose is much less. The relative ventilation distribution in the contralateral lung as the whole system would not change as much, thus the similarity of the sub-regions would be higher. This trend was similar in the SCC data analysis because of the same cause.

Ventilation distribution may vary due to body positioning [27, 28]. In this study, all the patients were in supine position when the 4DCT scans were performed, thus the difference due to body positioning should be minimal. However, slight ventilation variation may still exist due to variation in breath maneuver [16, 17, 29]. There was an interval of a few months between the 1st and 2nd 4DCT scans for each patient. Variation in breath maneuver may exist between the two scans, which could reduce the reproducibility of the ventilation distributions. Compared to the previous reproducibility studies [16, 17], with a larger cohort, the correlation and reproducibility values were slightly lower in this study. The possible causes for the lower values are 1. Longer intervals were involved in this study; 2. Lung cancer patients' clinical data were used in this study; 3. There was a course of radiation therapy between the two data sets in each case and possible other therapies. High radiation dose delivered in the radiotherapy reduces ventilation [23]. The ventilation change due to radiotherapy was minimized by using only low dose regions or contralateral lung data in the comparison. However, because of the long intervals and the healing/repairing process during the interval, physiological changes due to the radiotherapy course and other therapies may still exist even in the low dose regions. 4. A new lesion was present in the 2nd ventilation data set in each case. Because lung cancer can impair the lung function. Although the new gross tumor volume itself in each case was excluded from the ventilation comparison by the thresholding method, the potentially affected surrounding lung volumes were still included. Despite these factors, the correlation and reproducibility values in this study were good.

5   Conclusions

Based on the SCC and DSC values, we conclude that the two ventilation data sets in each case correlated and the reproducibility over time was fair when there were no obvious artifacts in the 4DCT. Ventilation data smoothing can reduce errors introduced in the DIR and thus improve the reproducibility. High quality 4DCT is essential for good reproducibility in ventilation distributions.

References

1. Harris, B., Bailey, D., Miles, S., Bailey, E., Rogers, K., et al.: Objective analysis of tomographic ventilation–perfusion scintigraphy in pulmonary embolism. Am. J. Resp. Crit. Care Med. **175**, 1173–1180 (2007)
2. Melo, M.F.V., Layfield, D., Harris, R.S., O'Neill, K., Musch, G., et al.: Quantification of regional ventilation-perfusion ratios with PET. J. Nucl. Med. **44**, 1982–1991 (2003)
3. Zhang, G.G., Huang, T.C., Dilling, T., Stevens, C., Forster, K.M.: Derivation of high-resolution pulmonary ventilation using local volume change in four-dimensional CT data. In: Dössel, O., Schlegel, W.C. (eds.) World Congress on Medical Physics and Biomedical Engineering, pp. 1834–1837. Springer, Heidelberg (2009)
4. Guerrero, T., Sanders, K., Noyola-Martinez, J., Castillo, E., Zhang, Y., et al.: Quantification of regional ventilation from treatment planning CT. Int. J. Radiat. Oncol. Biol. Phys. **62**, 630–634 (2005)
5. Reinhardt, J.M., Ding, K., Cao, K., Christensen, G.E., Hoffman, E.A., et al.: Registration-based estimates of local lung tissue expansion compared to xenon CT measures of specific ventilation. Med. Image Anal. **12**, 752–763 (2008)
6. Zhang, G., Dilling, T.J., Stevens, C.W., Forster, K.M.: Functional lung imaging in thoracic cancer radiotherapy. Cancer Control **15**, 112–119 (2008)
7. Ding, K., Cao, K., Fuld, M.K., Du, K., Christensen, G.E., et al.: Comparison of image registration based measures of regional lung ventilation from dynamic spiral CT with Xe-CT. Med. Phys. **39**, 5084–5098 (2012)
8. Kipritidis, J., Siva, S., Hofman, M.S., Callahan, J., Hicks, R.J., et al.: Validating and improving CT ventilation imaging by correlating with ventilation 4D-PET/CT using 68 Ga-labeled nanoparticles. Med. Phys. **41**, 011910 (2014)
9. Yamamoto, T., Kabus, S., Lorenz, C., Mittra, E., Hong, J.C., et al.: Pulmonary ventilation imaging based on 4-dimensional computed tomography: comparison with pulmonary function tests and SPECT ventilation images. Int. J. Radiat. Oncol. Biol. Phys. **90**, 414–422 (2014)
10. Zhang, G.G., Latifi, K., Du, K., Reinhardt, J.M., Christensen, G.E., et al.: Evaluation of the ΔV 4D CT ventilation calculation method using in vivo Xenon CT ventilation data and comparison to other methods. J. Appl. Clin. Med. Phys. **17**, 550–560 (2016)
11. Huang, T.-C., Hsiao, C.-Y., Chien, C.-R., Liang, J.-A., Shih, T.-C., et al.: IMRT treatment plans and functional planning with functional lung imaging from 4D-CT for thoracic cancer patients. Radiat. Oncol. **8**, 3 (2013)
12. Castillo, R., Castillo, E., McCurdy, M., Gomez, D.R., Block, A.M., et al.: Spatial correspondence of 4D CT ventilation and SPECT pulmonary perfusion defects in patients with malignant airway stenosis. Phys. Med. Biol. **57**, 1855–1871 (2012)
13. Siva, S., Thomas, R., Callahan, J., Hardcastle, N., Pham, D., et al.: High-resolution pulmonary ventilation and perfusion PET/CT allows for functionally adapted intensity modulated radiotherapy in lung cancer. Radiother. Oncol. **115**, 157–162 (2015)
14. Ding, K., Bayouth, J.E., Buatti, J.M., Christensen, G.E., Reinhardt, J.M.: 4DCT-based measurement of changes in pulmonary function following a course of radiation therapy. Med. Phys. **37**, 1261–1272 (2010)
15. Du, K., Bayouth, J.E., Cao, K., Christensen, G.E., Ding, K., et al.: Reproducibility of registration-based measures of lung tissue expansion. Med. Phys. **39**, 1595–1608 (2012)
16. Du, K., Bayouth, J.E., Ding, K., Christensen, G.E., Cao, K., et al.: Reproducibility of intensity-based estimates of lung ventilation. Med. Phys. **40**, 063504 (2013)

17. Yamamoto, T., Kabus, S., von Berg, J., Lorenz, C., Chung, M.P., et al.: Reproducibility of four-dimensional computed tomography-based lung ventilation imaging. Acad. Radiol. **19**, 1554–1565 (2012)
18. Latifi, K., Zhang, G., Stawicki, M., van Elmpt, W., Dekker, A., et al.: Validation of three deformable image registration algorithms for the thorax. J. Appl. Clin. Med. Phys. **14**, 19–30 (2013)
19. Janssens, G., de Xivry, J.O., Fekkes, S., Dekker, A., Macq, B., et al.: Evaluation of nonrigid registration models for interfraction dose accumulation in radiotherapy. Med. Phys. **36**, 4268–4276 (2009)
20. Zhang, G., Huang, T.-C., Dilling, T., Stevens, C., Forster, K.: Comments on 'Ventilation from four-dimensional computed tomography: density versus Jacobian methods'. Phys. Med. Biol. **56**, 3445–3446 (2011)
21. Simon, B.A.: Non-invasive imaging of regional lung function using X-ray computed tomography. J. Clin. Monit. Comput. **16**, 433–442 (2000)
22. Latifi, K., Feygelman, V., Moros, E.G., Dilling, T.J., Stevens, C.W., et al.: Normalization of ventilation data from 4D-CT to facilitate comparison between datasets acquired at different times. PLoS ONE **8**, e84083 (2013)
23. Latifi, K., Dilling, T., Feygelman, V., Moros, E., Stevens, C., et al.: Impact of dose on lung ventilation change calculated from 4D-CT using deformable image registration in lung cancer patients treated with SBRT. J. Radiat. Oncol. **4**, 265–270 (2015)
24. Latifi, K., Huang, T.-C., Feygelman, V., Budzevich, M.M., Moros, E.G., et al.: Effects of quantum noise in 4D-CT on deformable image registration and derived ventilation data. Phys. Med. Biol. **58**, 7661–7672 (2013)
25. Dice, L.R.: Measures of the amount of ecologic association between species. Ecology **26**, 297–302 (1945)
26. Lawrence, I.K.L.: A concordance correlation coefficient to evaluate reproducibility. Biometrics **45**, 255–268 (1989)
27. Kaneko, K., Milic-Emili, J., Dolovich, M.B., Dawson, A., Bates, D.V.: Regional distribution of ventilation and perfusion as a function of body position. J. Appl. Physiol. **21**, 767–777 (1966)
28. Yang, Q.H., Kaplowitz, M.R., Lai-Fook, S.J.: Regional variations in lung expansion in rabbits: prone vs. supine positions. J. Appl. Physiol. **67**, 1371–1376 (1989)
29. Roussos, C.S., Fixley, M., Genest, J., Cosio, M., Kelly, S., et al.: Voluntary factors influencing the distribution of inspired gas. Am. Rev. Respir. Dis. **116**, 457–467 (1977)

Modelling Strategies for the Advanced Design of Polymeric Orthodontic Aligners

Sandro Barone, Alessandro Paoli[✉], Armando Viviano Razionale,
and Roberto Savignano

Department of Civil and Industrial Engineering,
University of Pisa, Largo Lucio Lazzarino 1, 56126 Pisa, Italy
{s.barone,a.paoli,a.razionale}@ing.unipi.it,
roberto.savignano@for.unipi.it

Abstract. In the last decade, orthodontic removable thermoplastic aligners have become a common alternative to conventional fixed brackets and wires. However, the wide spread of this typology of orthodontic treatment was not followed by an adequate scientific investigation about its biomechanical effects onto the teeth. In the present work, a patient-specific framework has been developed with the aim of simulating orthodontic tooth movements by using plastic aligners. A maxillary and a mandibular dental arch were reconstructed by combining optical and radiographic imaging methods. A Finite Element (FE) model was then created to analyze two different aligner configurations. In particular, the effect of a non-uniform aligner's thickness and of a customized initial offset between the aligner and the patient dentition were studied. The force-moment systems delivered by the aligner to a mandibular central incisor during labiolingual tipping, and to a maxillary central incisor during rotation were analyzed and discussed.

Keywords: Orthodontic aligner design · Finite element model · Aligner thickness · Customized aligner shape

1 Introduction

Orthodontic devices are designed to provide the desired tooth arrangement by eliciting tooth movements. Traditionally, orthodontic systems exploit fixed appliances based on the combination of archwires and dental brackets, which deliver forces to any tooth in the arch. In the last decades, the interest for adult orthodontic corrections raised the requirement of aesthetic alternatives to conventional fixed devices since adult patients are very concerned with their appearance during the orthodontic treatment. For this reason, research activities in the orthodontic field focused not only on the effectiveness of the appliances on correcting tooth position, but also on the fulfilment of comfort and aesthetical issues during the treatment. In this context, the use of transparent tooth correction systems is becoming common for minimally invasive treatments. In particular, treatments based on clear removable thermoplastic appliances (aligners) are increasingly used [1]. This system consists of a set of thermoformed templates, made of transparent thermoplastic material, which are sequentially worn by the patient. The orthodontic three-dimensional force-moment system on each tooth is generated

© Springer International Publishing AG 2017
A. Fred and H. Gamboa (Eds.): BIOSTEC 2016, CCIS 690, pp. 67–83, 2017.
DOI: 10.1007/978-3-319-54717-6_5

by a pre-determined geometrical mismatch between the aligner shape and the dentition geometry. This condition is determined by using virtual 3D models of the patient's dentition and computer-aided design (CAD) methodologies [2]. Each single template, which corresponds to the new desired tooth position, is programmed to perform only a small part of the complete tooth movement. Therefore, a full treatment consists of multiple templates with varying shapes from the initial anatomical geometry to the target tooth position. The possibility to simulate and identify appropriate moment-to-force (M:F) ratios delivered to the target tooth is a key issue in order to predict and control tooth movements. At this purpose, the Finite Element Analysis (FEA) is one of the most used tools to evaluate the effectiveness of dental devices and has been widely used in dentistry since the 70's [3].

In this paper, a patient-specific framework has been developed in order to make feasible a customized simulation of orthodontic tooth movements by using thermoplastic aligners. A Finite Element (FE) model is created to design optimized appliances leading to more efficient orthodontic treatments. Even if the use of aligners is becoming an effective solution to treat malocclusion conditions, few attempts have been made to develop FE models describing the aligner's behavior in delivering forces [4–6].

The ability to make an accurate treatment prediction has long been a challenge for orthodontists, since tooth movements with aligners are more complex with respect to fixed appliances due to the lack of specific points of force application. Many parameters are certainly involved in determining the clinical outcome: tooth anatomy, aligner's material properties, amount of mismatch between aligner and dentition geometries, slipping motions between contact shapes. In particular, the aligner's thickness has demonstrated to have a great influence on the magnitude of the force produced by the appliance [7, 8]. In this work, the authors have used the developed FE model to study the influence of different aligner's design approaches on the treatment outcomes. First, the effect of non-uniform aligner's thickness has been investigated with respect to the amount and quality of the desired tooth movement. Then a different approach has been investigated by considering the aligner with a uniform thickness but introducing an initial non-uniform penetration between aligner and dentition geometries. The results of the two different aligner's modelling strategies have been compared by analyzing the force-moment system delivered to a mandibular central incisor during labiolingual tipping and to a maxillary central incisor during rotation.

2   Materials and Methods

The patient anatomical geometries as well as the aligner shape have been reconstructed by computer-aided scanning and digital imaging techniques.

In particular, multi-source data are used to obtain tooth anatomies including crown and root shapes. Ideal tooth movements can be achieved through orthodontic appliances that are designed by taking into account not only tooth crowns but also root geometrical features. For this reason, in this work, accurate crown geometries, obtained from high-precision optically scanned data, are merged with approximate representations of root geometries, which are derived from a raw and fast segmentation of Cone Beam Computed Tomography (CBCT) data.

The aligner geometry has then been modelled by exploiting CAD tools in order to create a layer, which closely mates the teeth crown surfaces except for the area corresponding to the tooth to be moved. In this region, a penetration between crown and aligner surfaces is introduced to generate the loading condition. In particular, the influence of both thickness and shape of the aligner has been investigated in order to optimize the effectiveness of the orthodontic treatment and reduce the patient discomfort. Figure 1 summarizes the overall framework.

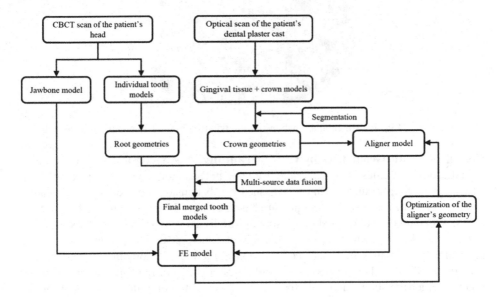

Fig. 1. Overall workflow.

2.1 Creation of the Patient Anatomical Model

The patient's anatomical model, composed of alveolar bone, teeth and periodontal ligaments (PDL), is obtained by exploiting information derived from a CBCT patient's scan and an optical scan of the relative dental plaster cast. The CBCT scan is used to obtain complete geometries of each individual tooth along with its relative spatial arrangement within the jawbone. An optical scanner, based on a coded structured light approach, has been used to acquire the plaster model created from the patient's mouth impression [9]. The aim of the optical scanning procedure is to reconstruct an accurate digital model composed of tooth crowns.

CBCT volumetric data are used to reconstruct the jawbone structure as well as the complete and individual tooth geometries. A CBCT scan yields a stack of slices corresponding to cross-sections through a maxillofacial volumetric region. CBCT data are stored in a sequence of Digital Imaging and Communications in Medicine (DICOM) images. An imaging slice is a 2D matrix of grey intensity values representing the x-ray attenuation of different anatomical tissues.

The three-dimensional model of the jawbone has been obtained by exploiting tools provided by an open-source software for medical image analysis [10]. A triangular mesh of the isosurface representing the bone shape (Fig. 2) has been obtained by segmenting the volumetric CBCT data set with a specific grey intensity value (isovalue).

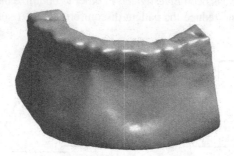

Fig. 2. Mandibular jawbone geometry used in the FE model.

Complete Tooth Geometries by CBCT Scanning. The reconstruction of the individual tooth anatomies is not so straightforward because tooth root regions cannot be easily separated from surrounding bone tissue by only considering pixel's grey-intensity values. Most of the existing techniques are based on slice-by-slice segmentation procedures, which involve the digital processing of hundreds of slices in order to reconstruct three-dimensional geometries, thus resulting in time-consuming procedures. In this paper, DICOM images are processed by adopting the methodology introduced in [11]. This method is based on processing a small number (four) of multi-planar reformation images, which are obtained for each tooth on the basis of anatomy-driven considerations. The reformation images greatly enhance the clearness of the target tooth contours, which are then extracted and used to automatically model the overall 3D tooth shape through a B-spline representation.

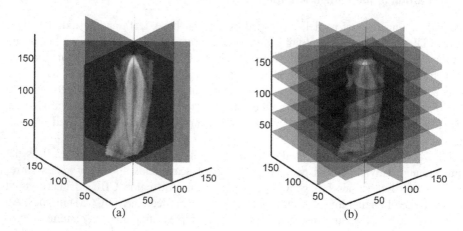

Fig. 3. (a) Four reference planar sections along with the 2D tooth contours, (b) B-spline curves computed for the transverse slices.

Practically, four reference planar sections are automatically extracted as passing from the tooth axis and oriented along the labiolingual direction, the mesiodistal direction and the two directions disposed at 45° with respect to these two meaningful clinical views. These reference sections are used to outline the tooth by interactively tracing four different 2D tooth contours (C_i) as shown in Fig. 3a. The four contours are used to automatically extract a B-spline curve. Each slice perpendicular to the tooth axis (transverse slice) intersect the C_i contours in eight points that are used as control points to compute a parametric B-spline curve of degree 2 (Fig. 3b). For each slice, 100 points are evaluated on the B-spline curve in order to obtain a point cloud representing the overall tooth shape. For further details, the reader can refer to [11].

Figure 4a shows, as example, the point clouds relative to the incisors, canine and premolar teeth of the inferior arch used in the present work. Figure 4b shows the respective StL models obtained by a tessellation of the respective point clouds.

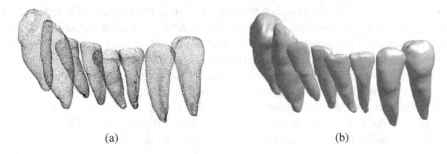

(a)   (b)

Fig. 4. Incisor, canine and premolar tooth models of an inferior arch as obtained by segmenting CBCT data. (a) Point clouds, (b) StL models.

The greatest benefit of this methodology consists in providing reliable approximations of individual tooth roots, by interactively contouring a few significant images created from the whole CBCT data set. The processing time is greatly reduced with respect to standard cumbersome slice-by-slice methods usually proposed within medical imaging software. However, the accuracies obtained for crown geometries, especially for multi-cusped shapes, cannot be considered adequate to simulate orthodontic treatments based on the use of customized appliances.

Crown Geometries by Optical Scanning. In this work, an optical scanner, based on a coded structured light approach, has been used to acquire the patient's plaster model. An accurate digital mouth reconstruction composed of both crown shapes and gingival tissue is then obtained as shown in Fig. 5a. The overall surface is then segmented into disconnected regions, representing the individual crown geometries and the gingiva (Fig. 5b) through a semi-automated procedure, which exploits the curvature of the digital mouth model [12].

Multi-source Data Fusion. For each tooth, the multi-source data obtained by using optical and tomographic scanning must be merged in order to create accurate multibody

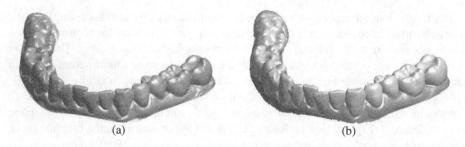

| (a) | (b) |

Fig. 5. (a) Digital mouth model as obtained by using the structured light scanner and (b) individual tooth crowns and gingival geometries as obtained by segmenting the model.

dental models. The crown surfaces obtained by optical scanning are aligned with the corresponding crown geometries segmented from the CBCT data set.

The meshes from the two sets of data are coarsely aligned into a common reference frame by manually selecting at least three common points. A refinement of the initial alignment is then performed by a fine registration procedure based on the Iterative Closest Point (ICP) technique. The crown geometries obtained by processing DICOM images are then removed by means of a disk vertex selection algorithm. Each vertex of the optic crown is projected into a point on the CBCT mesh. This point describes the center of a sphere, which is used to select the points of the CBCT mesh to remove. The final tooth models (Fig. 6) are then obtained by a Poisson surface reconstruction approach [13]. This allows for fully closed models composed of the most accurate representation for tooth crowns.

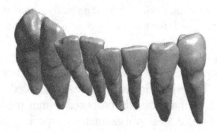

Fig. 6. Final merged tooth geometries (optical crowns + CBCT roots) used in the FE model.

PDL Modelling. PDL geometries cannot be easily visualized and reconstructed since usually the slice thickness is similar or even greater than the ligament space (about 0.2 mm) [14]. For this reason, in this work the PDL has been modelled for each tooth by detecting the interface area between bone and tooth models to which a 0.2 mm thick shell has been added. The volume shell is then subtracted from the alveolar bone in order to define the PDL volume [15]. The obtained PDL solid models are shown in Fig. 7.

2.2 Orthodontic Aligner Modelling Process

For each dental arch, the aligner geometry has been created by defining a layer completely congruent with the tooth crown surface. The individual teeth are firstly joined, root

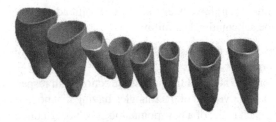

Fig. 7. PDL geometries used in the FE model.

geometries are deleted and undercut volumes manually removed in order to create a unique layer. The layer is thickened to create a 0.5 mm thick volume. Finally, the merged tooth geometries (shown in Fig. 6) are subtracted from the volume and the most external surface of the remaining geometry is removed with the aim at modelling the inner shape of the aligner. This procedure is carried out to guarantee an optimal fit between the mating surfaces of the tooth crowns and the appliance [16]. The standard aligner is supposed to have a uniform 0.7 mm thickness which originates from the mean thickness of the thermoplastic material disk (0.75 mm thick) before the thermoforming process [17]. For this reason, a shell has been created by thickening the inner shape of the aligner by 0.7 mm along the direction normal to the surface. Figure 8 shows the overall modelled geometries used to create the FE model.

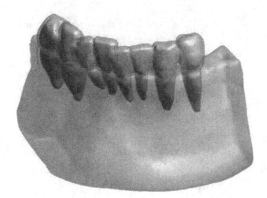

Fig. 8. Overall geometries used to create the FE model.

In this work, however, two further modelling strategies have been followed to test the influence of both a non-uniform aligner thickness and a customized initial penetration between the aligner and the target tooth on the effectiveness of the orthodontic treatment.

Non-uniform Aligner's Thickness Modelling. The idea is based on varying the appliance geometry by thickening the aligner in correspondence of highly deformed regions while thinning the model in correspondence of low deformed regions. This would allow optimizing the forces delivered to any tooth in the arch. In particular, the aligner displacement values obtained from a simulation performed with a uniform 0.7 mm thick aligner (reference simulation) have been used to pinpoint adjust the aligner's thickness. The procedure can be schematized as follows:

1. Evaluation of the displacement value (d) for each FE mesh node of the inner surface of the 0.7 mm uniform thick aligner;
2. Determination, for each node, of the normal direction with respect to the surface. The mean of the normal unit vectors of the triangles having that node as vertex is used;
3. Computation, for each node, of a new point along the normal direction having distance t from the node linearly defined as:

$$t = s_{min} + \frac{(d - d_{min})}{(d_{max} - d_{min})} \cdot (s_{max} - s_{min}) \tag{1}$$

where s_{min} and s_{max} are, respectively, the minimum and maximum values which define the aligner's thickness range, while d_{min} and d_{max} respectively represent the minimum and maximum displacement values computed for the 0.7 mm uniform thick aligner. The thickness range has been defined between 0.5 mm and 0.9 mm. Figure 9 shows a full-field map of the aligner thickness values (Fig. 9a) and three different cross sections (Fig. 9b).

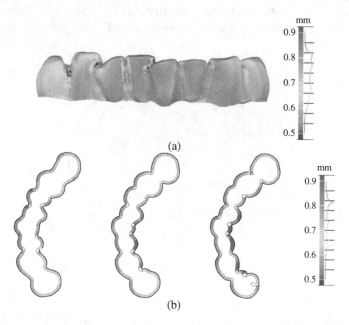

Fig. 9. (a) Full-field map of the aligner thickness values (expressed in mm) and three cross-sections (b).

Customized Initial Penetration Modelling. The second approach is based on varying the initial penetration between the aligner and the target tooth. The initial penetration is customized on the basis of the aligner deformation values obtained from the reference simulation. Each node of the aligner geometry is moved closer or farther with respect to the target tooth in the labiolingual direction (x-axis) proportionally to the node x-displacement in the reference simulation. A better tooth movement may be expected by counterbalancing the aligner deformation since the final tooth placement depends on the equilibrium between aligner and dentition. The procedure can be schematized as follows:

1. Evaluation of the x-displacement value (dx) for each FE mesh node of the inner surface of the 0.7 mm thick aligner;
2. Translation of each node along the x direction having distance t from the node linearly defined as:

$$t = -dx \qquad (2)$$

Figure 10 shows a full-field map of the aligner offset values, which create the initial penetration (Fig. 10a), and three different cross sections (Fig. 10b).

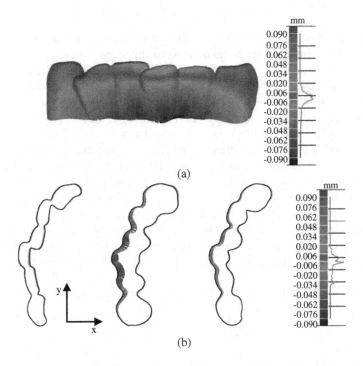

(a)

(b)

Fig. 10. (a) Full-field map of the aligner offset values (expressed in mm) and three cross sections (b).

2.3 Generation of the FE Model

The different bodies were imported in Ansys® 14. Each body was modeled with solid 10 nodes tetrahedrons. The approximate number of elements and nodes for each simulation was 134000 and 226000 respectively for the mandibular segment and 183000 and 267000 respectively for the maxillary segment. Figure 11 shows the maxillary and mandibular meshed models for the simulation performed with a uniform 0.7 mm thick aligner.

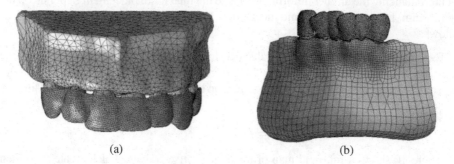

(a)   (b)

Fig. 11. Meshed models used for the simulation with 0.7 mm uniform thick aligner. (a) Maxillary arch, (b) Mandibular arch.

Material Properties. A linear elastic mechanical model was assigned to each body as shown in Table 1. Moreover, teeth and bone were supposed as made by a homogenous material, without discerning in enamel, pulp, dentin for the teeth and cortical and cancellous for the bone. This simplification does not affect the simulation results as shown in previous studies [18]. In technical literature, many are the biomechanical models that simulate the tooth ligament properties [19]. The investigation of the ligament in-vivo behavior is not a trivial task due to its small size (about 0.2 mm thickness). For this reason, most of the scientific literature has investigated the mechanical properties of the PDL through experimental analyses, thus developing five different models: linear elastic, bilinear elastic, viscoelastic, hyperelastic and multiphase [19]. However, the complex non-linear response of the PDL does not need to be addressed while performing an analysis about the first phase of the orthodontic reaction as in the present study [20]. The thermoplastic aligners are usually made from a polyethylene terephthalate glycol-modified (PETG) disc, whose mechanical properties can be retrieved from the manufacturer's datasheet. Its mechanical behavior has been approximated as linear elastic.

Table 1. Material properties used for the FE analyses.

	Young modulus (MPa)	Poisson's ratio
Bone	13000	0.3
Teeth	20000	0.3
PDL	0.059	0.49
Aligner (PETG)	2050	0.3

Loading and Boundary Conditions. Two orthodontic movements have been simulated:

- labiolingual tipping of a mandibular central incisor;
- rotation of a maxillary central incisor around its long axis.

The initial load configurations for the FE analysis were generated by the penetration between the aligner and the target tooth. The initial models do not present any penetration between teeth and aligner since the aligner is modelled onto the teeth surfaces. The target teeth must then be rotated around the Center of Resistance (C_{RES}) in order to create the initial penetration. The coordinates of the target teeth's C_{RES} were determined by using the method proposed by [21]. The tooth axes were defined according to [22]. The z-axis is associated with the lower inertia moment of the geometrical model and is obtained through the Principal Component Analysis (PCA) of the polyhedral surface by considering the masses associated with the barycenter of the triangles of the polyhedrons, which are proportional to its area. Two sections of the tooth were created and analyzed to identify the positive direction of the z-axis. The tooth was sliced by two different planes perpendicular to z-axis and 3 mm far from the tooth extremities. The section showing the worst approximation of a circle is considered as located upper (Γ_c). The mesiodistal (y-axis) direction and the labiolingual (x-axis) direction are orthogonal to the z-axis and are obtained by analyzing the principal component of inertia of the planar section Γ_c (Fig. 12a). Finally, the mandibular incisor is rotated around its C_{RES} along the y-axis, while the maxillary incisor is rotated around its C_{RES} along the z-axis.

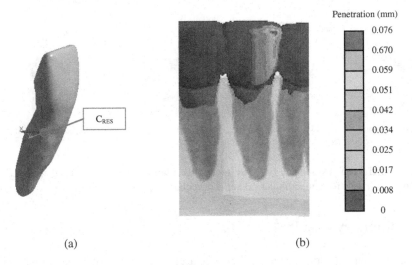

(a)   (b)

Fig. 12. Maxillary central incisor's Center of Resistance (C_{RES}) (a) and initial penetration between teeth and aligner (b).

The resulting maximum initial penetration turned out to be about 0.092 mm for the mandibular central incisor tipping and 0.076 mm for the maxillary central incisor rotation as shown in Fig. 12b. The bone extremities were fixed in all directions. An augmented Lagrangian formulation was used to simulate contact. Bonded contact surfaces were considered between bone and PDL and between PDL and teeth. Corresponding nodes cannot separate each other and a perfect adhesion between contact surfaces, without mutual sliding or separation, can be assumed. The aligner-teeth contact was set as frictionless, with a maximum allowed penetration of 0.01 mm, which provided the best accuracy-computational time ratio. An undesired initial penetration between the aligner geometry and non-target teeth could occur due to the meshing process. For this reason, the "*adjust to touch*" option was used for those contact couples in order to remove all the undesired initial penetrations.

Analysis Settings. Three different scenarios were simulated for each target tooth to compare the influence of the proposed aligner's modelling strategies onto the orthodontic movement:

- uniform 0.7 mm thickness;
- non-uniform thickness as obtained by applying Eq. (1);
- customized initial penetration as obtained by applying Eq. (2).

For each simulation, the resulting force-moment system delivered by the aligner to the target tooth was calculated at the C_{RES} (Fig. 13). Computational time resulted in about 2 h for each simulation, using a Workstation based on Intel Xeon CPU E3-1245 v3@3.40 GHz and 16 GB RAM.

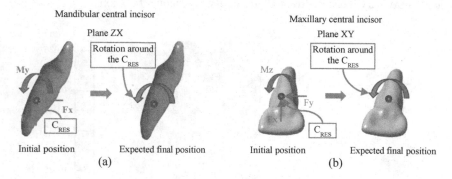

Fig. 13. Initial tooth position and Force System on the plane of interest (left) and expected final tooth position (right) for a mandibular (a) and a maxillary (b) central incisor.

3   Preliminary Results

The results obtained for each scenario were analyzed by comparing the moment along the y-axis and the resulting moment-to-force (M:F) ratio delivered to the tooth on the ZX plane and XY plane for the mandibular (Table 2) and maxillary incisor (Table 3), respectively. The M:F values describe the quality of the force system [23], while M_y and M_z define the amount of orthodontic movement.

The amount of moment delivered to the target tooth increased from the uniform to the non-uniform thick aligner by 75% and 133% for the mandibular and maxillary incisor, respectively. While the aligner with the customized initial penetration elicited a moment 176% higher than the 0.7 mm uniform thickness aligner for the mandibular incisor and 103% higher for the maxillary incisor. For the mandibular incisor, the same trend was also observed for the stress values in the PDL along the x-axis (Fig. 14). The magnitude of the PDL stress values directly affects the bone remodeling process, which is the main responsible of the orthodontic movement [18]. The PDL stress was not analyzed for the rotation of the maxillary incisor, because it has not the same significance. During a tipping movement the PDL shows distinct regions of tension and compression which can be easily analyzed. All different scenarios showed a positive stress value on the higher part of the anterior region and a negative stress value on the posterior, in agreement with the expected labiolingual movement. Figure 15 shows the full-field maps of the displacement values occurring on the target teeth for the different configurations.

Table 2. Force system delivered to the mandibular incisor for each scenario on the ZX plane.

Aligner	Uniform thickness	Non-uniform thickness	Initial offset
M_y (Nmm)	2.26	3.96	6.25
M_y/F_x (mm)	2.24	2.83	2.43
M_y/F_z (mm)	−3.42	−3.06	44.6

Table 3. Force system delivered to the maxillary incisor for each scenario on the XY plane.

Aligner	Uniform thickness	Non-uniform thickness	Initial offset
M_z (Nmm)	2.8	6.55	5.71
M_z/F_x (mm)	4.7	8.4	6.6
M_z/F_y (mm)	8.8	57.5	57.1

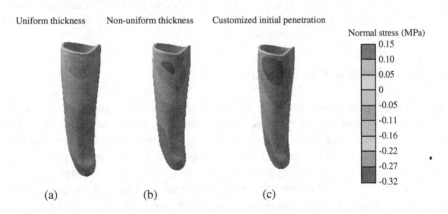

Uniform thickness   Non-uniform thickness   Customized initial penetration

Normal stress (MPa)

0.15
0.10
0.05
0
-0.05
-0.11
-0.16
-0.22
-0.27
-0.32

(a)   (b)   (c)

Fig. 14. PDL stress values along the x-axis of the mandibular target tooth for the different simulations: (a) uniform 0.7 mm thickness, (b) non-uniform thickness, (c) customized initial penetration.

Uniform thickness   Non-uniform thickness   Customized initial penetration

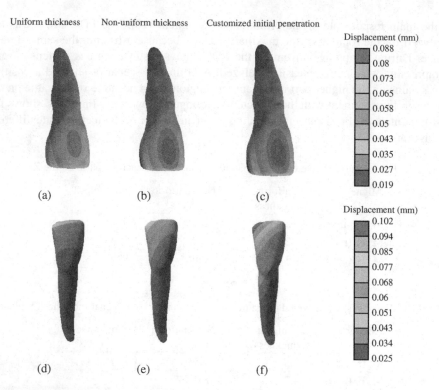

Fig. 15. Displacement values of the target tooth for the different simulations: (a) maxillary uniform 0.7 mm thickness, (b) maxillary non-uniform thickness, (c) maxillary customized initial penetration, (d) mandibular uniform 0.7 mm thickness, (e) mandibular non-uniform thickness, (e) mandibular customized initial penetration.

4 Discussion and Conclusions

Thermoformed plastic aligners have demonstrated limitations in exerting complex force-moment systems [24]. In particular, extrusion of central incisors and rotation and inclination of canine and premolar teeth are obtained in clinical practice by using auxiliary elements as attachments, bonded to the crowns surface, or divots and power ridges, which enhance the biomechanical effectiveness. However, the aligner thickness represents an additional critical element that should be optimized since the aligner material itself is the only element that imparts the force system. Minimal aligner thickness values would minimize patient discomfort. However, forces delivered by thick appliances are higher than those of thin materials are [7].

Typically, aligners are obtained by a vacuum thermoforming process performed onto 3D physical molds of the teeth manufactured by RP methodologies for each single step of the orthodontic treatment. A single thermoplastic polymer resin sheet (about 0.75 mm-thick) is stretched over each prototyped mold and trimmed to extract the final configuration. For this reason, a constant thickness is usually considered for the aligners.

In this paper, the influence of different aligner geometries, in terms of non-uniform thickness and modified initial penetration values, has been analyzed for a mandibular central incisor during labiolingual tipping and for a maxillary central incisor during rotation. Preliminary results have evidenced a more effective force-moment system delivered to the target tooth by pinpoint modulating the aligner thickness in order to vary its stiffness or the initial penetration between the aligner and the patient's dentition. The non-uniform thickness appliance elicited a higher magnitude of the desired moment M_y and a better quality of the movement as attested by the higher values obtained for the M:F parameter. These findings clearly call upon some considerations about the aligner's manufacturing process. Currently, the standard production processes are strictly constrained by the thermoforming procedures, which cannot provide aligners having non-uniform customized thickness values. An alternative method for the direct manufacturing of the aligner should be used. For instance, milling by CNC machines or layer-by-layer printing of a single or multiple polymeric materials, would allow to obtain non-uniform thin-walled polymeric orthodontic aligners. Nevertheless, essential aligner's properties are large spring-back, high stored energy, tolerance to mouth hostile environment, biocompatibility and low surface roughness in correspondence of the mating surfaces. These features should be taken into high consideration when considering an alternative production method. This topic certainly represents a challenging task, which should affect future research activities.

The second aligner's modelling strategy represents an alternative approach, which could be adopted to enhance the treatment effectiveness by using a uniform thickness aligner. This approach is based on the creation of an aligner with an initial penetration with respect to the dentition anatomy. The penetration values are customized by rapid prototyping a 3D physical mold of the teeth that have been moved on the virtual model from their actual placement by using the aligner deformation values obtained from the reference simulation. The physical aligner, obtained by a thermoforming process performed onto the modified 3D dental mold, is then supposed to maintain the same thickness of the usual thermoforming process guaranteeing, at the same time, a customized offset between appliance and dentition. This approach demonstrated to enhance the treatment effectiveness with respect to the standard uniform aligner.

A further parameter that influences the effectiveness of the orthodontic treatment, besides the aligner's thickness and the initial penetration, is represented by the mechanical properties of the thermoplastic materials. In the present study, the physical values indicated in material manufacturer's datasheets have been used. However, these values are given under standard atmospheric conditions. Temperature, humidity, and forming procedures may have marked effects on the actual mechanical properties, which may differ between the intraoral environment and room temperature [17]. For this reason, some experimental tests are currently being carried out by simulating intraoral environment in order to characterize the aligner's mechanical properties in working conditions.

References

1. Boyd, R.L.: Esthetic orthodontic treatment using the invisalign appliance for moderate to complex malocclusions. J. Dent. Educ. **72**(8), 948–967 (2008)
2. Beers, A.C., Choi, W., Pavlovskaia, E.: Computer-assisted treatment planning and analysis. Orthod. Craniofac. Res. **6**(Suppl. 1), 117–125 (2003)
3. Farah, J.W., Craig, R.G., Sikarski, D.L.: Photoelastic and finite-element stress analysis of a restored axisymmetric first molar. J. Biomech. **6**(5), 511–520 (1973). doi:10.1016/0021-9290(73)90009-2
4. Cai, Y.Q., Yang, X.X., He, B.W., Yao, J.: Finite element method analysis of the periodontal ligament in mandibular canine movement with transparent tooth correction treatment. BMC Oral Health **15**, 106 (2015). doi:10.1186/s12903-015-0091-x
5. Gomez, J.P., Pena, F.M., Martinez, V., Giraldo, D.C., Cardona, C.I.: Initial force systems during bodily tooth movement with plastic aligners and composite attachments: a three-dimensional finite element analysis. Angle Orthod. **85**(3), 454–460 (2015). doi:10.2319/050714-330.1
6. Savignano, R., Barone, S., Paoli, A., Razionale, A.V.: FEM analysis of bone-ligaments-tooth models for biomechanical simulation of individual orthodontic devices. In: 34th Computers and Information in Engineering Conference, Buffalo, New York, USA, 17–20 August 2014, p. V01AT02A081. ASME (2014)
7. Hahn, W., Dathe, H., Fialka-Fricke, J., Fricke-Zech, S., Zapf, A., Kubein-Meesenburg, D., Sadat-Khonsari, R.: Influence of thermoplastic appliance thickness on the magnitude of force delivered to a maxillary central incisor during tipping. Am. J. Orthod. Dentofac. Orthop. **136**(1), 12.e11–12.e17 (2009). doi:10.1016/j.ajodo.2008.12.015
8. Kwon, J.S., Lee, Y.K., Lim, B.S., Lim, Y.K.: Force delivery properties of thermoplastic orthodontic materials. Am. J. Orthod. Dentofac. Orthop. **133**(2), 228–234 (2008). doi:10.1016/j.ajodo.2006.03.034
9. Barone, S., Paoli, A., Razionale, A.V.: Computer-aided modelling of three-dimensional maxillofacial tissues through multi-modal imaging. Proc. Inst. Mech. Eng. H: J. Eng. Med. **227**(2), 89–104 (2013). doi:10.1177/0954411912463869
10. 3DSlicer: A multi-platform, free and open-source software package for visualization and medical image computing (2016). http://www.slicer.org/. Accessed 20 Apr 2016
11. Barone, S., Paoli, A., Razionale, A.V.: CT segmentation of dental shapes by anatomy-driven reformation imaging and B-spline modelling. Int. J. Numer. Meth. Biomed. Eng. **32**(6), e02747 (2016). doi:10.1002/cnm.2747
12. Barone, S., Paoli, A., Razionale, A.: Creation of 3D multi-body orthodontic models by using independent imaging sensors. Sensors **13**(2), 2033–2050 (2013). doi:10.3390/s130202033
13. Kazhdan, M., Bolitho, M., Hoppe, H.: Poisson surface reconstruction. In: Proceedings of the Fourth Eurographics Symposium on Geometry Processing, Cagliari, Sardinia, Italy, pp. 61–70. Eurographics Association (2006). http://dl.acm.org/citation.cfm?id=1281957.1281965
14. Dorow, C., Schneider, J., Sander, F.G.: Finite element simulation of in vivo tooth mobility in comparison with experimental results. J Mech. Med. Biol. **3**(1), 79–94 (2003). doi:10.1142/S0219519403000661
15. Liu, Y., Ru, N., Chen, J., Liu, S.S.-Y., Peng, W.: Finite element modeling for orthodontic biomechanical simulation based on reverse engineering: a case study. Res. J. Appl. Sci. Eng. Technol. **6**(17), 3267–3276 (2013)
16. Barone, S., Paoli, A., Razionale, A.V., Savignano, R.: Computer aided modelling to simulate the biomechanical behaviour of customised orthodontic removable appliances. Int. J. Interact. Des. Manuf. **10**(4), 387–400 (2016). doi:10.1007/s12008-014-0246-z

17. Ryokawa, H., Miyazaki, Y., Fujishima, A., Miyazaki, T., Maki, K.: The mechanical properties of dental thermoplastic materials in a simulated intraoral environment. Orthod. Waves **65**(2), 64–72 (2006). http://dx.doi.org/10.1016/j.odw.2006.03.003
18. Penedo, N.D., Elias, C.N., Pacheco, M.C.T., de Gouvêa, J.P.: 3D simulation of orthodontic tooth movement. Dent. Press J. Orthod. **15**, 98–108 (2010)
19. Fill, T.S., Toogood, R.W., Major, P.W., Carey, J.P.: Analytically determined mechanical properties of, and models for the periodontal ligament: critical review of literature. J. Biomech. **45**(1), 9–16 (2012). doi:10.1016/j.jbiomech.2011.09.020
20. Cattaneo, P.M., Dalstra, M., Melsen, B.: The finite element method: a tool to study orthodontic tooth movement. J. Dent. Res. **84**(5), 428–433 (2005)
21. Viecilli, R.F., Budiman, A., Burstone, C.J.: Axes of resistance for tooth movement: does the center of resistance exist in 3-dimensional space? Am. J. Orthod. Dentofac. Orthop. **143**(2), 163–172 (2013). doi:10.1016/j.ajodo.2012.09.010
22. Di Angelo, L., Di Stefano, P., Bernardi, S., Continenza, M.A.: A new computational method for automatic dental measurement: the case of maxillary central incisor. Comput. Biol. Med. **70**, 202–209 (2016). doi:10.1016/j.compbiomed.2016.01.018
23. Savignano, R., Viecilli, R., Paoli, A., Razionale, A.V., Barone, S.: Nonlinear dependency of tooth movement on force system directions. Am. J. Orthod. Dentofac. Orthop. **149**(6), 838–846 (2016). doi:10.1016/j.ajodo.2015.11.025
24. Kravitz, N.D., Kusnoto, B., BeGole, E., Obrez, A., Agran, B.: How well does invisalign work? A prospective clinical study evaluating the efficacy of tooth movement with invisalign. Am. J. Orthod. Dentofac. Orthop. **135**(1), 27–35 (2009). doi:10.1016/j.ajodo.2007.05.018

Bioinformatics Models, Methods and Algorithms

Implicitly Weighted Robust Classification Applied to Brain Activity Research

Jan Kalina[1,2(✉)] and Jaroslav Hlinka[1,2]

[1] Institute of Computer Science of the Czech Academy of Sciences,
Pod Vodárenskou věží 2, 182 07 Prague 8, Czech Republic
{kalina,hlinka}@cs.cas.cz
[2] National Institute of Mental Health, Topolová 748, 250 67 Klecany, Czech Republic

Abstract. In bioinformatics, regularized linear discriminant analysis is commonly used as a tool for supervised classification problems tailor-made for high-dimensional data with the number of variables exceeding the number of observations. However, its various available versions are too vulnerable to the presence of outlying measurements in the data. In this paper, we exploit principles of robust statistics to propose new versions of regularized linear discriminant analysis suitable for high-dimensional data contaminated by (more or less) severe outliers. The work exploits a regularized version of the minimum weighted covariance determinant estimator, which is one of highly robust estimators of multivariate location and scatter. The performance of the novel classification methods is illustrated on real data sets with a detailed analysis of data from brain activity research.

Keywords: High-dimensional data · Classification analysis · Robustness · Outliers · Regularization

1 Introduction

In bioinformatics, a common data analysis task is to learn a classification rule over high-dimensional data, i.e. data with the number of variables p exceeding the number of observations n ($n < p$ or even $n \ll p$) [2,8]. Thus, supervised classification methods (classifiers) represent important tools for the analysis of data observed in K different samples (groups) as

$$X_{11}, \dots, X_{1n_1}, \dots, X_{K1}, \dots, X_{Kn_K}, \tag{1}$$

while we assume $p > K \geq 2$ and denote $n = \sum_{k=1}^{K} n_k$. Sensitivity of various standard classification procedures to the presence of outlying measurements (outliers) in such high-dimensional data has been repeatedly reported as a serious problem in data mining as well as multivariate statistics [9,33].

The linear discriminant analysis (LDA) is well known to be too sensitive to the presence of outlying values in the data because it exploits the classical (non-robust) estimates in the form of means and empirical covariance matrix.

© Springer International Publishing AG 2017
A. Fred and H. Gamboa (Eds.): BIOSTEC 2016, CCIS 690, pp. 87–107, 2017.
DOI: 10.1007/978-3-319-54717-6_6

As an alternative, robust classification methods have been proposed which are resistant to the presence of outliers [4,16,35]. Highly robust methods are defined as methods with a high breakdown point, which measures the sensitivity of an estimator against noise or outliers in the data. Particularly, the finite-sample definition of the breakdown point corresponds to the maximal percentage of extremely severe outliers present in the data set, which still does not lead the method to a collapse, i.e. the estimators of the means and of the common scatter matrix are not shifted to infinity [6]. Nevertheless, robust classification methods are computationally feasible only for $n > p$ with a sufficiently small p.

In this paper, we propose new classification methods for high-dimensional data exploiting principles of robust statistics in a unique combination with the (Tikhonov) regularization. Section 2 recalls various existing approaches to regularized linear discriminant analysis for $n \ll p$. Section 3 recalls the minimum weighted covariance determinant estimator, which is one of highly robust estimators of multivariate location and scatter, and proposes its regularized version. Section 4 proposes four new robust regularized classification methods for high-dimensional data, exploiting the tools of Sect. 3. The following Sect. 5 illustrates the performance of the novel methods on several real data sets, while the largest attention is paid to data from brain activity research. A discussion follows in Sect. 6, where also the good comprehensibility of the newly proposed procedures is brought to attention, and finally Sect. 7 concludes the paper.

2   Regularized Linear Discriminant Analysis

Various available versions of a regularized LDA can be characterized as modifications of the standard LDA for the context of high-dimensional data. While these methods have found their applications in bioinformatics (e.g. [10,23,34]), they remain to be vulnerable to outliers because of the non-robustness of the empirical covariance matrix as well as means of each group.

Regularized LDA assumes a common covariance matrix Σ for each group. If $n < p$ or even $n \ll p$, the pooled estimator of Σ denoted by S is singular. Let us denote the mean of the observed values in the k-th group ($k = 1, \ldots, K$) by $\bar{X}_k$. Perhaps the simplest and most habitually used version of regularized LDA, which we denote by LDA* to avoid confusion, assigns a new observation $Z = (Z_1, \ldots, Z_p)^T$ to group k, if $l_k^* > l_j^*$ for every $j \neq k$, where the regularized linear discriminant score for the k-th group ($k = 1, \ldots, K$) has the form

$$l_k^* = \bar{X}_k^T (S^*)^{-1} Z - \frac{1}{2} \bar{X}_k^T (S^*)^{-1} \bar{X}_k + \log \pi_k. \tag{2}$$

Here, π_k is a prior probability of observing an observation from the k-th group,

$$S^* = (1 - \lambda)S + \lambda T \tag{3}$$

for $\lambda \in (0, 1)$ denotes a regularized estimator of Σ and the target matrix T is a given symmetric positive definite matrix of size $p \times p$. Its most common choices include the identity matrix $\mathcal{I}_p$ or a diagonal (non-identity) matrix $T = \bar{s}\,\mathcal{I}_p$, where $\bar{s} = \sum_{i=1}^p S_{ii}/p$. The regularized matrix (3) is guaranteed to be regular and positive definite even for $n \ll p$. A suitable value of λ is usually found by a cross validation.

Another important example of a regularized LDA is the shrunken centroid regularized discriminant analysis (SCRDA) [10], which performs also a regularization on the mean of each group, namely by shrinking each of the means towards the pooled mean in the L_1-norm.

Concerning the properties of regularized versions of LDA, it would be very difficult to assess them rigorously. Instead, they were rather investigated only by means of numerical simulations [10,34]. Basically, regularized LDA methods remain to be non-robust to the presence of severe outliers, although the regularization leads to their local insensitivity (robustness) to small measurement errors as advocated within the framework of robust optimization [1].

3 Robust Estimation of Multivariate Location and Scatter

This section devoted to robust estimation of multivariate location and scatter starts by recalling the highly robust minimum weighted covariance determinant estimator in Sect. 3.1. We will use the regularized M-estimator of multivariate data proposed in [3] to obtain a reliable initial estimator of the scatter matrix. This allows to define a regularized version of the minimum weighted covariance determinant estimator in Sect. 3.2. For the iterative computation of this scatter matrix estimator, an initial estimate will be necessary and we recommend to use Chen's estimator [3] for this purpose.

3.1 Minimum Weighted Covariance Determinant Estimator

The Minimum Covariance Determinant (MCD) and the Minimum Weighted Covariance Determinant (MWCD) estimators are highly robust affine-equivariant estimators of multivariate location and scatter. Let us consider a single sample of independent identically distributed p-variate random variables $X_1, \ldots, X_n$ (i.e. $K = 1$). The estimators are formulated for multivariate data coming from a unimodal elliptically symmetric distribution with a location parameter $\mu \in \mathbb{R}^p$ and a scatter matrix $\Sigma \in \mathbb{R}^{p \times p}$ (cf. [17]). The scatter matrix is a more general concept compared to the covariance matrix, which must not necessarily exist. Nevertheless, the two concepts are identical for Gaussian data. Standard estimators of μ and Σ are highly vulnerable to the presence of outliers in the data. On the other hand, the MCD and MWCD estimators, which will •
be now recalled, are more suitable for severely contaminated data compared to classical estimates and also compared to multivariate M-estimators [26,36].

The MCD estimator [31] is computed as the classical mean and the empirical covariance matrix taking into account however only the optimal subset of h observations, which yields the minimum determinant of the empirical covariance matrix over all such possible subsets of h observations. This corresponds to assigning weights equal to 1 or 0 to the observations, while the number of ones is a fixed value equal to h which must be specified by the user prior to the computations. Properties of the estimator were overviewed in [17].

The MWCD estimator [29] represents a generalization of MCD allowing to consider non-negative (possibly continuous) weights $w_1, \ldots, w_n$ to be assigned to the observations. The user specifies magnitudes of weights prior to computing the estimator but the weights themselves are assigned to individual observations only after a permutation, which is determined only during the computation of the estimator. Such implicitly given weights are based on the idea to down-weight outliers and to increase the influence of the majority of "good data".

Properties of the MWCD estimator were derived in [29] including the efficiency, Fisher consistency or influence function. The method attains the maximal breakdown point which is possible for an affine-equivariant estimator [25]; this is true if the outliers obtain weights exactly equal to 0 and the data are assumed in general position [30]. An approximative algorithm for computing the MWCD estimator may be obtained as a generalization of the MCD algorithm [31]. The advantage of the weighting scheme is its ability to reduce the local sensitivity compared to the MCD estimator; this is analogous to the experience with implicitly weighted methods in robust regression [19].

3.2   Regularized MWCD Estimator

We consider again the data in one group as in Sect. 3.1. Because the MWCD estimator cannot be computed for $n < p$, we define its regularized version which is computationally feasible also for $n \ll p$.

Let us first recall the work of Chen et al. [3], who proposed a regularized M-estimator of the scatter matrix of multivariate data based on a popular (Huber-type [15]) M-estimator of [36]. However, M-estimators of parameters in the multivariate model do not possess a high breakdown point [37]. In addition, the estimation does not yield any corresponding estimator of the mean.

Our proposal of a regularized MWCD estimator presented as Algorithm 1 exploits Chen's regularized M-estimator as an initial estimator of the scatter matrix. This depends on the value of a regularization parameter $\rho \in (0, 1)$. If there is no prior idea how to choose its suitable value, a reasonable recommendation is to choose a very small ρ. Nevertheless, its suitable value may be easily found by cross validation in specific tasks. This will be the case of classification problems in Sect. 4. In step 2, choosing robust rather than standard initial estimators is a common approach in a variety of iterative robust estimators [18].

*4   Robust Classification

Four novel robust versions of regularized LDA will be proposed in this section, together with algorithms for their efficient computation. The approach is suitable for multivariate data coming from a unimodal elliptically symmetric distribution as explained in Sect. 3.1.

4.1 MWCD-LDA*

We will propose a novel classification method denoted as MWCD-LDA*. We assume data (1) in K groups, while all the groups have the common scatter matrix $\Sigma \in \mathbb{R}^{p \times p}$. Our approach is based on estimating Σ as well as the means of the groups by the regularized MWCD estimator of Sect. 3.2.

Algorithm 1. Regularized MWCD estimator.

Input: p-dimensional observations $X_1, \ldots, X_n$, weights $w_1, \ldots, w_n$, positive definite symmetric matrix $T \in \mathbb{R}^{p \times p}$.

Output: $\bar{X}_{MWCD}, S_{MWCD}$.

1: **for** $i = 1$ to $10\,000$ **do**
2:     Randomly select an initial set of $n/2$ observations. Compute Chen's estimator, which will be denoted by B (for the mean) and C (for the scatter matrix).
3:     $j := 1$
4:     $L_{ij} := +\infty$
5:     **repeat**
6:         Compute

$$d(i; B, C) = \left[(X_i - B)^T C^{-1} (X_i - B) \right]^{1/2}, \quad i = 1, \ldots, n. \qquad (4)$$

Sort these values in ascending order and assign corresponding ranks to individual observations. This determines a permutation $\pi(1), \ldots, \pi(n)$ of the indexes $1, 2, \ldots, n$, which fulfills

$$d(\pi(1); B, C) \le \cdots \le d(\pi(n); B, C). \qquad (5)$$

7:         Assign the weights to each observation according to its rank evaluated in the previous step. In this way, e.g. the observation $X_{\pi(1)}$ obtains the weight w_1.
8:         Compute the weighted mean and weighted empirical covariance matrix S_w using these weights.
9:         $j := j + 1$
10:        $C := (1 - \lambda) S_w + \lambda T$
11:        $L_{ij} := \det(C)$
12:    **until** $L_{ij} \ge L_{i,j-1}$
13: **end for**
14: Determine the set of weights $\tilde{w}_1, \ldots, \tilde{w}_n$ minimizing L_{ij} over all considered i and j.

15: $\bar{X}_{MWCD} := \sum_{i=1}^n \tilde{w}_i X_i$
16:

$$S_{MWCD} := \sum_{i=1}^n \tilde{w}_i (X_i - \bar{X}_{MWCD})(X_i - \bar{X}_{MWCD})^T \qquad (6)$$

Concerning the estimator of the scatter matrix of data (1), we must be aware that Σ does not play the role of the covariance (nor scatter) matrix over all data but rather to the scatter matrix common for each of the groups. Therefore, we need to adapt Algorithm 1 to the situation with K groups.

Algorithm 1 in step 8 considers the empirical weighted covariance matrix, which has to be replaced for the data in K groups by

$$S^*_{MWCD} = \left(S^*_{ij}\right)^p_{i,j=1}, \tag{7}$$

where

$$S^*_{ij} = \sum_{k=1}^{K} \sum_{l=1}^{n_k} w_{kl} \left(X_{kli} - \bar{X}^k_{iw}\right) \left(X_{klj} - \bar{X}^k_{jw}\right). \tag{8}$$

Here, the summation over l runs over all observations $l = 1, \ldots, n$, which belong to the k-th group,

$$X_{kl} = (X_{kl1}, \ldots, X_{klp})^T \quad \text{for } k = 1, \ldots, K, \ l = 1, \ldots, n_k, \tag{9}$$

the weights are denoted as $w_{11}, \ldots, w_{1n_1}, \ldots, w_{K1}, \ldots, w_{Kn_K}$ and $\bar{X}^k_{iw}$ denotes the weighted mean of the k-th group with these weights. With this difference, Algorithm 1 yields S^*_{MWCD} as the estimator of Σ, which exploits the optimal weights denoted as $\tilde{w}_1, \ldots, \tilde{w}_n$.

The resulting weights will be now used also to define regularized MWCD-means of each group, which have the form

$$\bar{X}_{MWCD} = \sum_{l=1}^{n} \tilde{w}_l X_l \tag{10}$$

and

$$\bar{X}^k_{MWCD} = \sum_{l \in group\ k} \tilde{w}_l X_l, \quad k = 1, \ldots, K. \tag{11}$$

Now we come back to the original classification problem. Formally, MWCD-LDA* will assign a new observation $Z = (Z_1, \ldots, Z_p)^T$ to group k, if $\tilde{\ell}_k > \tilde{\ell}_j$ for every $j \neq k$, where

$$\tilde{\ell}_k = (\bar{X}_{k,MWCD})^T \left(S^*_{MWCD}\right)^{-1} Z$$
$$- \frac{1}{2}(\bar{X}_{k,MWCD})^T \left(S^*_{MWCD}\right)^{-1} \bar{X}_{k,MWCD} + \log \pi_k. \tag{12}$$

Equivalently, the classification rule can be also expressed exploiting a robust regularized Mahalanobis distance. In this respect, an observation Z is assigned to group k if

$$(\bar{X}_{j,MWCD} - Z)^T \left(S^*_{MWCD}\right)^{-1} (\bar{X}_{j,MWCD} - Z) + \log \pi_j \tag{13}$$

reaches its minimum over all $j = 1, \ldots, K$ exactly for k.

As both (12) and the group assignment (13) are rather obscure from the computational point of view, we recommend to avoid the expensive and numerically unstable computation of the Mahalanobis distance by solving a set of linear equations within Algorithm 2 for the task to classify an observation $Z = (Z_1, \ldots, Z_p)^T$.

The approach of Algorithm 2 is based on the eigendecomposition of S^*_{MWCD}. Its inversion is replaced is replaced by (16), which is based on expressing

$$
\begin{aligned}
(\bar{X}_{k,MWCD} - Z)^T (S^*_{MWCD})^{-1} (\bar{X}_{k,MWCD} - Z) \\
= (\bar{X}_{k,MWCD} - Z)^T Q D^{-1} Q^T (\bar{X}_{k,MWCD} - Z) \\
= \|D^{-1/2} Q^T (\bar{X}^{k,MWCD}_k - Z)\|^2.
\end{aligned} \tag{14}
$$

While S^*_{MWCD} depends on the parameter λ, its suitable value will be found by a cross validation in the form of a grid search over all possible values of $\lambda \in (0, 1)$.

Alternatively, the method can be computed using the Cholesky decomposition. Besides, if a specific choice $T = \mathcal{I}_p$ is considered, the computation of MWCD-LDA* can be performed by means of more efficient algorithms, which exceed the scope of this paper.

Algorithm 2. MWCD-LDA* for a general T based on the eigendecomposition.

Input: Data (1), $Z \in \mathbb{R}^p$, weights $w_1, \ldots, w_n$, positive definite symmetric matrix $T \in \mathbb{R}^{p \times p}$.

Output: Assignment of Z to one of the groups $1, \ldots, K$.

1: **for** $i = 1$ to 100 **do**
2:    $\lambda := i/100$
3:    Compute S^*_{MWCD} and $\bar{X}^k_{MWCD}$ for $k = 1, \ldots, K$ by a modification of Algorithm 1 using the given T and λ, replacing S_w by (7) with (8).
4:    Compute the matrix

$$
A = (\bar{X}_{1,MWCD} - Z, \ldots, \bar{X}_{K,MWCD} - Z) \tag{15}
$$

   of size $p \times K$.
5:    Compute the eigendecomposition $S^*_{MWCD} = QDQ^T$.
6:    Compute $B = D^{-1/2} Q^T A$.
7:    Assign Z to group k, if

$$
k = \arg\max_{l=1,\ldots,K} \left\{ \|B_l\|^2 + \log \pi_l \right\}, \tag{16}
$$

   where $\|B_l\|^2$ is the Euclidean norm of the l-th column of B.
8: **end for**
9: Determine the value of λ yielding the best classification performance and carry out steps 3 to 7 with it to find the final classification decision.

4.2 L_1-SCRRDA

Further, we propose to accompany regularizing the scatter matrix of Sect. 4.1 by regularizing the mean of each of the groups. The novel method represents a robustification of the SCRDA of [10] and will be denoted as L_1-SCRRDA, which abbreviates a shrunken centroid robust regularized discriminant analysis with the means regularized in the L_1-norm. We can also perceive the method to be based on an L_1-regularized robust Mahalanobis distance.

Let us consider the mean of the k-th group to be estimated by

$$\bar{X}^{(1)}_{k,MWCD} = \text{sgn}(\bar{X}_{k,MWCD}) \left(|\bar{X}_{k,MWCD}| - \Delta \right)_+$$
$$= \text{sgn}(\bar{X}_{k,MWCD}) \max \left\{ |\bar{X}_{k,MWCD}| - \Delta, 0 \right\}, \quad (17)$$

where $\Delta \in \mathbb{R}^p$ and $(x)_+$ denotes the positive part of $x \in \mathbb{R}^p$. In other words, the MWCD-mean is shrunken towards zero in the L_1-norm, which can be interpreted as a regularized (biased) version of the MWCD-mean.

The method L_1-SCRRDA exploits the matrix S^*_{MWCD} as in Sect. 4.1. It assigns an observation $Z = (Z_1, \ldots, Z_p)^T$ to group k, if $\ell_k^{(1)} > \ell_j^{(1)}$ for every $j \neq k$, where

$$\ell_k^{(1)} = (\bar{X}^{(1)}_{k,MWCD})^T (S^*_{MWCD})^{-1} Z$$
$$- \frac{1}{2} (\bar{X}^{(1)}_{k,MWCD})^T (S^*_{MWCD})^{-1} \bar{X}^{(1)}_{k,MWCD} + \log \pi_k. \quad (18)$$

Within the classification method, suitable values of both regularization parameters λ and Δ can be found by cross validation as in Algorithm 2, which can be adapted to the context of L_1-SCRRDA.

Remark 1. L_1-SCRRDA distinguished between two groups of variables as follows, using the notation $\bar{X}_{jk,MWCD}$ and $\bar{X}^{(1)}_{jk,MWCD}$ to evaluate the j-th coordinate of $\bar{X}_{k,MWCD}$ and $\bar{X}^{(1)}_{k,MWCD}$, respectively.

1. Major (more relevant) variables fulfilling $|\bar{X}_{jk,MWCD}| > \Delta$ for at least one k. Their values of $\bar{X}^{(1)}_{jk,MWCD}$ for these k are obtained by shrinking $\bar{X}_{jk,MWCD}$ towards zero by the amount of exactly Δ.
2. Minor (less relevant) variables fulfilling $|\bar{X}^{(1)}_{jk,MWCD}| \leq \Delta$ for each k. Their values of $\bar{X}^{(1)}_{jk,MWCD}$ are equal to 0.

Remark 2. L_1-regularization is generally understood to introduce sparseness and reduce the dimensionality. However, the universality of this property is rather a "golden legend" and holds e.g. in linear regression (lasso estimator) but not for (any) regularized LDA, although there have been misleading statements on variable selection and sparseness also in this context (cf. [10,34]). In the light of Remark 1, we stress that L_1-SCRRDA remains to depend also on the major variables, because $\bar{X}_{k,MWCD} - Z$ is the same for all k, but the variable influences the linear discriminant score through the scatter matrix. Also the computational complexity of L_1-SCRRDA is not reduced compared to a method regularizing in the L_2-norm instead, which will be proposed in the next subsection.

4.3 L_2-SCRRDA

An alternative regularized robust version of LDA denoted as L_2-SCRRDA is proposed, which combines the scatter matrix estimation of Sect. 4.1 with shrinking the means towards the pooled mean (across groups) in the L_2-norm. Thus, we denote this version of robust SCRDA as L_2-SCRRDA.

The pooled scatter matrix (across groups) is estimated again by S^*_{MWCD}. The classical mean of the k-th group is replaced by the MWCD-mean shrunken towards the overall MWCD-mean across groups $\bar{X}_{MWCD}$, i.e. we consider

$$\bar{X}^{(2)}_{k,MWCD} = \delta \bar{X}_{k,MWCD} + (1-\delta)\bar{X}_{MWCD} \tag{19}$$

for $k = 1,\ldots,K$ and a fixed $\delta \in (0,1)$.

The method L_2-SCRRDA assigns an observation Z to group k, if $\ell^{(2)}_k > \ell^{(2)}_j$ for every $j \neq k$, where

$$\ell^{(2)}_k = (\bar{X}^{(2)}_{k,MWCD})^T (S^*_{MWCD})^{-1} Z$$
$$- \frac{1}{2}(\bar{X}^{(2)}_{k,MWCD})^T (S^*_{MWCD})^{-1} \bar{X}^{(2)}_{k,MWCD} + \log \pi_k. \tag{20}$$

Suitable values of parameters λ and δ can be found again by cross validation. The method may be preferable to L_1-SCRRDA if the data contain a large number of variables with a small effect on the classification, but without any clearly dominant small subset of variables. The shrinkage in (19) performed in the L_2-norm is analogous to shrinking estimates of parameters in ridge regression.

4.4 M-LDA*

We can also define a version of robust LDA based on M-estimation. Assuming again the data (1) as denoted in Sect. 1, the M-estimator of the mean of the k-th group denoted as $\bar{X}^k_M$ for $k = 1,\ldots,K$ will be considered. As in Sect. 4.1, the Chen's regularized estimator is considered as the estimate of the scatter matrix Σ common for each of the groups. This matrix S^*_M will be rather denoted as $S^*_{M,\rho}$ to stress its dependence on the regularization parameter $\rho \in (0,1)$.

The robust regularized LDA based on M-estimation, which we denote as M-LDA*, may be performed by Algorithm 3, which is formulated for an observation $Z \in \mathbb{R}^p$ exploiting the Cholesky decomposition of the scatter matrix. Algorithm 3 can be also adapted to be suitable for previously mentioned classifiers (MWCD-LDA*, L_1-SCRRDA and L_2-SCRRDA) in a straightforward way.

5 Examples

5.1 Methods in the Computations

To illustrate the performance of the novel robust regularized versions of LDA, we analyze several different real data sets with $n < p$. Each of the examples learns a classification rule to two groups ($K = 2$).

Algorithm 3. M-LDA* based on the Cholesky decomposition.

Input: Data (1), $Z \in \mathbb{R}^p$.
Output: Assignment of Z to one of the groups $1, \ldots, K$.
1: Compute Huber's estimators $\bar{X}_M^1, \ldots, \bar{X}_M^K$.
2: Compute the matrix

$$A = \left(\bar{X}_M^1 - Z, \ldots, \bar{X}_M^K - Z \right) \tag{21}$$

of size $p \times K$.
3: **for** $i = 1$ to 100 **do**
4:    $\rho := i/100$
5:    Compute $S_{M,\rho}^*$ (Sect. 4.4).
6:    Compute the Cholesky decomposition $S_{M,\rho}^* = L_* L_*^T$, where L_* is a (regular) lower triangular matrix.
7:    Compute $B = L_*^{-T} A$.
8:    Assign Z to group k, if

$$k = \arg \max_{l=1,\ldots,K} \left\{ ||B_l||^2 + \log \pi_l \right\}, \tag{22}$$

    where $||B_l||^2$ is the Euclidean norm of the l-th column of B.
9: **end for**
10: Determine the value of ρ yielding the best classification performance and carry out steps 4 to 8 with them to find the final classification decision.

We performed the computations in R software. Each classification task for each of the data sets is analyzed by means of a 5-fold cross validation. Within such approach, the data set is randomly divided into 5 subsamples of (approximately) equal sizes. Among all possible partitions, we select randomly 100 of them and compute the average Youden's index I as a classification performance measure over them. The averaged values are presented in Table 1. We recall Youden's index to be defined as

$$I = \text{sensitivity} + \text{specificity} - 1, \tag{23}$$

i.e. it fulfils $I \in [-1, 1]$.

Standard classifiers used in the examples include also the lasso-regularized logistic regression denoted as lasso-LR or a support vector machine (SVM) with a radial basis function kernel. Concerning the choice of parameters of individual classifiers, all regularized versions of LDA use $T = \mathcal{I}_p$. For standard methods, default settings of parameters were used whenever appropriate. Concerning the choice of the implicit weights, the MWCD-LDA* and L_2-SCRRDA use linearly decreasing weights in the following form. Starting with the simple choice

$$w_i = 1 - \frac{i-1}{n}, \quad i = 1, \ldots, n, \tag{24}$$

we standardize them to $\sum_{i=1}^{n} w_i = 1$ to obtain the final formula for the weights

$$\tilde{w}_i = \frac{2\,(n-i+1)}{n(n+1)}, \quad i = 1, \ldots, n, \tag{25}$$

which are very small for outliers reducing considerably their influence.

To investigate the effect of dimensionality reduction, we also use the principal component analysis (PCA) and a robust Minimum Redundancy Maximum Relevance (MRMR) of [22]. The latter is a robust supervised variable selection method selecting a small set of a fixed (given) number of the most relevant variables while penalizing for redundancy [27].

Table 1. Youden's index (23) as a classification performance measure computed for a 5-fold cross validation study on various real data sets of Sect. 5. Particularly, data from Sect. 5.3 are considered not only raw but also after a contamination by normally distributed outliers $N(0, \sigma^2)$ for different values of σ.

	Section 5.3				Section		
	Raw	Contam. for $\sigma =$			5.5	5.6	5.7
		0.1	0.2	0.3			
n	168				48	42	32
p	4005				38 614	518	15
Regularized versions of LDA							
PAM	0.88	0.81	0.75	0.68	0.85	0.86	0.51
LDA*	1.00	0.95	0.94	0.77	1.00	0.89	0.71
SCRDA	1.00	1.00	1.00	0.99	1.00	0.91	0.80
MWCD-LDA*	1.00	1.00	1.00	1.00	1.00	0.91	0.79
L_2-SCRRDA	1.00	1.00	1.00	1.00	1.00	0.92	0.80
Other classification methods							
SVM	1.00	0.99	0.98	0.96	1.00	0.92	0.85
Classification tree	0.96	0.95	0.91	0.92	0.94	0.84	0.11
Lasso-LR	0.99	1.00	0.97	0.94	0.97	0.87	0.82
Number of principal components	10	10			10	20	4
PCA $\Longrightarrow$ LDA	1.00	0.94	0.93	0.88	0.15	0.70	0.59
PCA $\Longrightarrow$ LDA*	1.00	0.95	0.94	0.89	0.51	0.62	0.59
PCA $\Longrightarrow$ SCRDA	1.00	0.95	0.94	0.89	0.62	0.72	0.59
Number of selected genes	10	10			10	20	4
MRMR $\Longrightarrow$ LDA	1.00	0.94	0.93	0.89	0.90	0.88	0.72
MRMR $\Longrightarrow$ LDA*	1.00	0.96	0.93	0.89	0.96	0.88	0.76
MRMR $\Longrightarrow$ SCRDA	1.00	0.96	0.93	0.89	1.00	0.90	0.76

5.2   Description of the Brain Activity Study

We participate on a neuroscience research investigating the spontaneous activity of various parts of the brain by means of neuroimaging methods, motivated by a distant aim to investigate modifications of the resting-state brain networks in schizophrenic patients. Specific functions of individual parts of the brain have been already rigorously described [7], but spontaneous brain activity and especially connections between pairs of brain parts in the resting state (i.e. resting-state brain networks) are acknowledged as a hot topic in current neuroscience [14]. We will now describe the whole study leading to acquiring the original real data set of brain scans by functional magnetic resonance imaging (fMRI), while the performance of various classification methods will be compared in the following sections on several different classification tasks.

Our data are measured on $n = 24$ healthy probands (i.e. without a manifested psychiatric disease) participating in the study, who were examined under 7 different situations. One of them can be characterized as a resting state, i.e. rest without any stimulus. Besides, the probands were observing each of 6 different movies while measuring the brain activity in the same way. For the sake of the fMRI imaging, the brain is divided to 90 regions and we are interested only in values of correlation coefficients between a pair of brain regions.

The most widely spread method for measuring the (functional) connectivity between a pair of brain regions is the correlation (i.e. Pearson's correlation coefficient) of activity time series derived from these regions by e.g. simple spatial averaging across all the voxels in the brain regions [14]. In this spirit, we consider $p = 90 * 89/2 = 4005$ variables containing values of correlation coefficients for each of the 24 probands. The basic task is to classify the resting state from (any) movie, i.e. all movies together are considered to be one class. In general, fMRI measurements are commonly contaminated by measurement errors as well as outliers [38]. It is also true with our data, which makes the newly proposed robust methods appealing for their analysis.

We consider the classification task to separate between the resting state and (any) movie for healthy individuals in Sect. 5.3, while a more specific task to classify between the resting state and one particular movie is investigated in Sect. 5.4.

5.3   Brain Activity: Resting State vs. Movie

Let us now consider the task to learn a classification rule over the training data from Sect. 5.2 to distinguish between the resting state and a movie. Considering thus all six movies together to belong to one class, this is a classification task to $K = 2$ groups with $p = 4005$ variables. The resting state group contains 24 observations while the group of movies consists of $6 * 24 = 144$ observations.

Several classification methods yield the best result ($I = 1.00$) as shown in Table 1. This is true for standard (non-robust) methods as well as for MWCD-LDA* or L_2-SCRRDA. While the standard LDA is computationally infeasible, SCRDA as one of its available regularized version turns out to perform reliably. There seems no advantage of the robust regularized LDA over non-robust versions, which may be explained by the fact that the raw data are not contaminated by a remarkable percentage of severe outliers. Only PAM turns out to be heavily influenced by them, although it was originally presented as a denoised version of diagonalized LDA [34]. The classification rule of L_1-SCRRDA distinguishes between 81 major variables and the remaining minor variables in this example. An SVM formally gives a perfect classification result, while our critical evaluation of SVM will be presented in Sect. 6.1.

Additionally, we investigated the effect of dimensionality reduction on the classification performance. There seems no remarkable small group of genes responsible for a large portion of variability of the data and the first few principal components seem rather arbitrary. L_2-SCRRDA has a good classification ability if applied on principal components. Thus, the classification results after reducing the dimensionality bring other arguments in favor of the regularization approaches used in this paper.

We additionally performed an artificial contamination of the original data in order to investigate the performance of the novel robust classification methods. Each single measurement for each proband was contaminated by noise, which was generated as proband-independent following normal distribution $N(0, \sigma^2)$ for various values of σ. The noise was added to all measurements and classification rules are learned over this contaminated data set. We consider the noise with $\sigma = 0.1$ to be slight and with $\sigma = 0.3$ to be moderate, revealing already the advantage of robust methods compared to non-robust ones. Such contamination was repeated 100-times and the classification performance of various methods was evaluated for each case and finally averaged.

The results of the classification performance of various methods on data artificially contaminated by noise, as presented again in Table 1, show an evidence of a reasonable robustness of SCRDA as well the novel methods. The larger value of σ, the more influential outliers are present in the contaminated data set. Indeed, the reduction of the classification performance of the standard data mining methods is not caused by the noise itself, but rather by severe outliers. The robustness of SCRDA to (small) measurement errors has not however been systematically investigated [21] and we think that its ability to outperform the SVM has not been documented sufficiently in the literature. Still, the robustness of the new methods MWCD-LDA* and L_2-SCRRDA is even able to outperform the relatively robust SCRDA.

The MRMR variable selection allows to find a small set of variables with an ability to diagnose schizophrenic patients based only on the fMRI measurements of the brain in the resting state, which is an interesting result from the point of view of neuroscience research. Let us inspect the effect of dimensionality reduction performed by other approaches. If the variables are arranged according to values of the statistic of the two-sample t-test, the best performance with the Youden's index $I = 1.00$ can be obtained only if at least 21 variables are selected. If PAM is used to arrange the variable according to their contribution to separating the two groups, then at least 36 variables are need in order to reach $I = 1.00$.

Table 2. Example of Sect. 5.4. Youden's index (23) as a classification performance measure computed for a 5-fold cross validation study. PCA is used with a fixed number of 10 principal components. The number of major variables within L_1-SCRRDA is also shown (see Remark 1 in Sect. 4.2).

Classification method	Resting state vs. movie					
	#1	#2	#3	#4	#5	#6
SVM	1.00	1.00	1.00	1.00	1.00	1.00
L_2-SCRRDA	1.00	1.00	1.00	1.00	1.00	1.00
L_1-SCRRDA	1.00	1.00	1.00	1.00	1.00	1.00
Number of major variables	3	7	2	3	1	1
PCA $\implies$ LDA	1.00	1.00	1.00	1.00	1.00	1.00

5.4   Brain Activity: A Closer Look on Individual Movies

In addition, we solve more particular tasks to learn the classification rule allowing to distinguish between the resting state and only one given movie over the training data from Sect. 5.2. Such six tasks to classify between the resting state and the i-th movie ($i = 1, \ldots, 6$) always deal with $p = 4005$ variables and 24 observations for each of the two groups. A wide variety of classification procedures is able to reach $I = 1.00$ as shown in Table 2 for selected methods. By a robust variable selection method of [22], we additionally verified that small sets of variables can be found allowing to solve the classification tasks easily.

Finally, we considered other classification tasks with the aim to separate individuals pairs of movies, i.e. classifying between movie #1 and movie #2, between movie #1 and movie #3 etc. The results for each of these 15 tasks again show that $I = 1.00$ can be attained easily, even using a small number of variables. Particularly, we used again the robust variable selection of [22]. The minimal number of variables needed to obtain the $I = 1.00$ result turns out to be greater or equal to 2 and always less or equal to 30. Such small numbers can be explained by a small number of observations in each of the groups.

5.5   Cardiovascular Genetic Study

We illustrate the performance of the novel classifiers on data acquired within the cardiovascular genetic study of the Center of Biomedical Informatics in Prague. The data set was described and analyzed by standard methods in [20]. The aim of the study was to identify a small set of genes associated with excess genetic risk for the incidence of a cardiovascular disease among $p = 38\,590$ gene transcripts. The gene expressions were measured on $n = 48$ individuals, namely on 24 patients having a cerebrovascular stroke and 24 control persons.

Some methods reach the classification performance $I = 1.00$, which is true for some standard methods including SVM and LDA* and also for the novel methods MWCD-LDA* and L_2-SCRRDA. This can be explained by the very large p, compared to other data sets of this paper.

The dimensionality reduction by means of PCA has drastic consequences, which can be explained by its unsupervised nature ignoring the grouping structure of the data. Indeed, it is the MRMR variable selection which confirms this opinion. MRMR shows that there is a small number of variables responsible for the separation between the two groups and is able to yield much improved results, namely with only 10 most relevant genes allowing to separate both groups with $I = 1.00$.

5.6 Metabolomic Profiles Data

We analyze a publicly available benchmarking data set of prostate cancer metabolomic data [32] of $p = 518$ metabolites measured over two groups of patients, who are either those with a benign prostate cancer (16 patients) or with other cancer types (26 patients).

A detailed analysis of the data reveals that they are not contaminated by severe outliers. Still, MWCD-LDA* and L_2-SCRRDA are able to slightly outperform other regularized versions of LDA. Their result are comparable to the SVM classifier while other classifiers yield inferior results. The MRMR variable selection performs better compared to the unsupervised dimensionality reduction by means of PCA, while there is no small group of variables responsible for a large portion of variability of the data and the first few principal components seem rather arbitrary for the classification task.

5.7 Keystroke Dynamics Data

The last data set contains data from a biometric authentication study by means of keystroke dynamics, which was described and analyzed in [22]. Our aim is to illustrate the performance of the novel classifiers also on this data set with a small number of variables. A set of probands was asked to type the same short sequence (password) consisting of 8 characters repeatedly. The particular task now is to classify to 2 groups, i.e. to distinguish between two probands exploiting $p = 15$ variables (keystroke durations and latencies in milliseconds) available for each of them.

As Table 1 indicates, the best results are obtained with L_2-SCRRDA, while other robust regularized LDA versions together with SCRDA remain slightly inferior. Again, the SVM classifier is based on a large number of support vectors ($\geq 90\%$ of observations). Dimensionality reduction leads to a loss of information compared to methods using all variables. Our detailed analysis of the data reveals the percentage of severe outliers to be about 10%. Indeed, if we additionally performed a manual outlier detection and then ignored the outliers from the data set, MWCD-LDA* and L_2-SCRRDA still retain their performance which was not affected by the outliers. On the other hand, the performance of SVM and non-robust versions of LDA is suddenly improved.

6 Discussion

6.1 Advantages of Robust Regularized Classification

Regularized LDA has been advocated for its computational and statistical benefits, which may be revealed not only for $n < p$ but also for $n > p$ with a relatively small n [13]. Regularization is generally believed to ensures a robustness [12,34], although this does not hold as a universal principle. In the context of regularized LDA, only a robustness with respect to small (local) changes of the measured data is ensured as it has been observed empirically [1]. Nevertheless, regularized LDA is not robust to more severe noise or outliers, as revealed in our examples.

Appealing properties of regularized LDA have lead us to the idea of joining principles of suitable Tikhonov-type regularization with statistical robustness. Advantages of the newly proposed MWCD-LDA*, L_1-SCRRDA and L_2-SCRRDA include:

- High robustness to outliers thanks to a high breakdown point of MWCD (as a consequence of using the implicit weights similarly to the context of linear regression [19]);
- No assumption on the distribution of the outliers;
- Availability of efficient algorithms based on numerical linear algebra;
- No need for a prior dimensionality reduction;
- Comprehensibility.

While an SVM classifier yields the best classification performance in some of the examples, especially those with a relatively smaller p, we perceive also its drawbacks and try to summarize them.

- It depends on too many support vectors for $n < p$ (more than 90% of the observations play the role of support vectors in the examples);
- The necessity to optimize its parameters over a sufficiently large number of observations;
- A tendency to overfitting for $n < p$ [11];
- Internal structure not supposed to be understood (black box);
- Non-robustness to outliers.

6.2 Comprehensibility

Comprehensibility represents an important requirement in a wide variety of classification tasks in bioinformatics. Therefore, the discussion of comprehensibility of the newly proposed methods deserves to be presented as a separate subsection.

We consider the classical LDA itself to be comprehensible in the sense that it is based on the Mahalanobis distance of a given (new) measurement from each of the groups of data. The contribution of an individual observation to the final classification rule is only through the sufficient statistics, i.e. means of the corresponding groups and scatter matrix.

The classification rules of MWCD-LDA*, L_1-SCRRDA, L_2-SCRRDA and M-LDA* can be interpreted as based on a deformed (regularized) Mahalanobis distance between a new observation Z and the mean of each group. Let us discuss the particular situation of MWCD-LDA* and consider the singular value decomposition (SVD) of S^*_{MWCD} in the form $S^*_{MWCD} = Q\Lambda Q^T$. The aforementioned deformed Mahalanobis distance can be interpreted as the Euclidean distance applied on $\Lambda^{-1/2}Q^T Z$. More specifically, if we assume Z to come from one of the groups with the covariance matrix Σ, we obtain in a straightforward way

$$\text{var } \Lambda^{-1/2}Q^T Z = \Lambda^{-1/2}Q^T \cdot \text{var } Z \cdot Q\Lambda^{-1/2}$$
$$= \Lambda^{-1/2}Q^T \cdot Q\Lambda Q^T Q\Lambda^{-1/2} = \mathcal{I}_p. \tag{26}$$

The deformed Mahalanobis distance of L_1-SCRRDA and L_2-SCRRDA takes additionally into account a regularization of the means.

Regularizing the means can be theoretically justified as exploiting Stein's statistical estimation [13, 28], extending Stein's shrinkage estimator originally proposed for the mean of multivariate normal data. The regularization of the means within L_1-SCRRDA and L_2-SCRRDA replaces (unbiased) arithmetic means by their (biased) shrunken counterparts, allowing to reduce their mean square error if the regularization parameters are sufficiently small.

Also the implicit weights assigned to individual observations in methods of Sects. 4.1, 4.2 and 4.3 allow a clear interpretation. Less reliable observations (potential outliers) obtain small or negligible weights. Such permutation of the weights is used which minimizes the determinant of a weighted empirical covariance matrix. The weights are used to compute the weighted mean and weighted empirical covariance matrix. In the numerical examples, we have verified that outlying measurements obtain small weights, which ensures the robustness of the method.

6.3   Limitations

Let us mention also the limitations of the newly proposed classification methods.

- Suitability of all the novel methods for data following an elliptically symmetric unimodal multivariate distribution.
- All the novel methods require an intensive computation.
- The implicit weights in methods of Sects. 4.1, 4.2 and 4.3 are assigned to individual observations (rather than perhaps to individual variables).
- The variability not substantially different across variables is unexpressedly assumed for all regularized LDA methods. Still, the novel methods seem to yield reliable results on the data sets of Sect. 5, although this implicit assumption of homogeneous variances of all variables is violated in them.
- L_1-SCRRDA and L_2-SCRRDA are more computationally demanding compared to MWCD-LDA*, but yield comparable results, i.e. there seems no major added value of regularizing the means in contrary to the experience of e.g. [10].

Finally, we need to recall that the regularization itself may be a too radical modification of the original problem of rank n, which is replaced by a problem of rank p, which may be much larger. Such increase of the dimensionality of the scatter matrix may cause the new problem to be very distant from the original problem even if an extremely small λ is used and if the regularized problem was solved in an arbitrary-precision arithmetic [5, 24].

7 Conclusions and Future Work

The analysis of high-dimensional data with the number of variables p largely exceeding the number of observations n becomes an important task in numerous tasks of bioinformatics. While numerous available algorithms for the regularized LDA are popular for the analysis of high-dimensional data [21], regularized LDA turns out to be vulnerable to the presence of outliers, because it is based on the same maximum likelihood estimation principle as the standard LDA. It is the maximum likelihood estimation which causes the high sensitivity of the standard as well as of various regularized versions of LDA to outliers.

In this paper, we combine robustness to the presence of outliers with regularized estimation of the scatter matrix of the multivariate data in a unique way. As a result, four new robust classification methods for high-dimensional observations are proposed in Sect. 4. Three of the methods are based on implicit weighting of individual observations, while M-LDA* is based on M-estimation. In addition, newly proposed methods L_1-SCRRDA and L_2-SCRRDA replace also the sample mean of each group by a regularized (shrunken) robust estimator.

We analyzed several real data sets fulfilling $n < p$ in Sect. 5. These data sets coming from various problems of (bioinformatics) research can be characterized as high-dimensional in sense of $n < p$ or even $n \ll p$.

Particularly, we pay the largest attention to the analysis of an original brain activity data set from a neuroscience research study investigating connections among brain parts during a resting state. Results of various classification methods show distinct differences between the resting and non-resting state. At the same time, different movies shown to the set of 24 probands turn out to activate different connections between pairs of brain parts.

To investigate the performance of individual methods on data contaminated by noise, we also introduced an artificial contamination to the brain activity data. Indeed, robustness to moderate or severe outliers is an important requirement in the analysis of high-dimensional data, especially if the number of observations is small. The results reveal a regularized LDA in a standard form to be sensitive to outliers. SCRDA turns out to be moderately robust, which is an effect of the regularizing also the means, which yields denoised versions of standard means. The novel methods turn out to be even more robust also against severely outlying measurements. It is an artificial contamination of the data which reveals the robustness of the novel methods as their strength and the whole study with artificial contamination reveals the advantage of robust methods compared to non-robust ones. However, regularizing the means applied on the robust methods does not bring any major additional benefit compared to MWCD-LDA*, while it requires a high increase in computational complexity.

Open problems concerning the newly proposed methods as well as more general ideas for a future research in the area of robust analysis of high-dimensional data contain the following tasks.

- Formulating more efficient algorithms tailor-made for important specific choices of the target matrix T as alternatives to Algorithms 2 or 3.
- Comparing various approaches to regularizing the means (i.e. for various norms or various shrinkage targets) in a large simulation study.
- Comparing the performance and robustness of the new methods with approaches based on a robust PCA.
- Investigating the non-robustness of other standard regularized classification methods (e.g. of PAM).
- Extending the combination of regularization and robustness to other methods based on the Mahalanobis distance, such as classification trees, entropy estimators, k-means clustering, or dimensionality reduction.
- Combining regularization and robustness to other methods, including neural networks or SVM or even linear regression (e.g. robust lasso estimator).
- Developing other multivariate methods based on the regularized MWCD estimators, e.g. robust PAM or robust regularized PCA.

From the point of view of the neuroscience research, future investigations are planned to search for a small set of variables allowing to distinguish schizophrenic patients from control individuals based only on the fMRI measurements of the brain measured in the resting state.

Acknowledgments. Preliminary results were first presented at the BIOSTEC/ BIOINFORMATICS 2016 conference (21–23 February 2016 in Rome), where they were published in the proceedings.

The work was supported by the project Nr. LO1611 with a financial support from the MEYS under the NPU I program. The work of J. Kalina was financially supported by the Neuron Fund for Support of Science. The work of J. Hlinka was supported by the Czech Science Foundation project No. 13-23940S.

References

1. Ben-Tal, A., El Ghaoui, L., Nemirovski, A.: Robust Optimization. Princeton University Press, Princeton (2009)
2. Bühlmann, P., van de Geer, S.: Statistics for High-dimensional Data. Springer, New York (2011)
3. Chen, Y., Wiesel, A., Hero, A.O.: Robust shrinkage estimation of high dimensional covariance matrices. IEEE Trans. Sig. Process. **59**, 4097–4107 (2011)
4. Croux, C., Dehon, C.: Robust linear discriminant analysis using S-estimators. Can. J. Stat. **29**, 473–493 (2001)
5. Davies, P.: Data Analysis and Approximate Models: Model Choice, Location-Scale, Analysis of Variance, Nonparametric Regression and Image Analysis. Chapman & Hall/CRC, Boca Raton (2014)

6. Davies, P.L., Gather, U.: Breakdown and groups. Ann. Stat. **33**, 977–1035 (2005)
7. Duffau, H.: Brain Mapping: From Neural Basis of Cognition to Surgical Applications. Springer, Vienna (2011)
8. Dziuda, D.M.: Data Mining for Genomics and Proteomics: Analysis of Gene and Protein Expression Data. Wiley, New York (2010)
9. Filzmoser, P., Todorov, V.: Review of robust multivariate statistical methods in high dimension. Analytica Chinica Acta **705**, 2–14 (2011)
10. Guo, Y., Hastie, T., Tibshirani, R.: Regularized discriminant analysis and its application in microarrays. Biostatistics **8**, 86–100 (2007)
11. Han, H., Jiang, X.: Overcome support vector machine diagnosis overfitting. Cancer Inf. **13**, 145–148 (2014)
12. Hansen, P.C.: Rank-deficient and Discrete Ill-posed Problems: Numerical Aspects of Linear Inversion. SIAM, Philadelphia (1998)
13. Hastie, T., Tibshirani, R., Friedman, J.: The Elements of Statistical Learning, 2nd edn. Springer, New York (2008)
14. Hlinka, J., Paluš, M., Vejmelka, M., Mantini, D., Corbetta, M.: Functional connectivity in resting-state fMRI: is linear correlation sufficient? NeuroImage **54**, 2218–2225 (2011)
15. Huber, P.J., Ronchetti, E.M.: Robust Statistics, 2nd edn. Wiley, New York (2009)
16. Hubert, M., Rousseeuw, P.J., van Aelst, S.: High-breakdown robust multivariate methods. Stat. Sci. **23**, 92–119 (2008)
17. Hubert, M., Debruyne, M.: Minimal covariance determinant. Wiley Interdisc. Rev. Comput. Stat. **2**, 36–43 (2010)
18. Jurečková, J., Portnoy, S.: Asymptotics for one-step M-estimators in regression with application to combining efficiency and high breakdown point. Commun. Stat. Theor. Methods **16**, 2187–2199 (1987)
19. Kalina, J.: Implicitly weighted methods in robust image analysis. J. Math. Imag. Vis. **44**, 449–462 (2012)
20. Kalina, J., Seidl, L., Zvára, K., Grünfeldová, H., Slovák, D., Zvárová, J.: System for selecting relevant information for decision support. Stud. Health Technol. Inf. **183**, 83–87 (2013)
21. Kalina, J.: Classification analysis methods for high-dimensional genetic data. Biocybern. Biomed. Eng. **34**, 10–18 (2014)
22. Kalina, J., Schlenker, A.: A robust and regularized supervised variable selection. BioMed Res. Int. (2015). Article no. 320385
23. Kindermans, P.-J., Schreuder, M., Schrauwen, B., Müller, K.-R., Tangermann, M.: True zero-training brain-computer interfacing-an online study. PLoS One **9** (2014). Article no. 102504
24. Kůrková, V., Sanguineti, M.: Learning with generalization capability by kernel methods of bounded complexity. J. Complex. **21**, 350–367 (2005)
25. Lopuhaä, H.P., Rousseeuw, P.J.: Breakdown points of affine equivariant estimators of multivariate location and covariance matrices. Ann. Stat. **19**, 229–248 (1991)
26. Maronna, R.A., Martin, D.R., Yohai, V.J.: Robust Statistics: Theory and Methods. Wiley, New York (2006)
27. Peng, H., Long, F., Ding, C.: Feature selection based on mutual information: criteria of max-dependency, max-relevance, and min-redundancy. IEEE Trans. Pattern Anal. Mach. Intell. **27**, 1226–1238 (2005)
28. Pourahmadi, M.: High-dimensional Covariance Estimation. Wiley, New York (2013)
29. Roelant, E., van Aelst, S., Willems, G.: The minimum weighted covariance determinant estimator. Metrika **70**, 177–204 (2009)

30. Rousseeuw, P.J., Leroy, A.M.: Robust Regression and Outlier Detection. Wiley, New York (1987)
31. Rousseeuw, P.J., van Driessen, K.: A fast algorithm for the minimum covariance determinant estimator. Technometrics **41**, 212–223 (1999)
32. Sreekumar, A., et al.: Metabolomic profiles delineate potential role for sarcosine in prostate cancer progression. Nature **457**, 910–914 (2009)
33. Steinwart, I., Christmann, A.: Support Vector Machines. Springer, New York (2008)
34. Tibshirani, R., Narasimhan, B.: Class prediction by nearest shrunken centroids, with applications to DNA microarrays. Stat. Sci. **18**, 104–117 (2003)
35. Todorov, V., Filzmoser, P.: An object-oriented framework for robust multivariate analysis. J. Stat. Softw. **32**(3), 1–47 (2009)
36. Tyler, D.E.: A distribution-free M-estimator of multivariate scatter. Ann. Stat. **15**, 234–251 (1987)
37. Tyler, D.E.: Breakdown properties of the M-estimators of multivariate scatter (2014). http://arxiv.org/pdf/1406.4904v1.pdf
38. Wager, T.D., Keller, M.C., Lacey, S.C., Jonides, J.: Increased sensitivity in neuroimaging analyses using robust regression. NeuroImage **26**, 99–113 (2005)

Finding Median and Center Strings for a Probability Distribution on a Set of Strings Under Levenshtein Distance Based on Integer Linear Programming

Morihiro Hayashida[1]([✉]) and Hitoshi Koyano[2]

[1] Bioinformatics Center, Institute for Chemical Research, Kyoto University,
Gokasho, Uji, Kyoto 611-0011, Japan
morihiro@kuicr.kyoto-u.ac.jp
[2] Graduate School of Medicine, Kyoto University, 54 Kawahara-cho, Shogoin,
Sakyo-ku, Kyoto 606-8397, Japan

Abstract. For a data set composed of numbers or numerical vectors, a mean is the most fundamental measure for capturing the center of the data. However, for a data set of strings, a mean of the data cannot be defined, and therefore, median and center strings are frequently used as a measure of the center of the data. In contrast to calculating a mean of numerical data, constructing median and center strings of string data is not easy, and no algorithm is found that is guaranteed to construct exact solutions of center strings. In this study, we first generalize the definitions of median and center strings of string data into those of a probability distribution on a set of all strings composed of letters in a given alphabet. This generalization corresponds to that of a mean of numerical data into an expected value of a probability distribution on a set of numbers or numerical vectors. Next, we develop methods for constructing exact solutions of median and center strings for a probability distribution on a set of strings, applying integer linear programming. These methods are improved into faster ones by using the triangle inequality on the Levenshtein distance in the case where a set of strings is a metric space with the Levenshtein distance. Furthermore, we also develop methods for constructing approximate solutions of median and center strings very rapidly if the probability of a subset composed of similar strings is close to one. Lastly, we perform simulation experiments to examine the usefulness of our proposed methods in practical applications.

1 Introduction

Taking an average is a fundamental statistical method for understanding a data set. In this paper, we focus on a set of strings. For instance, nucleotide sequences of DNAs and RNAs are represented by strings as well as protein amino acid sequences. The number of such sequences has rapidly increased, and analytical methods are required. In the field of evolutionary studies of organisms, it would be an aim to find the DNA nucleotide sequence of common ancestors. In the field

© Springer International Publishing AG 2017
A. Fred and H. Gamboa (Eds.): BIOSTEC 2016, CCIS 690, pp. 108–121, 2017.
DOI: 10.1007/978-3-319-54717-6_7

of protein science, it is essential to detect functional motifs in protein amino acid sequences. Also in the field of image recognition, there are several applications such as post-processing of optical character recognition (OCR) results [2] and shape recognition [4]. Furthermore, it can be applied to classification and clustering of strings and biological sequences [20].

Several measures have been proposed for representing a "center" of a data set of strings because a mean cannot be defined for a string data set. One is a *median string*, which is defined as a string that minimizes the sum of distances with strings included in a set [15]. Another is a *center string*, which is defined as a string that minimizes the maximum of distances with strings [8]. Several dissimilarities such as Levenshtein distance [18], Hamming distance [9], and Jaro-Winkler distance [26] have been proposed, where the Jaro-Winkler distance is known not to obey the triangle inequality. The Levenshtein distance between two given strings s and t allows three types of edit operations, insertion, deletion, and substitution, and can be calculated in polynomial time $O(|s||t|)$ using dynamic programming, where $|s|$ denotes the length of s. The Hamming distance has been also used for closest strings and related problems [6,7]. Data reduction techniques that reduce candidates of a center string under the Hamming distance were developed [11]. However, they mentioned that their parameterized methods would be not applicable for finding center strings under the Levenshtein distance. A genetic algorithm for finding closest strings under rank distance was developed [5], where the rank distance has been applied in biology, natural language processing, and authorship attribution.

The problems of finding the median and center strings for a finite set of strings under the Levenshtein distance have been proved to be NP-complete for an unbounded alphabet [10], and even for a binary alphabet [21,22]. It has been proved that a related problem, the consensus string problem with consensus error (CSCE), is also NP-complete if the distance function is the weighted edit distance with a metric penalty matrix [24]. An exact algorithm for finding the median string under the Levenshtein distance using dynamic programming was proposed [17], which requires an N-dimensional array and $O(n^N)$ time and space for a set of N strings with length n. Therefore, for example, it requires $10^{10} \cdot 4$ bytes $= 40\,\text{GB}$ memory for $n = N = 10$. For computing approximate median strings in practical time, several methods have been proposed. If given strings are similar, the path by the optimal dynamic programming should be close to the main diagonal. Hence, the method to restrict candidate paths to a region near the diagonal was proposed [19]. A greedy algorithm starts from an empty string, and selects a letter that minimizes the exact consensus error [3]. An online algorithm takes the current approximate median string and a new string, and calculates a weighted mean of these strings [12]. In a stochastic approach, some conditional probability from a string to another was defined, and an approximate median string was obtained by expectation maximization technique [23]. An iterative algorithm applies the edit operation with some highest score to the current string until a better solution is not found [1]. These methods output approximate median strings, and there are a few methods to output optimal median strings. Methods for finding optimal center strings have not been developed.

In this paper, hence, we approach the problem of finding optimal median and center strings by applying efficient solvers for integer linear programming (ILP) problems and propose novel ILP-based methods named ILPMed and ILP-Cen for generalized definitions of median and center strings to a probability distribution on a set of strings [16]. Furthermore, making use of the triangle inequality on the Levenshtein distance, we improve ILPMed and ILPCen into faster methods, ILPMedTri and ILPCenTri. By restricting ILPs to a region near the diagonal and forcing each distance between two strings within some threshold, we propose ILP-based methods, ILPMedDiag, ILPCenDiag, ILPMedFar, and ILPCenFar, for finding approximate median and center strings. We perform several computational experiments and verify the efficiency of our methods.

2   Methods

We use the Levenshtein distance because it is often used and a fundamental edit distance. In this section, we briefly review the computation of the Levenshtein distance, median, center strings, and propose integer linear programming formulations for exact and approximate median and center strings. Let $\mathcal{A} = \{a_1, \ldots, a_z\}$ be an alphabet composed of z letters, for instance, $\mathcal{A} = \{A, T, G, C\}$ for DNA nucleotide sequences. We define $\mathcal{A}^*$ to be the set of all strings on $\mathcal{A}$ with varying lengths, and for a string $s \in \mathcal{A}^*$, $|s|$ denotes the length of s.

2.1   Levenshtein Distance

The Levenshtein distance $d(s,t)$ between two strings s and t is defined as the minimum cost of sequences of edit operations transforming $s = s_1 \cdots s_n$ into $t = t_1 \cdots t_m$, and can be calculated by the following dynamic programming [25].

$$D[0,0] = 0, \tag{1}$$

$$D[i,j] = \min \begin{cases} D[i-1, j-1] + \gamma(s_i \to t_j) \\ D[i-1, j] + \gamma(s_i \to \epsilon) \\ D[i, j-1] + \gamma(\epsilon \to t_j) \end{cases} \tag{2}$$

where ϵ denotes an empty letter, $\gamma(s_i \to t_j)$, $\gamma(s_i \to \epsilon)$, and $\gamma(\epsilon \to t_j)$ denote the costs of substitution, deletion, and insertion, respectively. Then, $D[n, m]$ is the Levenshtein distance $d(s, t)$.

2.2   Median and Center Strings

Given N strings $s^{(k)}$ with length n_k $(k = 1, \ldots, N)$ on $\mathcal{A}^*$, the median string is defined by

$$\mathrm{argmin}_{t \in \mathcal{A}^*} \sum_{k=1}^{N} d(t, s^{(k)}). \tag{3}$$

Similarly, the center string is defined by

$$\operatorname{argmin}_{t \in \mathcal{A}^*} \max_{k \in \{1,\ldots,N\}} d(t, s^{(k)}). \tag{4}$$

For a given probability distribution $p(s)$ on $\mathcal{A}^*$, we define median and center strings by

$$\operatorname{argmin}_{t \in \mathcal{A}^*} \sum_{s \in \mathcal{A}^*} p(s)d(t, s), \tag{5}$$

$$\operatorname{argmin}_{t \in \mathcal{A}^*} \max_{s \in \mathcal{A}^*} p(s)d(t, s), \tag{6}$$

respectively. If $p(s^{(k)}) = \frac{1}{N}$ for $k = 1,\ldots,N$ and $p(s) = 0$ for all $s \notin \{s^{(k)}\}$, Eqs. (3) and (4) are equivalent to Eqs. (5) and (6), respectively.

2.3   Integer Linear Programming Formulation

Since it is known that problems of finding median and center strings under the Levenshtein distance are NP-hard [21,22], we make use of integer linear programming which efficient solvers have been developed. We can find a median string t by integer linear programming if the Levenshtein distance $d(t, s^{(k)})$ between t and $s^{(k)}$ can be calculated in linear formulas. It, however, is difficult to directly represent the array $D[i, j]$ in the dynamic programming by integer linear programming because it includes the selection of the minimum value in Eq. (2).

Suppose that a probability distribution $p(s)$ on $\mathcal{A}^*$ is given, where the number of strings s satisfying $p(s) > 0$ is finite, N, that is, $p(s^{(k)}) > 0$ ($k = 1,\ldots,N$). We use integer numbers $1,\ldots,|\mathcal{A}|$ instead of letters in $\mathcal{A}$ because a variable takes a value in linear programming. $s_i^{(k)}$ ($i = 1,\ldots,n_k$) is given as a constant value of $1,\ldots,|\mathcal{A}|$, and represents the i-th letter in $s^{(k)}$. t_j ($j = 1,\ldots,m$) is a variable taking a value of $1,\ldots,|\mathcal{A}|$, and represents the j-th letter in the median string. Then, we propose the following integer linear programming formulation, called ILPMed, for finding the median string for $p(s)$ under the Levenshtein distance with costs C_{sub}, C_{del}, C_{ins} of substitution, deletion, insertion.

$$\min \ \sum_{k=1}^{N} p(s^{(k)}) \Big\{ \sum_{i=1}^{n_k} C_{del} x_{k,i,0} + \sum_{j=1}^{m} C_{ins} y_{k,0,j}$$

$$+ \sum_{i=1}^{n_k} \sum_{j=1}^{m} (C_{del} x_{kij} + C_{ins} y_{kij} + C_{sub} h_{kij}) \Big\} - C_{ins}(m - l)$$

subject to
  for all $k = 1,\ldots,N$,

$$1 = x_{k,1,0} + y_{k,0,1} + z_{k,1,1}, \tag{a1}$$

$$x_{k,i,0} = x_{k,i+1,0} + y_{k,i,1} + z_{k,i+1,1} \quad \text{for all } i < n_k, \tag{a2}$$

$$x_{k,n_k,0} = y_{k,n_k,1}, \tag{a3}$$

$$y_{k,0,j} = x_{k,1,j} + y_{k,0,j+1} + z_{k,1,j+1} \quad \text{for all } j < m, \tag{a4}$$

$$y_{k,0,m} = x_{k,1,m}, \tag{a5}$$

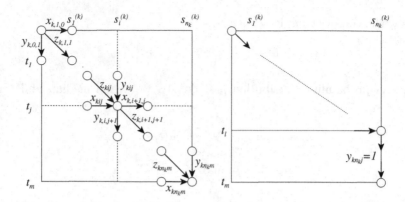

Fig. 1. Illustration on variables appeared in our integer linear programming formulation. (Left) Variables x_{kij}, y_{kij}, and z_{kij} represent a path in dynamic programming for calculating the Levenshtein distance if the value of the variable is equal to 1. (Right) Variable l represents the length of string t. For all $j > l$, y_{kn_kj} is forced to be 1.

$$x_{kij} + y_{kij} + z_{kij} = x_{k,i+1,j} + y_{k,i,j+1} + z_{k,i+1,j+1}$$
$$\text{for all } i < n_k, \ j < m, \quad \text{(a6)}$$
$$x_{kn_kj} + y_{kn_kj} + z_{kn_kj} = y_{k,n_k,j+1} \quad \text{for all } j < m, \quad \text{(a7)}$$
$$x_{kim} + y_{kim} + z_{kim} = x_{k,i+1,m} \quad \text{for all } i < n_k, \quad \text{(a8)}$$
$$x_{kn_km} + y_{kn_km} + z_{kn_km} = 1, \quad \text{(a9)}$$
$$y_{kn_kj} \geq \tfrac{1}{m}(j - l) \quad \text{for all } j, \quad \text{(b)}$$
$$\text{for all } k, \ i, \ j,$$
$$s_i^{(k)} - t_j \leq |\mathcal{A}| g_{kij}, \quad \text{(c1)}$$
$$t_j - s_i^{(k)} \leq |\mathcal{A}| g_{kij}, \quad \text{(c2)}$$
$$h_{kij} \geq z_{kij} + g_{kij} - 1, \quad \text{(d1)}$$
$$h_{kij} \leq \tfrac{1}{2}(z_{kij} + g_{kij}), \quad \text{(d2)}$$
$$x_{kij}, y_{kij}, z_{kij}, g_{kij}, h_{kij} \in \{0, 1\},$$
$$t_j \in \{1, \ldots, |\mathcal{A}|\}, \ 0 \leq l \leq m,$$

where m is a sufficient large constant integer (e.g., the sum of n_k), and l is the variable representing the length of median string.

In the formulation, variable x_{kij} takes 1 if $s_i^{(k)}$ is deleted, otherwise 0 (see Fig. 1). y_{kij} takes 1 if t_j is inserted, otherwise 0. z_{kij} takes 1 if $s_i^{(k)}$ is substituted with t_j, otherwise 0. There must be exactly one path from the upper left to the lower right for each string $s^{(k)}$. If either of x_{kij}, y_{kij}, and z_{kij} is 1, either of $x_{k,i+1,j}$, $y_{k,i,j+1}$, and $z_{k,i+1,j+1}$ must be 1, which is represented by Eq. (a6). According to the position (i, j), Eqs. (a1–9) are constructed. Equation (b) represents the constraint that the length of median string t is l, and y_{kn_kj} is forced to be 1 if $j > l$. It is difficult to represent the Levenshtein distance $d(t, s^{(k)}) = \sum_{i=1}^{n_k} C_{del} x_{k,i,0} + \sum_{j=1}^{l} C_{ins} y_{k,0,j} + \sum_{i=1}^{n_k} \sum_{j=1}^{l} (C_{del} x_{kij} + C_{ins} y_{kij} + C_{sub} h_{kij})$ in the formulation because l is also a variable to be found. Hence, we use a constant integer m instead of l. Then, the sum includes the extra cost of $C_{ins}(m - l)$.

We reduce the cost such that the objective function represents the sum in Eq. (5). It should be noted that for all $j > l$, y_{kn_kj} is forced to be 1 (see Fig. 1). Eqs. (c1–2) represent that g_{kij} becomes 1 if $s_i^{(k)}$ is the same as t_j. Eqs. (d1–2) represent that h_{kij} becomes 1 if and only if both of z_{kij} and g_{kij} are 1. It means that the substitution cost from $s_i^{(k)}$ to t_j is C_{sub} if $s_i^{(k)} \neq t_j$, otherwise 0.

It is guaranteed that we can find the median string for $p(s)$ under the Levenshtein distance by solving this integer linear programming formulation because the objective function is equivalent to the sum in Eq. (5), t can be any string with length up to $m = \sum_{k=1}^{N} n_k$, and the sum for a string with length more than m is larger than that for the concatenated string of all $s^{(k)}$.

In a similar way to median strings, we propose the following integer linear programming formulation, called ILPCen, for finding the center string for a probability distribution $p(s)$.

min $\quad d$

subject to

$\quad$ for all $k = 1, \ldots, N$,

$$p(s^{(k)})\Big\{\sum_{i=1}^{n_k} C_{del}x_{k,i,0} + \sum_{j=1}^{m} C_{ins}y_{k,0,j}$$

$$+ \sum_{i=1}^{n_k}\sum_{j=1}^{m}(C_{del}x_{kij} + C_{ins}y_{kij} + C_{sub}h_{kij}) - C_{ins}(m - l)\Big\} \leq d,$$

$1 = x_{k,1,0} + y_{k,0,1} + z_{k,1,1},$

$x_{k,i,0} = x_{k,i+1,0} + y_{k,i,1} + z_{k,i+1,1} \qquad$ for all $i < n_k$,

$x_{k,n_k,0} = y_{k,n_k,1},$

$y_{k,0,j} = x_{k,1,j} + y_{k,0,j+1} + z_{k,1,j+1} \qquad$ for all $j < m$,

$y_{k,0,m} = x_{k,1,m},$

$x_{kij} + y_{kij} + z_{kij} = x_{k,i+1,j} + y_{k,i,j+1} + z_{k,i+1,j+1}$

$\qquad\qquad\qquad\qquad\qquad\qquad$ for all $i < n_k,\ j < m$,

$x_{kn_kj} + y_{kn_kj} + z_{kn_kj} = y_{k,n_k,j+1} \qquad$ for all $j < m$,

$x_{kim} + y_{kim} + z_{kim} = x_{k,i+1,m} \qquad$ for all $i < n_k$,

$x_{kn_km} + y_{kn_km} + z_{kn_km} = 1,$

$y_{kn_kj} \geq \frac{1}{m}(j - l) \qquad$ for all j,

$\quad$ for all k, i, j,

$s_i^{(k)} - t_j \leq |\mathcal{A}|g_{kij},$

$t_j - s_i^{(k)} \leq |\mathcal{A}|g_{kij},$

$h_{kij} \geq z_{kij} + g_{kij} - 1,$

$h_{kij} \leq \frac{1}{2}(z_{kij} + g_{kij}),$

$x_{kij}, y_{kij}, z_{kij}, g_{kij}, h_{kij} \in \{0, 1\},$

$t_j \in \{1, \ldots, |\mathcal{A}|\},$

$0 \leq l \leq m,\ d \geq 0.$

Here, d is a variable that represents the maximum in Eq. (6).

2.4   Constraints for Efficient Execution

In order to find median and center strings more efficiently, it is necessary to add some constraints to ILPMed and ILPCen. Here, we propose three types of constraints utilizing the triangle inequality, estimation of Levenshtein distance, and alignment width.

Constraints on Triangle Inequality. Since the Levenshtein distance $d(s,t)$ for strings s and t satisfies the triangle inequality, we add the following constraints to ILPMed and ILPCen, respectively. We call the resulted ILPs, ILPMedTri and ILPCenTri, respectively.

$$d(s^{(k_1)}, s^{(k_2)}) + d(s^{(k_1)}, t) \geq d(s^{(k_2)}, t) \qquad \text{for all } k_1, k_2 (k_1 \neq k_2),$$
$$d(s^{(k_1)}, t) + d(s^{(k_2)}, t) \geq d(s^{(k_1)}, s^{(k_2)}) \qquad \text{for all } k_1, k_2 (k_1 < k_2),$$

where $d(s^{(k_1)}, s^{(k_2)})$ is a constant value calculated from given strings $s^{(k_1)}, s^{(k_2)}$, and $d(s^{(k)}, t)$ is the expression with variables $x_{kij}, y_{kij}, h_{kij}$, and l, represented by

$$\sum_{i=1}^{n_k} C_{del} x_{k,i,0} + \sum_{j=1}^{m} C_{ins} y_{k,0,j}$$
$$+ \sum_{i=1}^{n_k} \sum_{j=1}^{m} (C_{del} x_{kij} + C_{ins} y_{kij} + C_{sub} h_{kij}) - C_{ins}(m - l). \tag{7}$$

It should be noted that the constraints do not avoid ILPMed and ILPCen from finding optimal solutions, and ILPMedTri and ILPCenTri are able to find exact median and center strings. Jiang et al. also utilized the triangle inequality, and proposed a linear program for finding a lower bound for a median string [13,14]. Their method, however, cannot provide a median string.

Constraints on Paths Near Diagonal. If strings $s^{(k)}$ are similar to each other, we can restrict candidate paths to a region near the diagonal without loss of optimality. We introduce a constant positive integer w, and propose integer linear programming formulations, called ILPMedDiag and ILPCenDiag, by reducing variables, $x_{kij}, y_{kij}, z_{kij}, g_{kij}$, and h_{kij} with $|i - j| > w$ from ILPMed and ILPCen, respectively.

Constraints by Farthest String. We examine another restriction for paths in an alignment between given and desired strings. For a median string t and a string $s^{(k)}$, we have

$$p(s^{(k)})d(t, s^{(k)}) \leq \sum_{k'=1}^{N} p(s^{(k')})d(t, s^{(k')}) \tag{8}$$

$$\leq \sum_{k'=1}^{N} p(s^{(k')})d(s^{(k)}, s^{(k')}) \tag{9}$$

$$\leq \max_{k'=1,...,N} d(s^{(k)}, s^{(k')}). \tag{10}$$

Then, we have $d(t, s^{(k)}) \leq \frac{1}{p(s^{(k)})} \max_{k'=1,...,N} d(s^{(k)}, s^{(k')})$. If $p(s^{(k)})$ is too small, the right-hand side becomes large, and the constraint is not effective. Although it is difficult to prove that

$$d(t, s^{(k)}) \leq \max_{k'=1,...,N} d(s^{(k)}, s^{(k')}) \tag{11}$$

for each k and median or center string t, we propose integer linear programming formulations, called ILPMedFar and ILPCenFar, by adding Eq. (11) as constraints to ILPMed and ILPCen, respectively. As well as the constraints on the triangle inequality, $d(t, s^{(k)})$ is replaced with Eq. (7), and $d(s^{(k)}, s^{(k')})$ is a constant value.

3   Computational Experiments

For the evaluation of our methods, we performed several computational experiments. We used $C_{del} = C_{ins} = C_{sub} = 1$ to calculate the Levenshtein distance, and used an alphabet $\mathcal{A}$ with 4 letters as DNA and RNA nucleotide sequences. We randomly generated two types of probability distributions, $p_1(s)$ and $p_2(s)$, on $\mathcal{A}^*$. In $p_1(s)$, N strings $s^{(k)}$ with length n_k were generated as strings satisfying $p_1(s) > 0$ while varying $N = 2, \ldots, 10$ and $n_k = 2, \ldots, 10$, where n_k was the same for all $k = 1, \ldots, N$. Each $s_i^{(k)}$ was generated as $\min(1 + \lfloor |\alpha| \rfloor, |\mathcal{A}|)$, where α followed the normal distribution with mean 0 and variance 1, and $\lfloor \alpha \rfloor$ is the largest integer not greater than α. The probability of $p_1(s^{(k)})$ was generated uniformly at random such that $\sum_{k=1}^{N} p_1(s^{(k)}) = 1$ holds. In $p_2(s)$, N strings $s^{(k)}$ were generated from a string of $a_1 \cdots a_1$ ($a_1 \in \mathcal{A}$) with length n by applying randomly selected edit operations of substitution, insertion, and deletion, three times, where we examined $N = 2, \ldots, 10$ and $n = 5, \ldots, 10$, and the length n_k of $s^{(k)}$ could be different for different k. The probability of $p_2(s^{(k)})$ was generated uniformly at random such that $\sum_{k=1}^{N} p_2(s^{(k)}) = 1$ holds. For each case of $p_1(s)$, $p_2(s)$, n_k, and N, we generated a set of N strings $s^{(k)}$ with $p_1(s^{(k)})$ or $p_2(s^{(k)})$ ten times, and took the average of execution times. We used CPLEX (version 12.5) as the integer linear programming solver under a linux operating system with Xeon 2.9 GHz processor and 35 GB memory. It should be noted that we cannot run the existing method [17] based on dynamic programming for 10 strings with length 10 on the computer for finding the median string.

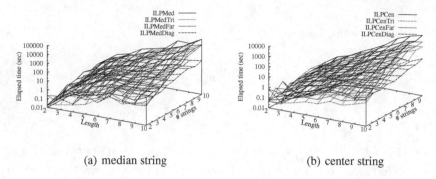

(a) median string   (b) center string

Fig. 2. Results on the average execution time in seconds on a log scale for probability distributions $p_1(s)$ for $N = 2, \ldots, 10$ and $n_k = 2, \ldots, 10$ (a) by ILPMed, ILPMedTri, ILPMedFar, and ILPMedDiag with $w = 2$ for finding median strings, and (b) by ILP-Cen, ILPCenTri, ILPCenFar, and ILPCenDiag with $w = 2$ for finding center strings.

Figure 2 shows results on the average execution time in seconds on a log scale by ILPMed, ILPMedTri, ILPMedFar, and ILPMedDiag with $w = 2$ and by ILPCen, ILPCenTri, ILPCenFar, and ILPCenDiag with $w = 2$ for probability distributions $p_1(s)$ for $N = 2, \ldots, 10$ and $n_k = 2, \ldots, 10$. Tables 1 and 2 show the detailed average and standard deviation of execution time by ILPMed, ILPMedTri, ILPMedFar, and ILPMedDiag and by ILPCen, ILPCenTri, ILP-CenFar, and ILPCenDiag for $n = 10$. We can see from these that the average execution times by ILPMed and ILPCen rapidly, almost exponentially, increased with both of the number N of strings and the length n_k because the problems are NP-hard. On the other hand, the average execution times by ILPMedDiag and ILPCenDiag were smaller than those by other methods in almost all cases because candidate solutions for the problems were restricted to the region near the diagonal. The execution times by ILPMedFar and ILPCenFar were often larger than those by ILPMedDiag, ILPCenDiag, ILPMedTri, ILPCenTri, and the constraints for each distance between two strings were not effective. The constraints, however, concerning the triangle inequality on the Levenshtein distance in ILPMedTri and ILPCenTri substantially reduced the execution times by ILPMed and ILPCen.

Figure 3 shows results on the average execution time in seconds on a log scale by ILPMed, ILPMedTri, ILPMedFar, and ILPMedDiag with $w = 2$ and by ILPCen, ILPCenTri, ILPCenFar, and ILPCenDiag with $w = 2$ for probability distributions $p_2(s)$ for $N = 2, \ldots, 10$ and $n = 5, \ldots, 10$. Tables 3 and 4 show the detailed average and standard deviation of execution time by ILPMed, ILPMedTri, ILPMedFar, and ILPMedDiag and by ILPCen, ILPCenTri, ILP-CenFar, and ILPCenDiag for $n = 10$. Also for $p_2(s)$, the average execution times by ILPMedDiag and ILPCenDiag were smaller than those by other methods,

Table 1. Results on the average and standard deviation of execution time in seconds by ILPMed, ILPMedTri, ILPMedFar, and ILPMedDiag with $w = 2$ for probability distributions $p_1(s)$ for $n = 10$ and $N = 2, \ldots, 10$.

N	ILPMed		ILPMedTri		ILPMedFar		ILPMedDiag	
	Average	s.d.	Average	s.d.	Average	s.d.	Average	s.d.
2	14.4	6.8	2.4	1.7	4.8	3.1	1.7	1.8
3	21.5	10.5	5.3	1.8	6.4	1.8	7.0	4.4
4	23.7	15.3	22.5	18.7	24.0	23.5	7.0	4.5
5	168.0	69.0	68.6	61.1	77.7	69.7	15.7	3.6
6	798.6	280.4	283.7	183.1	239.6	59.2	21.7	8.3
7	1400.6	773.3	542.5	326.0	548.6	196.0	41.6	20.5
8	11269.8	9074.5	1570.0	1238.5	2925.1	2251.8	61.0	39.2
9	22438.1	16417.6	2589.5	1528.4	4474.2	3989.5	37.7	17.9
10	18467.8	16625.5	4104.7	3825.7	4313.1	1987.4	67.4	22.8

Table 2. Results on the average and standard deviation of execution time in seconds by ILPCen, ILPCenTri, ILPCenFar, and ILPCenDiag with $w = 2$ for probability distributions $p_1(s)$ for $n = 10$ and $N = 2, \ldots, 10$.

N	ILPCen		ILPCenTri		ILPCenFar		ILPCenDiag	
	Average	s.d.	Average	s.d.	Average	s.d.	Average	s.d.
2	6.4	1.9	1.9	1.9	3.4	2.0	3.8	2.5
3	18.1	7.4	6.3	4.7	16.9	9.4	7.0	5.1
4	72.1	27.2	7.9	6.4	17.8	12.7	5.5	4.0
5	464.8	312.8	20.0	10.6	76.6	57.6	8.8	4.8
6	1015.6	580.6	72.7	46.3	278.2	244.4	6.9	3.0
7	2810.3	1218.8	173.4	148.6	484.6	252.6	10.7	4.4
8	5936.2	3522.8	142.9	82.6	1509.4	997.3	20.4	9.5
9	5086.0	1358.6	659.6	447.2	1485.8	825.0	23.9	10.5
10	8120.3	4336.0	597.5	545.8	1777.4	1221.1	41.1	29.5

and those by ILPMedTri and ILPCenTri were smaller than those by ILPMed and ILPCen, respectively. The slopes of ILPMed and ILPCen along the length for $p_2(s)$ were smaller than those for $p_1(s)$, respectively. In addition, the average execution times for $p_2(s)$ were smaller than those for $p_1(s)$. Figure 4 shows results on the average objective value by ILPMed, ILPMedFar, and ILPMedDiag with $w = 2$ and by ILPCen, ILPCenFar, and ILPCenDiag with $w = 2$ for probability distributions $p_2(s)$ for $N = 2, \ldots, 10$ and $n = 5, \ldots, 10$. It is noted that the objective values by ILPMed and ILPCen for $p_1(s)$ were almost the same as those by ILPMedDiag and ILPCenDiag, respectively. In $p_2(s)$, three edit operations were applied to strings. Differences of objective values between

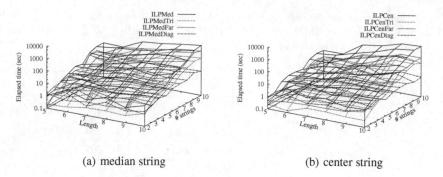

(a) median string   (b) center string

Fig. 3. Results on the average execution time in seconds on a log scale for probability distributions $p_2(s)$ for $N = 2, \ldots, 10$ and $n = 5, \ldots, 10$ (a) by ILPMed, ILPMedTri, ILPMedFar, and ILPMedDiag with $w = 2$ for finding median strings, and (b) by ILP-Cen, ILPCenTri, ILPCenFar, and ILPCenDiag with $w = 2$ for finding center strings.

Table 3. Results on the average and standard deviation of execution time in seconds by ILPMed, ILPMedTri, ILPMedFar, and ILPMedDiag with $w = 2$ for probability distributions $p_2(s)$ for $n = 10$ and $N = 2, \ldots, 10$.

N	ILPMed		ILPMedTri		ILPMedFar		ILPMedDiag	
	Average	s.d.	Average	s.d.	Average	s.d.	Average	s.d.
2	4.7	3.2	1.1	1.5	0.9	0.9	1.6	2.4
3	12.4	7.8	2.5	1.1	4.5	3.4	2.1	2.5
4	18.9	11.5	4.1	2.4	10.7	5.1	4.8	5.3
5	23.3	23.2	8.9	4.2	13.8	8.1	3.2	2.5
6	98.6	76.1	11.8	7.0	51.3	43.2	6.4	3.9
7	154.3	120.0	38.5	37.2	100.0	72.9	7.5	4.0
8	293.7	170.0	71.2	44.9	243.0	321.9	7.1	6.5
9	943.4	747.9	117.3	85.1	606.2	595.7	14.1	8.4
10	1160.4	276.1	108.1	77.9	414.8	180.7	9.4	5.5

ILPMed and ILPMedDiag with $w = 2$, and between ILPCen and ILPCenDiag with $w = 2$, occurred. The objective values by ILPMedFar and ILPCenFar were almost the same as those by ILPMed and ILPCen, respectively, as well as in $p_1(s)$. We can obtain optimal strings using ILPMedDiag and ILPCenDiag by increasing the width w of diagonal.

Table 4. Results on the average and standard deviation of execution time in seconds by ILPCen, ILPCenTri, ILPCenFar, and ILPCenDiag with $w = 2$ for probability distributions $p_2(s)$ for $n = 10$ and $N = 2, \ldots, 10$.

N	ILPCen		ILPCenTri		ILPCenFar		ILPCenDiag	
	Average	s.d.	Average	s.d.	Average	s.d.	Average	s.d.
2	2.7	2.8	0.4	0.4	1.8	1.8	2.4	2.8
3	7.5	4.5	1.4	0.7	3.8	1.3	2.3	2.6
4	19.5	13.9	3.4	1.7	7.4	3.4	4.9	3.1
5	26.6	20.0	5.0	3.0	9.6	4.1	2.1	2.4
6	157.0	110.7	18.1	18.1	23.3	14.6	3.8	2.6
7	320.4	257.5	20.3	11.0	90.6	137.2	4.9	4.1
8	641.3	425.0	55.8	62.4	98.5	99.9	4.8	4.4
9	1259.2	676.6	94.8	40.6	152.6	82.6	3.1	3.3
10	1266.7	488.1	53.0	21.6	145.0	97.3	3.9	3.6

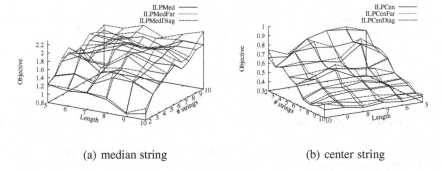

(a) median string   (b) center string

Fig. 4. Results on the average objective value for probability distributions $p_2(s)$ for $N = 2, \ldots, 10$ and $n = 5, \ldots, 10$ (a) by ILPMed, ILPMedFar, and ILPMedDiag with $w = 2$ for finding median strings, and (b) by ILPCen, ILPCenFar, and ILPCenDiag with $w = 2$ for finding center strings.

4   Conclusion

We extended the definitions of median and center strings for a data set of strings into those for a probability distribution $p(s)$ on a set $\mathcal{A}^*$ of all strings on an alphabet $\mathcal{A}$. We then proposed novel ILP-based methods, ILPMed, ILPCen, ILPMedTri, and ILPCenTri, for finding optimal median and center strings for $p(s)$ on $\mathcal{A}^*$. After that, we developed methods, ILPMedDiag, ILPCenDiag, ILPMedFar, and ILPCenFar for finding approximate median and center strings for $p(s)$ on $\mathcal{A}^*$ very rapidly in the case where similar strings are observed with high probability. We performed computational experiments, and confirmed that the execution times by ILPMedDiag and ILPCenDiag were smaller than those by other ILP-based methods, respectively, and ILPMedDiag and ILPCenDiag considerably reduced the execution times. The triangle inequality worked as effective constraints in finding optimal solutions of median and center strings, and consequently, ILPMedTri and ILPCenTri substantially reduced the execution times. However, for applications to large-scale data sets, it is desired that ILPMedTri and ILPCenTri are improved into faster ones. The Levenshtein distance between two strings can be calculated in polynomial time. Nevertheless, the number of candidate paths in ILPMedTri and ILPCenTri is enormous, and should be reduced. On the other hand, ILPMedDiag and ILPCenDiag are considered to be useful if given strings are similar to each other because the number of candidate paths in ILPMedDiag and ILPCenDiag is small. As future work, we need to analyze computational time and space complexities for our proposed methods. Furthermore, we would like to improve our methods by introducing other types of restriction to the variables than those in this paper. In addition, we will improve ILPMedTri and ILPCenTri into methods that find optimal solutions more rapidly based on approximate solutions obtained by ILPMedDiag and ILPCenDiag, and consider decomposition of strings and linear programming relaxation.

Acknowledgements. This work was partially supported by Grants-in-Aid #24500361 and #26610037 from MEXT, Japan.

References

1. Abreu, J., Rico-Juan, J.: A new iterative algorithm for computing a quality approximate median of strings based on edit operations. Pattern Recogn. Lett. **36**, 74–80 (2014)
2. Bunke, H., Jiang, X., Abegglen, K., Kandel, A.: On the weighted mean of a pair of strings. Pattern Anal. Appl. **5**, 23–30 (2002)
3. Casacuberta, F., de Antoni, M.: A greedy algorithm for computing approximate median strings. In: Proceedings of National Symposium on Pattern Recognition and Image Analysis, pp. 193–198 (1997)
4. Chen, S., Tung, S., Fang, C., Cherng, S., Jain, A.: Extended attributed string matching for shape recognition. Comput. Vis. Image Underst. **70**, 36–50 (1998)
5. Dinu, L., Ionescu, R.: An efficient rank based based approach for closest string and closest substring. PLoS ONE **7**(6), e37576 (2012)
6. Gramm, J.: Fixed-parameter algorithms for the consensus analysis of genomic data. Ph.D. thesis, Universität Tübingen (2003)
7. Gramm, J., Niedermeier, R., Rossmanith, P.: Fixed-parameter algorithms for closest string and related problems. Algorithmica **37**, 25–42 (2003)
8. Gusfield, D.: Algorithms on Strings, Trees and Sequences. Cambridge University Press, New York (1997)
9. Hamming, R.: Error detecting and error correcting codes. Bell Syst. Tech. J. **29**(2), 147–160 (1950)
10. de la Higuera, C., Casacuberta, F.: Topology of strings: median string is NP-complete. Theoret. Comput. Sci. **230**, 39–48 (2000)
11. Hufsky, F., Kuchenbecker, L., Jahn, K., Stoye, J., Böcker, S.: Swiftly computing center strings. BMC Bioinform. **12**, 106 (2011)

12. Jiang, X., Abegglen, K., Bunke, H., Csirik, J.: Dynamic computation of generalised median strings. Pattern Anal. Appl. **6**, 185–193 (2003)
13. Jiang, X., Bunke, H.: Optimal lower bound for generalized median problems in metric space. In: Caelli, T., Amin, A., Duin, R.P.W., Ridder, D., Kamel, M. (eds.) SSPR/SPR 2002. LNCS, vol. 2396, pp. 143–151. Springer, Heidelberg (2002). doi:10.1007/3-540-70659-3_14
14. Jiang, X., Wentker, J., Ferrer, M.: Generalized median string computation by means of string embedding in vector spaces. Pattern Recogn. Lett. **33**, 842–852 (2012)
15. Kohonen, T.: Median strings. Pattern Recogn. Lett. **3**, 309–313 (1985)
16. Koyano, H., Kishino, H.: Quantifying biodiversity and asymptotics for a sequence of random strings. Phys. Rev. E **81**(6), 061912 (2010)
17. Kruskal, J.: An overview of sequence comparison: time warps, string edits, and macromolecules. SIAM Rev. **25**(2), 201–237 (1983)
18. Levenshtein, V.: Binary codes capable of correcting deletions, insertions and reversals. Doklady Adademii Nauk SSSR **163**(4), 845–848 (1965)
19. Lopresti, D., Zhou, J.: Using consensus sequence voting to correct OCR errors. Comput. Vis. Image Underst. **67**(1), 39–47 (1997)
20. Martínez-Hinarejos, C., Juan, A., Casacuberta, F.: Median strings for k-nearest neighbour classification. Pattern Recogn. Lett. **24**, 173–181 (2003)
21. Nicolas, F., Rivals, E.: Complexities of the centre and median string problems. In: Baeza-Yates, R., Chávez, E., Crochemore, M. (eds.) CPM 2003. LNCS, vol. 2676, pp. 315–327. Springer, Heidelberg (2003). doi:10.1007/3-540-44888-8_23
22. Nicolas, F., Rivals, E.: Hardness results for the center and median string problems under the weighted and unweighted edit distances. J. Discrete Algorithms **3**, 390–415 (2005)
23. Olivares-Rodríguez, C., Oncina, J.: A stochastic approach to median string computation. In: da Vitoria Lobo, N., Kasparis, T., Roli, F., Kwok, J.T., Georgiopoulos, M., Anagnostopoulos, G.C., Loog, M. (eds.) SSPR/SPR 2008. LNCS, vol. 5342, pp. 431–440. Springer, Heidelberg (2008). doi:10.1007/978-3-540-89689-0_47
24. Sim, J.S., Park, K.: The consensus string problem for a metric is NP-complete. J. Discrete Algorithms **1**, 111–117 (2003)
25. Wagner, R., Fischer, M.: The string-to-string correction problem. J. ACM **21**(1), 168–173 (1974)
26. Winkler, W.: String comparator metrics and enhanced decision rules in the Fellegi-Sunter model of record linkage. In: Proceedings of the Section on Survey Research Methods, pp. 354–359 (1990)

Accelerating the Exploitation of (bio)medical Knowledge Using Linked Data

Mohammad Shafahi[✉], Hamideh Afsarmanesh, and Hayo Bart

Faculty of Science, Informatics Institute, University of Amsterdam,
Science Park 904, Amsterdam, The Netherlands
{m.shafahi,h.afsarmanesh}@uva.nl, hayojay@live.nl

Abstract. Early identification and treatment of a diseases, especially when chronic, can reduce severe complications for the patients, doctors, and the society as a whole. Therefore, becoming aware and having insight about the state of the art findings on diseases, if communicated properly to different stakeholders, will benefit all. The medical research field, however, is vast and dynamically evolves with new discoveries. Additionally, new results are being continuously generated. The new discoveries on diseases address their diagnosis, prognosis, and possible treatment pathways for each disease, which are typically published in medical articles. Research results, however, are not reflected in practice by practitioners, unless they are officially verified by governments and authoritative health institutes, and appear in medical guidelines. Developing the medical guidelines requires identifying every relevant medical article, traversing through and validating it, as well as gathering and inter-relating that data to the information from other relevant sources, such as the drug interaction databases.

Therefore, optimal exploitation of medical advances and research results by all its stakeholders, being researchers, practitioners, and patients, is essential. This, however, is hindered due to both the lack of integration of their typically disparate information, and the lack of facilities for coherent, up-to-date, and personalized access by their stakeholders. The few researches that address these issues do not sufficiently address the needed dynamism in data, lack intuitiveness in their use, and present a rather limited amount of information, which is usually obtained from a single source. This research aims to address these gaps through the development of BioMed Xplorer, presenting a model and a tool that enables researchers to rapidly query and explore biomedical knowledge from multiple sources, while preserving provenance data, and presenting all inter-linked information through an intuitive and personalized user interface. Results are further validated by some domain experts, through contrasting it against the state of the art, and with a task-based validation experimenting with the real case of updating medical guidelines.

Keywords: BioMed Xplorer · Disease related information · Semantic Web · Knowledge base ontology · Visualization · Provenance data · Medical knowledge · External data source · RDF · Graph · Knowledge exploration

© Springer International Publishing AG 2017
A. Fred and H. Gamboa (Eds.): BIOSTEC 2016, CCIS 690, pp. 122–144, 2017.
DOI: 10.1007/978-3-319-54717-6_8

1   Introduction and Research Approach

The (bio)medical field is vast and dynamic, with knowledge developing rapidly as a result of continuously ongoing research. Within this field extensive research is conducted into identifying different aspects of diseases such as diagnosis, prognosis, and possible treatment pathways. The available knowledge from such research enables governments and authoritative health institutes to develop medical guidelines, which are used by practitioners for improving patients health by treating a disease or injury, or making a diagnosis. Conventional methods for developing medical guidelines would involve identifying the risk factors and their effects on the disease from the ever-evolving body of (bio)medical knowledge. Achieving this aim would thus involve checking a vast amount of scientific publications for relevant statements regarding factors that might affect a disease. This, however, is a cumbersome and costly activity, especially when considering the fact that the U.S. National Library of Medicine's (NLM) bibliographic database MEDLINE, as of today, contains over 22 million citations, over 750,000 of which were added in 2014 [31], and that these numbers have grown exponentially [17]. As a result of the sheer size and continuous growth of the body of (bio)medical knowledge, exploration of this body of knowledge, as well as finding the relevant knowledge for inclusion in medical guidelines, becomes increasingly challenging for researchers, potentially causing an information overload [17,22]. As a result, publishing and/or updating versions of medical guidelines is a process that normally takes long, about half a decade, during which new discoveries are achieved. These new discoveries, however, will not be practiced and thus cannot improve the health and well being of the target population, as they are not yet published.

Numerous researchers have reckoned this problem and have attempted to address it from different perspectives [9,22], for example through the development of comprehensive visualizations that represent knowledge extracted from (bio)medical publications [8,20,25,26,29]. Even though the visual nature of these knowledge representation and visualization tools provides practitioners and researchers with great expressive power, four common shortcomings can be identified among them, being: (*i*) their restricted scope, focusing just on a particular sub-domain of the (bio)medical field, (*ii*) their lack of intuitiveness and rather sharp learning curve, (*iii*) the scarce amount of information represented, solely limited to names and identifiers, lacking descriptions or definitions, while these are available externally, and (*iv*) the fact that they are either no longer active (AliBaba, PGviewer), or do not work properly (EBIMed). From these shortcomings it thus becomes clear that there is a need for a meaningful representation of the available (bio)medical knowledge that: (*a*) is intuitive, and (*b*) represents information from multiple sources. As such the following research question can be conceived:

Can we develop a model of the (bio)medical knowledge that is available from large, disperse, heterogeneous, and dynamic sources across the web?

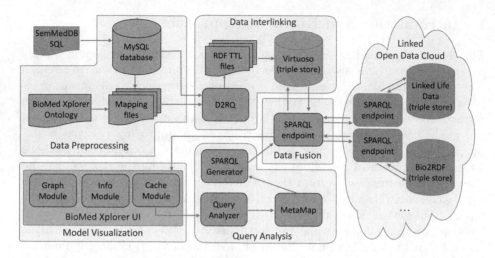

Fig. 1. The system architecture of BioMed Xplorer.

In order to address this research question a research approach has been designed that consists of the following seven phases: (*1*) State of the Art Assessment, (*2*) Data Source Characterization and Selection, (*3*) Data Preprocessing and Ontology Design, (*4*) Data Interlinking and Fusion with external sources, (*5*) Stakeholder Customization and Query Analysis, (*6*) Model Visualization, and (*7*) Validation. Completion of these seven phases delivers a system with an architecture that is shown in Fig. 1. As one might notice, the architecture consists of five core modules, each of which corresponds to one of the major design and development stages. The components of these modules will be gradually defined in the corresponding sections, as such fully describing the system architecture.

The remainder of this paper is structured according to the seven phases that were outlined above, with each section elaborately discussing a particular phase of the research. In Sect. 2, the characterization and selection of data sources for inclusion in the model is described. This is followed by a discussion of the data preprocessing and ontology design in Sect. 3, whereas Sect. 4 covers the fusion and interlinking of the data. In Sect. 5 the stakeholders of the system and how their queries to the system are analyzed is described. The visualization of the model is subsequently discussed in Sect. 6, while the work is validated in Sect. 7. Finally, Sect. 8 concludes the paper.

2   Data Source Characterization and Selection

Central to the development of a model is the data that eventually will be represented in the model and thus needs to be utilized for building and populating the model. With the research question in mind the identification, and subsequent selection, of data sources that provide disease related information, pertaining to, for example, symptoms, inheritability, and genetics of a disease, thus are the first key steps in the development process of the disease related information model. A search for disease related information results in a wide variety of structured (i.e. standardized terminologies or vocabularies, ontologies, and databases) and unstructured (e.g. websites [32,35]) data sources. Data from unstructured sources requires conversion to a structured format, for example using Natural Language Processing (NLP) techniques, and thus cannot be directly incorporated into the disease related information model. As a result we have decided to only incorporate structured data sources.

A medical knowledge model that is represented in a network-like format consists of two components, namely concepts and relationships among these concepts. Concepts can be sourced from standardized terminologies, or from ontologies. Some well-known terminologies in the biomedical field are the *International Classification of Diseases (ICD)*, *Medical Subjects Headings (MeSH)*, and *Systematized Nomenclature of Medicine Clinical Terms (SNOMED CT)*, whereas the *National Cancer Institute Thesaurus (NCIt)*, the *Disease Ontology*, and the *Gene Ontology* are among the frequently used ontologies within the biomedical field. Instead of sourcing concepts from one or multiple individual terminologies, one can source the concepts from the *Unified Medical Language System (UMLS) Metathesaurus* [34] or the *National Cancer Institute Metathesaurus (NCIm)* [30], both of which integrate, among many others, the aforementioned sources into a single terminology. Using these metathesauri provides the opportunity of broadening the scope of the concepts that are covered and, as such, expanding the knowledge base of the model by using concepts represented in the majority of separate terminologies. Therefore, the use of either the UMLS or NCIm to define concepts in the model is preferred over the use of separate terminologies. Relationships can also be sourced from these two sources. More extensive relationships, however, can be obtained from the *Online Mendelian Inheritance in Man (OMIM)* database [19], MalaCards [37], or SemMedDB [21].

It is essential to designate a primary data source for the development of the medical knowledge model as the available structured data sources for disease related information overlap in terms of covering the same information in different formats and presentations. Considering the overarching aim of this research in aiding (bio)medical stakeholders in their knowledge explorations efforts, and due to the fact that (bio)medical knowledge originating from peer-reviewed literature is considered trustworthy and rich, relationships directly derived from (bio)medical literature are selected as the primary relationships in the model. To this end, SemMedDB is thus selected as the primary source, presenting disease related information, for incorporation into the developed model. This choice is further motivated by the fact that SemMedDB is considerably larger (containing over 70 million statements) than the other identified sources containing disease related information. Finally, the broad scope, covering terms across the entire biomedical domain, also played a role in the choice for SemMedDB.

3   Data Preprocessing

Due to the large amounts of heterogeneous and dynamic information that is nowadays available across a multitude of sources, relational databases are considered to be less than ideal for storing and instantiating knowledge representations of information of such nature [14]. Linked data, on the other hand, provides a promising solution to this issue as it is able to cope with such large amounts of dynamic and heterogeneous information [5]. To this end we therefore aim to develop our model using Semantic Web technologies. According to [1,5] the Semantic Web consists of three main components, being (*i*) *labeled graphs* that encode meaning by representing concepts and the relations among them, and are usually expressed as (subject-predicate-object) triples in RDF; (*ii*) *Uniform Resource Identifiers (URIs)* to uniquely identify the items in the datasets as well as to assert meaning, which is reflected in the design of RDF; and (*iii*) *ontologies* to formally define the relations that can exist among data items. In order to develop our model using the Semantic Web, the existence of these three components needs to be ensured. Processing the data in SemMedDB such that these three components exist, is therefore the main aim of the preprocessing stage.

3.1   Ontology Design

In order to generate labeled graphs from a relational database, such as SemMedDB, and to ensure the use of URIs, an ontology needs to be developed that represents the desired data structure of these graphs. This ontology should define the data items, as well as the relations among them, that are aimed to be represented. Considering that the planned model should represent the statements, and their provenance data, in SemMedDB as a RDF graph, it is key for the ontology to closely resemble SemMedDB's database design. Prior work has been conducted in this area by Tao et al. [28]. In their work, Tao et al. aimed to optimize the organization and representation of Semantic MEDLINE data (SemMedDB) for translational science studies by reducing redundancy through the application of Semantic Web technologies. This is achieved by representing the concepts and associations in SemMedDB as RDF. Despite successfully decreasing the redundancy of the information in SemMedDB, two shortcomings can be identified in the ontology that was developed by Tao et al. First of all, the ontology represents a limited amount of information compared to the information that is available in SemMedDB. This, in turn, impedes the ability to incorporate external resources into the model since unique identifiers, which are required to retrieve the appropriate entities from these external sources, are among the information from SemMedDB that is omitted. The second shortcoming is the lack of reuse of terms defined in existing vocabularies, which is one of the founding principles of the Semantic Web [27]. The developed ontology by Tao et al. defines all terms used, whereas equivalent classes might already exist in other vocabularies in the Web of Data. Such reuse would facilitate the linking of data to a Web of Data, which is an overarching goal of the Semantic Web [5].

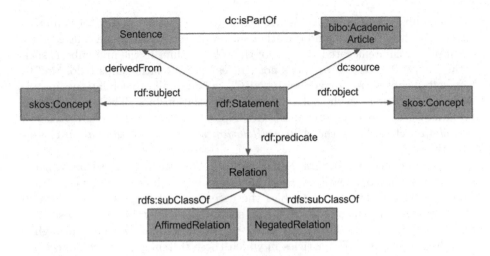

Fig. 2. BioMed Xplorer Ontology, only showing the object properties.

Despite the limitations of the ontology developed by Tao et al., this ontology is considered as a starting point as well as an opportunity to improve on and extend upon it. To this extent, the BioMed Xplorer Ontology is developed that addresses the identified shortcomings by representing most of the information contained within SemMedDB, as well as by reusing as much terms from existing vocabularies or ontologies as possible. The BioMed Xplorer Ontology is developed in the Web Ontology Language (OWL2) and is published on a Persistent Uniform Resource Locator (PURL) [36] domain[1]. Such a locator allows the underlying Web address of a resource to change while not affecting the availability of the systems that depend on this resource. The BioMed Xplorer Ontology is shown in Fig. 2.

RDF Reification. The provenance data in SemMedDB can be considered as meta-statements as they apply to statements as a whole. Such meta-statements, however, are not natively supported by RDF. Therefore RDF Reification needs to be applied in order to represent this provenance data in the ontology. Tao et al. also recognized this need, however, they did not use the RDF Reification vocabulary as outlined by W3C [39]. The BioMed Xplorer Ontology, on the other hand, implements the RDF reification vocabulary.

As the statements contained in SemMedDB relate two UMLS concepts to each other, both the subject and object of an *rdf:Statement* instance are modelled as instances of a *Concept* class. The concepts are related to each other through one, of 58, relationships that are identified by SemRep [33]. The predicate of an *rdf:Statement* instance therefore is modelled as one of 58 instances of the *Relation* class. This set of relationships consists of two disjunctive subsets,

---

[1] http://purl.org/net/fcnmed.

with one subset containing 31 relationships derived from the UMLS Semantic Network, such as "causes", and the other subset containing the remaining 27 relationships, which are negated versions of the relationships in the first subset, such as "neg_causes", referring to "does not causes" [21]. Relationships belonging to the negated subset are prefixed with "NEG", whereas all other relationships are considered to belong to the subset of affirmed relationships. These two subsets of relations are represented in the BioMed Xplorer Ontology as two subclasses of the *Relation* class, being the *AffirmedRelation* and the *NegatedRelation* classes respectively.

The provenance data in SemMedDB includes both the sentences from which a statement is derived, as well as the publications in which these sentences occur. Reification of the statements enables the assertion of this provenance data to their respective statements. To this end, sentences are represented as instances of the *Sentence* class, which are related to the *rdf:Statement* class through a *derivedFrom* property. The articles in which these statements and sentences are contained, are represented as instances of an *Articles* class, which are related to the *rdf:Statement* class through a *source* property. Furthermore, sentences are related to articles through the *partOf* property, indicating that a sentence is part of an academic article. In addition to the object properties, relating classes to each other, discussed in this section, a number of datatype properties, associating data values (such as identifiers) to classes, are asserted to each of the classes in the BioMed Xplorer Ontology as well. Collectively, these properties aim to represent as much information from SemMedDB in the ontology as possible.

Vocabulary Reuse. The BioMed Xplorer Ontology aims to reuse as much existing classes and properties as possible. To this extent all elements of the ontology, which include the classes and both the object and datatype properties, except elements from the RDF or RDFS namespaces, have been checked for the presence of already defined equivalent concepts or properties in existing vocabularies. This has been accomplished by making use of the online RDF vocabulary search and lookup tool vocab.cc [18] that allows one to enter any term, and retrieve any classes and properties that (partially) match the term. In general the highest ranked term that corresponds to the role of the term in the BioMed Xplorer Ontology (e.g. class or property) is selected for reuse in the BioMed Xplorer. In the end, the search for existing terms lead to the incorporation of terms from three existing vocabularies, being (*i*) the Bibliographic Ontology [10], (*ii*) the Dublin Core Metadata Terms [12], and (*iii*) the Simple Knowledge Organization System [38].

4   Data Interlinking and Fusion

The ontology developed in Sect. 3.1 defines the desired data structure for the developed model. Generating the labeled graphs from the SQL in SemMedDB, however, requires a mapping that specifies how the data in the database is matched and converted to the appropriate class instances, properties, and property values specified in the ontology. Such a mapping can be developed using D2RQ, a declarative language for describing mappings between relational databases, RDF(S), and OWL ontologies [7]. The developed D2RQ mapping files have been made available online[2]. A mapping file enables RDF applications to access relational databases as virtual RDF graphs through the companion tool D2R Server [6]. These virtual RDF graphs can subsequently be queried using the SPARQL protocol, with the D2RQ mapping translating the SPARQL queries to SQL queries, and translating the query results back to RDF. Both D2RQ and D2R are jointly available in the D2RQ Platform [7]. With the developed mapping file, the data in SemMedDB can be interlinked as RDF triples according to the specified ontology, as such surfacing and populating the actual disease related information model. Furthermore, the combination of the ontology and the use of RDF ensures the ability to link to the data in the information model from external datasets, this can be achieved through the URIs assigned to instances and properties.

In order to achieve complete data fusion with external sources, as such creating a truly Linked Data model, the data in the disease related information model should be linked to related entities or instances in external (RDF) data sources. This can be achieved by setting RDF links between the data in the model and these external data sources [4]. One common way of setting such links between data sets is through the *owl:sameAs* property, which indicates that two linked individuals refer to the same thing [11]. Establishing these links subsequently enables the incorporation of data from the external data sources into the disease related information model. Key to this data fusion process is the identification of external data sources containing instances that are equivalent to the instances in the developed model. The search for these data sources containing equivalent instances has been facilitated by searching the Linked Open Data cloud[3] for unique standardized instance identifiers. Among these identifiers in SemMedDB are the UMLS Concept Unique Identifier (CUI), the Entrez-Gene ID, and the OMIM identifier for concepts, as well as the PubMed Identifier (PMID) for publications. The search of the Linked Open Data cloud for (bio)medical RDF data sources that represent either (bio)medical concepts, identified by one of the aforementioned identifiers, or publications, identified by the PubMed identifier, returned two main external data sources that could be fused with the data in SemMedDB: Linked Life Data [23], and Bio2RDF [3].

5   Stakeholder Customization and Query Analysis

The (bio)medical knowledge graph developed in the previous sections can be queried using a SPARQL endpoint, but the medical stakeholders would prefer to query the knowledge graph using customized queries that most suites their purpose, or would prefer to traverse the knowledge graph in a visualized manner.

---

[2] The mapping files are available online from: https://goo.gl/1yD0WO.
[3] For details see http://lod-cloud.net/.

While the latter will be discussed in Sect. 6, this section will discuss how three main groups of (bio)medical stakeholders, namely patients, practitioners, and researchers, can benefit from BioMed Xplorer.

5.1   Researchers

Researchers can benefit most from using BioMed Xplorer to identify the state of the art related to their specific study within the medical domain. BioMed Xplorer can support them in the identification of the most relevant literature related to their study case and/or hypothesis, e.g. about the risk factors, medications etc. for a disease. This in turn can facilitate the researcher's task of validating their hypotheses related to a disease meta-model, by checking it against the state of the art. In order to query the system researchers can provide a hypothesis statement in natural language. This statement in combination with the researchers user profile will then be used by the Query Analyzer to build a graph representation of the hypothesis. In order to build the hypothesis graph, MetaMaps [2] is used to translate the natural language statement into medical concepts. These concepts are then further analyzed by the Query Analyzer in order to be placed within the hypothesis triples (i.e. as the subject, object, or predicate) that creates the graph. The hypothesis graph is finally translated into a SPARQL query by the SPARQL generator. As a result of this query the system provides the researcher with an extended graph having the subject and object as nodes of the graph and the predicates and/or the negation of the predicates as the edges of the graph. This graph is dynamic and can be further traversed by the researcher in order to investigate the state of the art related to his/her hypothesis. As such, BioMed Xplorer provides a user-friendly graphical Explore-Interface, supporting medical researchers in exploring medical information gathered from various sources in one comprehensive information graph. The graphical user interface is further discussed in Sect. 6.

5.2   Practitioners

Practitioners can use BioMed Xplorer to discover and apply state of the art in research to patient cases without the need to go through a large number of pages of medical guidelines, which, usually, are also not up to date with current research. In order to query the system, practitioners can provide one or more patient symptoms/condition statements in natural language or as an alternative provide the Electronic Health Record (EHR) of a patient. These statements in combination with the practitioners user profile and the patient profile, if available, are then used by the Query Analyzer to build a graph representation of the patients medical condition. In order to build the patients condition graph, MetaMaps [2] is used to translate the natural language statement into medical concepts. These concepts are then further analyzed by the Query Analyzer in order to be grouped based on their location in the condition statements creating a graph without edges but with clustered nodes (see Fig. 3). The graph is finally translated into a SPARQL query by the SPARQL generator. As a result of this

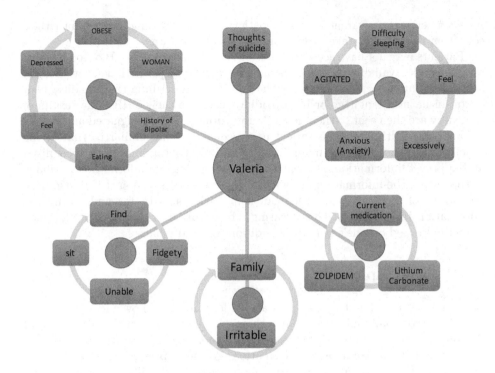

Fig. 3. Graph representation of a patient's (i.e. Valeria) condition.

query the system provides the practitioner with an extended and fully connected graph of the patients condition including a collection of information gathered from a variety of sources. Similar to the result of the researchers, this graph is dynamic and can be further traversed by the researcher in order to investigate the state of the art related to a patients condition.

5.3   Patients

At their early stages, many diseases do not present noticeable symptoms. Even when some simple symptoms are shown, these are either not properly identified by the patients, as a result of their lack of medical know-how, or are not given sufficient attention by the practitioners, due to not having all the information related to the entire picture of the patients health at once. Therefore, people can be at risk of gradually developing serious diseases, such as diabetes type 2 or heart disease, without being aware.

According to the World Health Organization, in 2012, 1.5 million deaths were directly caused by diabetes. Many of the risk factors that can be associated with these diseases are known, varying from the lack of exercise and consuming unhealthy food, to gene inheritance. It is also known that early detection and treatment of these diseases can reduce the burden of severe future complications.

As such, creating awareness among patients and providing insight into their condition benefits all stakeholders and the society at large.

Patients may not have access to BioMed Xplorer directly, but BioMed Xplorer can be made available to them though their general practitioners or medical authority. Patients can either query the system by describing their medical condition in natural language or by providing access to their EHR. The results of the query are then sent to their general practitioner or the designated authorities for approval. Ultimately the patient can receive a result similar to that of the practitioners, though much more limited and only containing highly recognized and approved information. By continuously extending the information available in the graph, the information provided to the patients is also enriched, e.g. with definitions of medical terms. This can help patients to better understand the information in the graph. Thus creating a customized graph prepared for each individual based on one's personal condition and daily habits.

6   BioMed Xplorer UI

Further assistance for (bio)medical stakeholders in their knowledge exploration efforts can be achieved by enabling them to intuitively explore the body of (bio)medical knowledge. To this end it is therefore imperative to visualize the developed disease related information graph, representing this body of knowledge, that incorporates other disease related information gathered and aggregated from disperse sources across the Web. With this in mind three key requirements for the BioMed Xplorer UI can be imagined, being that it should: (*i*) be usable and intuitive (e.g. supported by an appropriate visualization paradigm), (*ii*) concisely represent provenance data (e.g. the publications as well as sentences from which statements are derived), and (*iii*) represent information from multiple sources (e.g. concept summaries and definitions). Based on these identified requirements, BioMed Xplorer has been developed and made available[4] on the Web. The BioMed Xplorer UI supports the visualization of the information and has been developed in JavaScript in combination with the d3.js[5] and jQuery[6] libraries.

The data visualized by the BioMed Xplorer UI is obtained from BioMed Xplorer's back-end, which consists of a Virtuoso triple store containing the disease related information model in RDF. This triple store provides a built-in SPARQL endpoint that can be queried by BioMed Xplorer using the SPARQL [13] protocol. Efficient query handling has been achieved by the development of a caching mechanism.

The developed user interface has three key features being: (*i*) a graph-based visualization, (*ii*) the exploration of concept information, and (*iii*) the exploration and assessment of relationships. Each of these three features will be briefly discussed in the remainder of this section.

---

[4] BioMed Xplorer is available http://goo.gl/qeuW5k (best viewed in Firefox).
[5] For details see http://d3js.org/.
[6] For details see http://jquery.com/.

6.1   Graph Visualization

The BioMed Xplorer UI employs a graph-based visualization of (bio)medical knowledge as shown in Fig. 4. Exploration and traversal of the knowledge graph is supported through the expansion of concepts (by double clicking on concepts) and collapsing of concepts (by right clicking on concepts). Additionally, panning (by click and drag) and zooming (by scrolling) is supported as well.

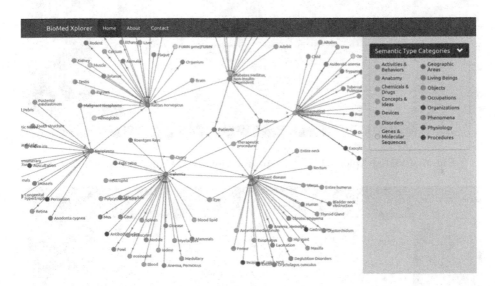

Fig. 4. A screenshot from BioMed Xplorer UI and its graph-based visualization of (bio)medical knowledge, representing (bio)medical concepts as nodes and their inter-relationships as edges.

6.2   Exploring Concept Information

Within the BioMed Xplorer UI, concept information can be explored through concept summaries, which can be opened by clicking on concepts, and concept overviews (as shown in Fig. 5), which can be opened by choosing to show details in a concept summary. Concept information includes a wide range of information available from within the model as well as from external sources, such as Linked Life Data and Bio2RDF.

6.3   Exploring and Assessing Relationships

Relationships between concepts can be explored in the BioMed Xplorer UI through statement summaries, which can be opened by clicking on an edge, and statement overview, which can be opened by choosing to show details in a relationship summary. Within statement overviews, a wide range of statement information is available, as is shown in Fig. 6. Among the available information is:

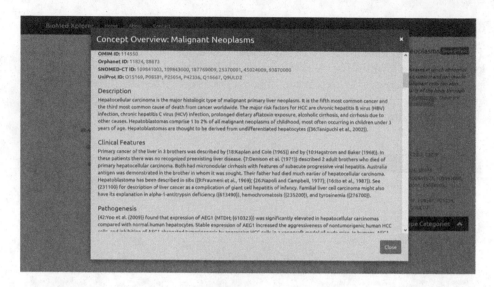

Fig. 5. BioMed Xplorer UI's concept overview for "Malignant Neoplasms".

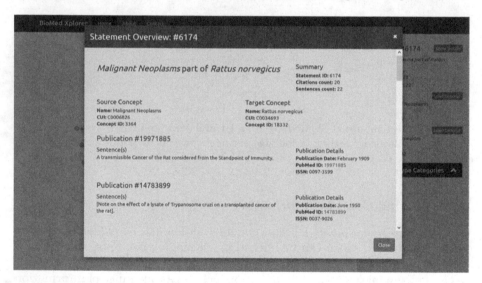

Fig. 6. BioMed Xplorer UI's statement overview for "Malignant Neoplasms" part of "Rattus Norvegicus".

the complete statement, two aggregates of the available provenance data, a brief overview of the source and target concepts of the relationship, the sentences from which the statement is derived at a publication level, as well as the details of the publications.

7 Validation

Keeping the key roles of both the BioMed Xplorer Ontology, as the foundation for the knowledge base, and the BioMed Xplorer UI, as the visualization of this knowledge base in mind, the validation of these two outcomes is imperative. This validation aims to assess whether the proposed solutions successfully address the already identified gap as well as how the proposed solutions measure against existing work. To this end, a threefold validation of both the ontology and the visualization has been conducted. In this regard a comparison to prior work has been conducted first, the results of which are shown in Tables 1 and 2 respectively. Secondly, an evaluation has been performed by 6 experts in the field. Results of

Table 1. Comparison of BioMed Xplorer Ontology with the ontology developed by Tao et al. [28].

Characteristic	Tao et al. (2012)	BioMed Xplorer Ontology
RDF reification	No	Yes
Vocabulary reuse	No	Yes
Links to external data sources	No	Yes
Number of related data-items captured	4	17
Provenance data captured	Publications	Publications and sentences

Table 2. Comparison of BioMed Xplorer UI with five knowledge visualization tools developed in prior work.

Characteristic	AliBaba	EBIMed	PG-viewer	Semantic MED-LINE	Semantic Navigator	BioMed Xplorer
Scope	Limited	Limited	Limited	Biomedical	Biomedical	Biomedical
Available	No	No	No	Yes	Yes	Yes
Visualization paradigm	Graph	Tabular	Tree	Graph	Graph	Graph
Concept categorization	Yes	Yes	Yes	Yes	No	Yes
Incorporation of links to external sources	Yes	Yes	No	Yes	No	Yes
Incorporation of data from external sources	Yes	Yes	Yes	No	No	Yes
Presentation of concept related information	Yes	No	Yes	Yes	No	Yes
Incorporation of provenance data	Yes	Yes	Yes	Yes	No	Yes

this expert evaluation showed that both the BioMed Xplorer Ontology and the BioMed Xplorer UI successfully satisfy the identified requirements, with average grades of a 7.8 and 7.6 out of 10 respectively. Details of the expert evaluation of both the BioMed Xplorer Ontology and the BioMed Xplorer UI are provided in Tables 3 and 4. Finally BioMed Xplorers strengths in preforming tasks required by its stakeholders is assessed using task-based validation approach.

Table 3. Frequency distribution of the five point Likert-scale scores for evaluating the BioMed Xplorer Ontology. A score of 1 indicates disagreement and 5 indicates agreement.

Statement	1	2	3	4	5
The ontology is capable of representing (statements of) biomedical knowledge	0	0	1	4	1
The ontology models (statements of) biomedical knowledge appropriately	0	1	1	3	1
The ontology is capable of representing the provenance data associated with (statements of) biomedical knowledge	0	0	2	2	2
The ontology models provenance data associated with (statements of) biomedical knowledge appropriately	0	1	1	2	2
The ontology globally fits its purpose	0	0	2	2	2

Table 4. Frequency distribution of the five point Likert-scale scores for evaluating the BioMed Xplorer UI. A score of 1 indicates disagreement and 5 indicates agreement.

Statement	1	2	3	4	5
The implemented functionalities support and facilitate the exploration of biomedical knowledge	0	1	1	1	3
Color coding of the nodes is helpful	0	0	1	2	3
The information in the summaries has a clear structure	0	1	2	1	2
The information in the summaries is relevant	0	1	1	3	1
The information in the extended details has a clear structure	0	1	1	3	1
The information in the extended details is relevant	0	0	1	2	3
The interface is well structured/organized	0	0	2	3	1
The graphical user interface has an adequate look and feel	0	0	2	2	2
The tool behaves as expected	0	2	2	1	1
The visualization is intuitive in its use	0	2	1	2	1
The visualization of information is simple and smooth	0	1	0	3	2
The system globally fits its purpose	0	2	0	1	3

7.1   Comparison to Related Work

Five main knowledge representation and visualization tools are identified that attempt to address similar challenges associated with exploring the body of (bio)medical knowledge through the representation and visualization of the knowledge contained within scientific publications. Among these tools are: AliBaba [25], EBIMed [26], PGviewer [29], Semantic MEDLINE [20], and the Semantic Navigator [8]. Due to the close correspondence between the aims of these tools and the aims of our research, these five tools are considered as the base for comparison to BioMed Xplorer. A comparison of the characteristics of these five selected tools is provided in Table 2.

The five tools identified from prior work have two main shortcomings. On the one hand they focus on a particular subdomain of the (bio)medical field (AliBaba, EBIMed, PGviewer) and, as such, inhibit the exploration of the body of (bio)medical knowledge. On the other hand they employ an alternative visualization paradigm (EBIMed and PGviewer) that is less focused on the visual representation of knowledge. The BioMed Xplorer UI overcomes these shortcomings, as such improving over most tools developed in prior work, through its broad scope, aiming to cover the complete (bio)medical domain, and its graph-based visualization. More specifically, BioMed Xplorer can be considered to be on par with Semantic MEDLINE and the Semantic Navigator as these two tools both focus on the entire (bio)medical field as well as employ a graph-based paradigm for visualizing (bio)medical knowledge. The three aforementioned tools are furthermore available on the Web, whereas AliBaba, EBIMed, and PGviewer are no longer available. The position of BioMed Xplorer is further reinforced by the fact that it is the only tool that is based on RDF, which improves its ability to handle large amounts of heterogeneous data from disperse sources. The other tools are based on traditional relational databases, as such inhibiting their ability to incorporate data from additional external sources into these tools.

In addition to representing (bio)medical knowledge through statements that relate two (bio)medical concepts to each other, the presentation of concept and statement related information is also of great importance, as it provides background knowledge about the concepts involved in the statements or about the statements themselves. To this end, BioMed Xplorer is on par with all of the other tools considering the presentation of statement related information. This information typically includes the complete statement itself, including its source and target concepts, the type of the statement, as well as the provenance data associated with the statements in terms of the abstract or sentences, and publications from which the statements were derived. Such provenance data is provided by all the tools included in the comparison, except for the Semantic Navigator as this tool solely represents the relationships stored in the UMLS. Occasionally, the statement related information might also include aggregates of the provenance data, such as the number of sentences and publications from which a particular statement is derived, as is the case for Semantic MEDLINE and BioMed Xplorer. Whereas the BioMed Xplorer is on par with the other tools in relation to the presentation of statement related information and the incorporation of provenance data, it in fact improves over these tools on the presentation of concept related information. The concept related information presented in BioMed Xplorer UI extends well beyond the conventional information that is incorporated. While, tools such as EBIMed and the Semantic Navigator do not present any of such concept related information at all, data items such as (semantic) types, synonyms, and parts of publications that mention the particular concept are presented by AliBaba, PGviewer, and Semantic MEDLINE. BioMed Xplorer extends this further through the incorporation of a wide range of cross-identifiers of concepts, a definition, and a range of data items pertaining to the clinical features, diagnosis, inheritance, pathogenesis, and genetics of a disease from OMIM, if available. The presentation of this wide range of concept related information in BioMed Xplorer is partially facilitated through the incorporation of data from external (structured) data sources, including Linked Life Data and Bio2RDF, which demonstrates its superiority compared to the other tools developed in prior work. Among these other tools, the incorporation of information from such external sources is either largely absent (such as in Semantic MEDLINE and Semantic Navigator), or limited to the inclusion of information from PubMed (such as in AliBaba, EBIMed, and PGviewer). Links to external data sources, usually in the form of cross-references to standardized terminologies, on the other hand, are commonly used by the tools developed in prior work, with only PGviewer and the Semantic Network lacking such cross references.

For the validation of the ontology, the ontology developed by Tao et al. [28] is considered as the base to which our developed ontology is compared. A comparison of the characteristics of the two ontologies is provided in Table 1. As is clear from this table, the ontology developed in this research improves the ontology developed by Tao et al. on a number of aspects, which will be further discussed below, as such contributing to the validation of the ontology developed in this research. As was discussed in Sect. 3.1, reification has been applied in both ontologies to allow triples to involve a particular (bio)medical statement, as a whole, into another statement, and thus enable meta-statements: *statements about statements*. This can be achieved by treating a statement, relating two (bio)medical concepts to each other through a relation, as a separate entity to which the subject, the predicate, and the object of the original statement are assigned using an object property. The ontology developed by Tao et al. performs this by making use of the *Association* class in combination with the *hass_name*, *has_predicate*, and *haso_name* properties. BioMed Xplorer Ontology, on the other hand, makes use of the official RDF reification vocabulary that uses the *rdf:Statement* class in combination with the *rdf:subject*, *rdf:predicate*, and *rdf:object* properties. Additionally, the use of this official RDF reification vocabulary also contributes to the reuse of existing vocabularies, one of the key principles of Semantic Web [27]. To further promote this base principle of the Semantic Web, the developed ontology, in addition to the use of the RDF reification vocabulary, makes extensive use of existing classes and properties from other vocabularies. This is a considerable improvement over the ontology developed by Tao et al., as the reuse of existing vocabularies, aside from the RDF vocabulary, is not present in their ontology.

Since the purpose of the developed model is to enable researchers to explore the body of (bio)medical knowledge as well as its (disease) related information, the amount of information captured by the ontology is of great importance. To this extent, the ontology developed by Tao et al. can be considered as rather limited due to the fact that there is no direct evidence of the incorporation of any (disease) related information beyond the three datatype properties assigning names to concepts and relations, as well as identifiers to publications. The ontology developed in our research improves on this point by associating 17 data-type properties to the ontology classes that can be used to capture a wide range of (disease) related information. This is further facilitated by the incorporation of RDF links to the equivalent resources in external data sources, including Linked Life Data and Bio2RDF, which contain a wealth of (disease) related information. No such links are incorporated in the ontology developed by Tao et al.

Finally, the developed ontology extends the ontology developed by Tao et al. by incorporating the sentences from which the represented statements are derived, in addition to the publications from which these sentences are a part, as a component of the provenance data that is associated to the statements. The incorporation of these sentences provides additional value to the disease related information model as it enables the presentation of the direct source of a particular statement, as opposed to the presentation of solely the publication from which a statement is derived.

7.2   Task-Based Validation

We can further evaluate BioMed Xplorer by assessing its strength in preforming tasks required by its stakeholders. As such, a task belonging to researchers has been selected. More specifically, we experimented with the task of finding medical publications which are relevant for updating a given guideline, as our benchmark.

A medical guideline usually exceeds hundred pages of text and tables and consists of numerous conclusions, each of them in the form of a short paragraph, for example:

"Adjuvant therapy with tamoxifen in breast-conserving treatment of DCIS, removed with tumour-free excision margins, results in limited improvement of local tumour control and no survival benefit."
(3rd conclusion in Section 1.1.2 (on pages 14-16), from the Dutch National Breast Cancer Guideline, 2004 - conclusion nr. 2 in our experiments)

Each conclusion is annotated with 1 to 10 citations from the medical literature that provide the evidence for that conclusion. For example, the above recommendation was supported by three citations to the literature, from the years 1991 and 2002.

Considering the above, we can evaluate BioMed Xplorer by investigating how well it can perform in finding all the recent medical publications, which are relevant for updating an conclusion, for each conclusion in a guideline. In order to preform this evaluation, we use the 2012 revision of the Dutch National Breast Cancer guideline that is an update to the 2004 guideline [24] and create a gold standard. This gold standard consists of 5 corresponding pairs of example conclusions from 2004 and 2012 that have distinct revised statements in each version of the guideline, while addressing the same or similar subject. For these 5 conclusions, the new publication evidences listed in the 2012 version is considered the gold standard for our search. Ideally BioMed Xplorer should suggest all of the gold standard evidences when presented with the 2004 statement as a hypothesis.

As the metrics we consider the *hits*, being the number of papers suggested by BioMex Xplorer that are in the gold standard for a given conclusion of the guideline. We would like the *hits/goal* to approach 1, where *goal* is the total number of papers in the gold standard for a given conclusion.

Aiming at assessing BioMed Xplorer, we preform two experiments. In the first experiment the query that the researcher provides to BioMed Xplorer (i.e. the hypothesis statement(s)) is just based on the guideline statement. In this experiment for each conclusion a hypothesis graph is represented to the researcher by BioMed Xplorer. Each edge of this hypothesis graph (i.e. each hypothesis statement) has a statement overview (see Fig. 6) containing the citations for the statement. These citations from the statement overview are then compared against the evidence items for the conclusions in the guideline, giving us the score of the query in terms of the number of *hits*. We repeat this procedure for each of the 5 conclusion from the 2004 guideline. The results of this experiment are reported in Table 5.

Table 5. Results of querying the conclusion text.

Conclusion	Goal	Edges of the hypothesis graph	Hit	%	Count
C1	2	Excision TREATS Noninfiltrating Intraductal Carcinoma	1	50	60
C2	4	Adjuvant therapy USES Tamoxifen	1	25	339
		Breast LOCATION_OF Tamoxifen	0	0	37
		Therapeutic procedure TREATS Neoplasm	1	25	3799
C3	14	Modified radical mastectomy PRECEDES Radiation therapy	0	0	8
C4	2	Primary Carcinoma PART_OF Breast	0	0	1425
C5	3	Mammaplasty TREATS Woman	2	67	179
		Reconstructive Surgical Procedures TREATS Woman	0	0	251

Table 5 suggests that BioMed Xplorer achieves comparable if not better precision than other existing competitive methods in literature for updating medical guidelines [15,16].

So far, queries to BioMed Xplorer are just one or two sentences of the guideline conclusions. To see if a more elaborate hypothesis statement will enhance the results of BioMed Xplorer, the second experiment involves the researcher also using the abstracts provided for each set of the conclusions in the hypothesis statement, instead of only using the guideline conclusions. In relation to the 5 conclusions there are 3 abstracts in the guideline that can be used to build a more elaborate hypothesis statement. Also based on the results in Table 5, we have decided to focus only on "TREATS" and "USES" Edges of the hypothesis graph for our evaluation.

When evaluating the results of this experiment in Table 6 one can notice a substantial improvement in the results compared to the first experiment and even in one case a 100% success. Comparing abstracts A1 (i.e. for Conclusions 1–2) and A3 (i.e. for Conclusions 5) in Table 6, also suggests that the more elaborate a hypothesis statement is the better the results will be.

Table 6. Results of querying the conclusion and abstract text.

Abstract	Conclusions	Goal	Hit	%	Edges of the hypothesis graph	Hit	Count
A1	C1–C2	6	6	100	Excision TREATS Noninfiltrating Intraductal Carcinoma	1	60
					Adjuvant therapy USES Tamoxifen	1	339
					Therapeutic procedure TREATS Neoplasm	1	3799
					Radiation therapy TREATS Noninfiltrating Intraductal Carcinoma	3	145
					Therapeutic procedure TREATS Invasive Carcinoma	0	59
					Therapeutic procedure TREATS Noninfiltrating Intraductal Carcinoma	2	156
A2	C3–C4	5	4	80	Pharmacotherapy TREATS Woman	0	802
					Pharmacotherapy TREATS Malignant neoplasm of breast	1	2420
					Reexcision TREATS Neoplasm	0	17
					Radiation therapy TREATS Woman	3	664
					Clinical Research USES Clinical Trials, Phase II	0	50
A3	C5	3	2	67	Mammaplasty TREATS Woman	2	179
					Reconstructive Surgical Procedures TREATS Woman	0	251
					Mammaplasty TREATS Patients	1	710

8   Conclusion

The medical field of research is vast and dynamically evolving with new discoveries and investigation of hypotheses on diseases and medications being published. The U.S. National Library of Medicine's bibliographic database MEDLINE, contains over 22 million publication citations, over 750,000 of which were added in

2014, and its citation numbers have been growing exponentially. As a result of the sheer size and growth of this body of knowledge, extraction and provision of the relevant information for its stakeholders is increasingly challenging, causing either an information overload or starvation on their targets. To both enhance the utilization of the vast amount of existing medical knowledge by different stakeholders, and simultaneously reduce the time and cost of providing awareness to public about their individual health conditions, we propose BioMed Xplorer. BioMed Xplorer gathers, clusters, and integrates knowledge in the (bio)medical field, such as knowledge about diseases, risk factors and their semantic inter-relationships, from a variety of authoritative sources. This knowledge base can then be queried with a customized query for each type of stakeholder. Researchers can query BioMed Xplorer with their hypothesis statements in order to identify the state of the art related to their hypothesis, in turn facilitating the researcher's task of validating their hypotheses related to a disease meta-model, by checking it against the state of the art. Practitioners can query BioMed Xplorer with one or more patient condition statements in order to discover and apply the state of the art research to patient cases, without the need to go through pages and pages of medical guidelines, in turn facilitating the decision making process of the practitioners. Patients can query BioMed Xplorer with their medical condition in order to gain awareness and insight about their health condition based on daily conditions and habits.

Future work will focus on implementing key indicators, representing the importance of instances, to more efficiently regulate which concepts and statements are presented to the user. To this end, indicators such as the degree of concepts, or number of sentences or publications from which a statement is derived, might be used. A second point of future work will focus on extending BioMed Xplorer's functionality with extensive filtering options, as such enabling the user to view important, or less important, concepts and statements based on key indicators. Finally we plan to not only provide customized queries to different types of stakeholders but also provide customized user interfaces for interacting with the knowledge base[7].

Acknowledgements. This work was carried out on the Dutch national e-infrastructure with the support of SURF Foundation. We also like to thank the School of Medicine at Democritus University of Trace for helping with some requirements identification and validation.

References

1. Antoniou, G., Van Harmelen, F.: A Semantic Web Primer. MIT Press, Cambridge (2004)
2. Aronson, A.: Effective mapping of biomedical text to the UMLS metathesaurus: the MetaMap program. In: Proceedings of AMIA Symposium, pp. 17–21 (2001)

---

[7] For details visit: https://www.surf.nl/en/services-and-products/hpc-cloud/index.html.

3. Belleau, F., Nolin, M.A., Tourigny, N., Rigault, P., Morissette, J.: Bio2RDF: towards a mashup to build bioinformatics knowledge systems. J. Biomed. Inform. **41**(5), 706–716 (2008)
4. Berners-Lee, T., Bizer, C., Heath, T.: Linked data-the story so far. Int. J. Seman. Web Inf. Syst. **5**(3), 1–22 (2009)
5. Berners-Lee, T., Hendler, J., Lassila, O., et al.: The semantic web. Sci. Am. **284**(5), 28–37 (2001)
6. Bizer, C., Cyganiak, R.: D2R server-publishing relational databases on the semantic web. In: Poster at the 5th International Semantic Web Conference, pp. 294–309 (2006)
7. Bizer, C., Seaborne, A.: D2RQ-treating non-RDF databases as virtual RDF graphs. In: Proceedings of the 3rd International Semantic Web Conference (ISWC 2004), vol. 2004. Citeseer, Hiroshima (2004)
8. Bodenreider, O.: A semantic navigation tool for the UMLS. In: Proceedings of the AMIA Symposium, pp. 971. American Medical Informatics Association (2000)
9. Cohen, A.M., Hersh, W.R.: A survey of current work in biomedical text mining. Briefings Bioinform. **6**(1), 57–71 (2005)
10. D'Arcus, B., Giasson, F.: Bibliographic Ontology Specification (BIBO) (2015). http://bibliontology.com/specification. Accessed 28 Aug 2015
11. Dean, M., Schreiber, G., Bechhofer, S., van Harmelen, F., Hendler, J., Horrocks, I., McGuinness, D.L., Patel-Schneider, P.F., Stein, L.A.: Owl web ontology language reference. W3C Recommendation, 10 February 2004
12. Dublin Core Metadata Initiative (DCMI): Dublin Core (DC) (2015). http://dublincore.org/. Accessed 28 Aug 2015
13. Harris, S., Seaborne, A., Prudhommeaux, E.: SPARQL 1.1 query language. W3C Recommendation 21 (2013)
14. Hendler, J.: Data integration for heterogenous datasets. Big Data **2**(4), 205–215 (2014)
15. Hu, Q., Huang, Z., den Teije, A., van Harmelen, F.: Detecting new evidence for evidence-based guidelines using a semantic distance method. In: Proceedings of the 15th Conference on Artificial Intelligence in Medicine (AIME 2015) (2015)
16. Hu, Q., Huang, Z., ten Teije, A., van Harmelen, F., Marshall, M., Dekker, A.: A topic-centric approach to detecting new evidences for evidence-based medical guidelines. In: Proceedings of the 9th International Joint Conference on Biomedical Engineering Systems and Technologies (HealthInf 2016) (2016)
17. Hunter, L., Cohen, K.B.: Biomedical language processing: what's beyond PubMed? Mol. Cell **21**(5), 589–594 (2006)
18. Institute of Applied Informatics and Formal Description Methods - Karlsruhe Research Institute: Vocab.cc. (2015). http://www.vocab.cc/. Accessed 28 Aug 2015
19. Johns Hopkins University: Online Mendelian Inheritance in Man (OMIM) (2015). http://www.omim.org/. Accessed 28 Aug 2015
20. Kilicoglu, H., Fiszman, M., Rodriguez, A., Shin, D., Ripple, A., Rindflesch, T.C.: Semantic MEDLINE: a web application for managing the results of PubMed searches. In: Proceedings of the Third International Symposium for Semantic Mining in Biomedicine, vol. 2008, pp. 69–76. Citeseer (2008)
21. Kilicoglu, H., Shin, D., Fiszman, M., Rosemblat, G., Rindflesch, T.C.: SemMedDB: a PubMed-scale repository of biomedical semantic predications. Bioinformatics **28**(23), 3158–3160 (2012)
22. Lu, Z.: PubMed and beyond: a survey of web tools for searching biomedical literature. Database 2011, baq036 (2011)

23. Momtchev, V., Peychev, D., Primov, T., Georgiev, G.: Expanding the pathway and interaction knowledge in linked life data. In: Proceedings of International Semantic Web Challenge (2009)

24. NABON: Breast cancer, dutch guideline, version 2.0. Tech. rep., Integraal kankercentrum Netherland, Nationaal Borstkanker Overleg Nederland (2012)

25. Plake, C., Schiemann, T., Pankalla, M., Hakenberg, J., Leser, U.: AliBaba: PubMed as a graph. Bioinformatics **22**(19), 2444–2445 (2006)

26. Rebholz-Schuhmann, D., Kirsch, H., Arregui, M., Gaudan, S., Riethoven, M., Stoehr, P.: EBIMed - text crunching to gather facts for proteins from MEDLINE. Bioinformatics **23**(2), e237–e244 (2007)

27. Shadbolt, N., Hall, W., Berners-Lee, T.: The semantic web revisited. IEEE Intell. Syst. **21**(3), 96–101 (2006)

28. Tao, C., Zhang, Y., Jiang, G., Bouamrane, M.M., Chute, C.G.: Optimizing semantic MEDLINE for translational science studies using semantic web technologies. In: Proceedings of the 2nd International Workshop on Managing Interoperability and compleXity in Health Systems, pp. 53–58. ACM (2012)

29. Tao, Y., Friedman, C., Lussier, Y.A.: Visualizing information across multidimensional post-genomic structured and textual databases. Bioinformatics **21**(8), 1659–1667 (2005)

30. U.S. National Cancer Institute: NCI Metathesaurus (NCIm) (2015). https://ncim. nci.nih.gov. Accessed 28 Aug 2015

31. U.S. National Library of Medicine: Medline fact sheet (2015). http://www.nlm. nih.gov/pubs/factsheets/medline.html. Accessed 28 Aug 2015

32. U.S. National Library of Medicine: Medline Plus (2015). http://www.nlm.nih.gov/ medlineplus/. Accessed 28 Aug 2015

33. U.S. National Library of Medicine: SemRep (2015). http://semrep.nlm.nih.gov. Accessed 28 Aug 2015

34. U.S. National Library of Medicine: Unified Medical Language System (UMLS) (2015). http://www.nlm.nih.gov/research/umls/. Accessed 28 Aug 2015

35. WebMD, LLC: WebMD (2015). http://www.webmd.com/. Accessed 28 Aug 2015

36. Weibel, S.L., Jul, E., Shafer, K.E.: PURLs: Persistent Uniform Resource Locators. OCLC Online Computer Library Center (1996)

37. Weizmann Institute of Science: MalaCards (2015). http://www.malacards.org/. Accessed 28 Aug 2015

38. World Wide Web Consortium: Simple Knowledge Organization System (SKOS) (2015). http://www.w3.org/2004/02/skos/. Accessed 28 Aug 2015

39. World Wide Web Consortium: RDF 1.1 Semantics. World Wide Web Consortium (2014)

A Stochastic Framework for Neuronal Morphological Comparison: Application to the Study of *imp* Knockdown Effects in Drosophila Gamma Neurons

A. Razetti[1(✉)], X. Descombes[2], C. Medioni[3], and F. Besse[3]

[1] University of Nice Sophia Antipolis, I3S, 2000 Route des Lucioles, Sophia Antipolis, France
arazetti@unice.fr
[2] Inria, CRISAM, 2003 Route des Lucioles, Sophia Antipolis, France
xavier.descombes@inria.fr
[3] Institute of Biology Valrose, University of Nice Sophia Antipolis, Parc Valrose, Nice, France
{caroline.medioni,florence.besse}@unice.fr

Abstract. In order to reach their final adult morphology, Gamma neurons in Drosophila brain undergo a process of pruning followed by regrowth of their main axons and branches called remodelling. The mRNA binding protein Imp was identified to play a fundamental role in this process. One of Imp targets, *profilin* mRNA, encodes for an actin regulator that has been shown to be involved in axon remodelling. In this paper we intend to further understand the role of Imp and the importance of *profilin* mRNA expression regulation during remodelling. To do so, we propose a stochastic framework to exhaustively compare the adult morphology between wild type (WT), *imp* knockdown (Imp) and *imp* knockdown rescued by Profilin (Prof Rescue) neurons. Our framework consists in (i) the selection of the main neuron morphological features, (ii) their stochastic modelling and parameter estimation from data and (iii) a maximum likelihood analysis for each individual neuron to quantitatively assess the similarity or difference between groups. Thanks to this framework we show that *imp* mutant neurons can be divided in two phenotypical groups with a different aberrancy degree, and that *profilin* overexpression partially rescues the main axon and branch development thereby it reduces the proportion of neurons with the strongest remodelling phenotype.

Keywords: Gamma neurons · Remodelling · Stochastic models · Maximum likelihood analysis

1 Introduction

Gamma neurons in Drosophila brain mushroom body are in charge of high functions such as olfactory learning and memory [1]. Mutations affecting their adult shape cause several behavioral dysfunctions [2].

During metamorphosis, gamma neurons go through a process of pruning –where the main part of their axons and dendrites is lost– followed by regrowth of the main axon and branches, resulting in the establishment of the adult shape [3]. The correct development of this process gives rise to well-formed and functional adult neurons.

© Springer International Publishing AG 2017
A. Fred and H. Gamboa (Eds.): BIOSTEC 2016, CCIS 690, pp. 145–166, 2017.
DOI: 10.1007/978-3-319-54717-6_9

This study is focused on the role of the mRNA binding protein Imp in the remodelling process. Medioni et al. [4] have shown that even though Imp is not essential during the initial axonal growth of gamma neurons, it is necessary during their remodelling. They report that, in adults, ~50% of *imp* mutants display shorter axons than wild types (WT) and fail to reach their target. *imp* mutants also exhibit an overall loss of branch number and complexity [4].

Molecular and genetic analysis have further shown that *profilin* mRNA, which encodes an actin cytoskeleton regulator [5–7], is a direct and functional target of Imp and both are key regulators of the Drosophila gamma neuron axonal remodelling process, acting on the same molecular pathway. Interestingly, the overexpression of *profilin* in *imp* mutants seems to partially rescue the main axon length, but not the branch complexity. These results suggest that Imp controls axonal extension during remodelling at least partly by regulating *profilin* mRNA expression. However they also suggest that the branching process may be dependent on the regulation of other *imp* mRNA targets, yet to be identified.

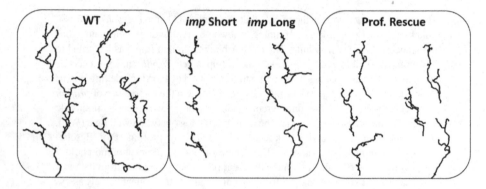

Fig. 1. Representation of the morphology of each one of the groups (in order: Wild type, *imp* mutant and *imp* mutant rescued by Profilin). *imp* mutants are divided into short and long species (named Imp Sh and Imp L respectively) as both phenotypes are equally observed [4].

In this paper, we intend to further understand the role of Imp and the importance of *profilin* mRNA regulation during remodelling, based on a deep analysis of the impact of *imp* knock-down, and its rescue by *profilin* overexpression, in neuron adult morphology (Fig. 1). To do so, an exhaustive morphological comparison of the main features between WT and the mutated axons is needed. We propose in this work a stochastic framework to accurately achieve this comparison, that can be summarized in three steps: (i) selection of relevant morphological features describing the data, (ii) stochastically model the behavior followed by each of the chosen morphological features and estimate the associated parameters to each model from the data (for WT and mutant groups) and (iii) perform a maximum likelihood analysis for each individual neuron (classification) –considering the features separately and altogether– to quantitatively assess the global similarity or difference between groups (through the classification performance). We also developed statistical tests under null hypothesis between the neuron groups for each morphological feature to enrich the analysis. This approach provides both a biological interpretation and a quantification of resemblance between biological samples, detecting differences as well as similarities between the groups. This framework is general and can be applied to model and characterize any kind of neurons (Fig. 2).

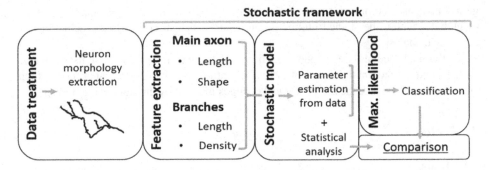

Fig. 2. Scheme of the proposed stochastic framework for the comparison of neuron morphologies between groups.

Because effects of *imp* knockdown and rescue with Profilin can be identified in the main axon as well as in the branch development or separately, we considered both structures separately. The four chosen main morphological features are: "main axon length", "main axon shape", "first order branch distribution along the main axon" and "branch length distribution". To measure these features, we segmented a set of images corresponding to each neuron group (wild type or mutated) to obtain a numeric 3D tree-shaped skeleton representing the morphology of each neuron. We then measured the feature values using homemade software. The image segmentation as well as the measurement of each feature are described in the following sections (Fig. 2).

Neuron morphological automatic classification has already been addressed in the bibliography. Kong et al. [8] proposed an unsupervised clustering of ganglion cells in the mouse retina by the k-means algorithm in order to define cell types. They initially disposed of 26 morphological parameters and found out that clustering with only three of them was the most effective way. Guerra et al. [9] established the advantage of applying supervised classification methods regarding morphological feature based classification to distinguish between interneurons and pyramidal cells. They also conclude that reducing the number of features to an optimal number outperforms the classical approach of using all the available information. Lopez-Cruz et al. [10] built a consensus Bayesian multinet representing the opinions of a set of experts regarding the classification of a pool of neurons. The morphological parameters chosen by each expert to make their decisions are not considered. A different approach was proposed by Mottini et al. [11] which consists on classifying different neuron types by reducing them to trees and calculating a distance, combining geometrical and topological information.

The different published approaches intend to accurately discriminate between different types of neurons, considering misclassification as a methodological error and consequently developing techniques to avoid these cases. However, similarities between populations are not necessarily to be excluded as they may reflect the properties of biological samples and provide useful information for their characterization. Furthermore, these methods do not intend to understand which morphological characteristic is discriminant between different species and at which level. A deeper multi-criteria statistical analysis is thus required. Our approach allows to assess the similarities and dissimilarities between the populations for each chosen morphological feature separately as well as considering them all together for a global analysis. Neurons are treated individually through the resemblance analysis as well as globally within the studied groups through the statistical comparisons, achieving an exhaustive analysis.

In the next sections, we introduce each one of the features followed by its stochastic model. Next we present the results of the classification combining different criteria, which allows to finally deduce the morphological changes induced by the studied mutations.

2 Data

2.1 Images

We used 3D images taken with a confocal microscope. Each set of images shows the distal part of an axonal tree at adult stage (Fig. 3). Single axons are labelled by GFP using the MARCM technique [12], which allows to image a single mutated (or wild type) neuron in a wild type environment. The database we used for this study consists in 46 wild type images, 48 *imp* mutants and 15 *imp* mutants rescued by *profilin* overexpression.

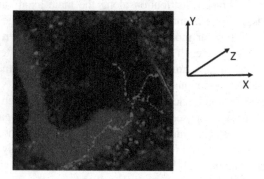

Fig. 3. Maximum intensity Z projection showing a wild type axon and the morphology of the mushroom body.

The voxel size varies among the images and is anisotropic in the Z axis. The voxel length in Z is between 5 and 12 times its length in X and Y, which varies from 0.09 to 0.15 µm (Fig. 3).

2.2 Segmentation

To avoid artificial jumps along the Z axis due to image anisotropy, we applied a simple quadratic interpolation algorithm included in FIJI (the open source image analysis software developed by NIH, Maryland, USA) [13].

An automatic segmentation of these images is still not available regarding our needs due to a noisy background and poorly defined and non-continuous neuron trace. Therefore we segmented the images with the open software Neuromantic [14], specially developed to segment 2 or 3D neurons manually or semi-automatically. As output we obtained a set of points along the main axon and branches that we connected using a Bresenham-inspired 6-connectivity algorithm. We chose this connectivity to keep further measurements simple. After this process we obtained a tree-like set of numeric 3D curves that describe the morphology of each neuron (Fig. 4).

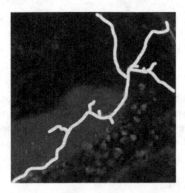

Fig. 4. Zoom of the Z projected image showed in Fig. 3, where the neuron has been segmented to obtain a tree-like set of numeric 3D curves.

To ensure all the neurons to be similarly oriented we rotated the images and aligned them with respect to reference structures (i.e. medial and dorsal lobes of the mushroom body, Fig. 3). We only considered a rigid transformation to avoid axon deformation. Conserved morphology was preferred rather than more accurate spatial location.

2.3   Tree Hierarchy

When studying their morphology it is necessary to understand how neurons are structured i.e. main axon and first, second, third (etc.) order branches (the neuron cell body and dendrites are not considered in this study). To accurately label each path of the tree that represents each neuron, we developed an automatic recursive algorithm capable of processing trees of any order. It starts by taking the whole tree and selects the main axon. After this process a number of independent subtrees is formed. During the following steps each of the subtrees is analyzed to assign their main path and so on, until no more untagged segments are left. In each step, the main path of the subtree is assigned following the same criteria used by experts when done visually: total length, directionality and sense coherence. To achieve this, we consider the points for each path between the root and the leaves of the tree (i.e. the whole neuron) or subtree and calculate a linear regression obtaining a straight guideline, which will determine directionality and sense coherence. For each path in the analyzed tree/subtree, a cost function is computed that depends on the distance between each point in the path and the guideline (directionality), the parallelism between them (accounting for the sense coherence of the path) and the path total length. Finally the path that minimizes this cost function is selected as "main axon" in the case of the whole tree (first step), or "main branch" in the case of the different subtrees (Fig. 5).

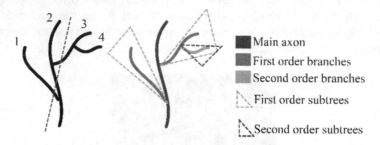

Fig. 5. Scheme of the three-hierarchy algorithm. For a given tree, the guideline is calculated as the linear regression of all the points in the tree, followed by the cost function for each possible path (1–4). The one that minimizes is assigned as "main axon" (here path 2). The algorithm is applied recursively to each subtree resulting in the hierarchy of the entire tree.

3   Model Development

After the processes described in the previous section, our data is composed of 3D skeletons of neurons where each unitary 6-connected segment is described by its round coordinates or pixels. Taking this simple geometrical description of the neuron into account, we defined the main morphological features that may best describe and discriminate the individuals: the main axon length and sinuosity, as well as the branch density and length distribution. In the following sections we develop the probabilistic models chosen to describe each feature, estimate their parameters and compute associated statistical tests under null hypothesis between the different groups (WT: wild type neurons that are used as controls, Imp: *imp* knockdown neurons, reported to be morphologically aberrant in the literature, and Prof Rescue: *imp* mutants with an overexpression of *profilin*, known to partially suppress the *imp* mutant phenotype) (Fig. 1). Besides, we derive the likelihood of each model that is used in Sect. 4.

3.1   Main Axon Length

The main axon length was measured taking the total amount of pixels in the corresponding path and multiplying it by the pixel size (μm). The length distribution was modelled as Gaussian, and the mean and standard deviation for each group X ($\mu_{m.a}^{X}$, $\sigma_{m.a}^{X}$) were calculated from data. We observed the bimodal behaviour in the Imp group reported by Medioni et al. [4] (Fig. 6). Therefore, in order to make a more accurate modelling of this parameter, we separated *imp* mutant neurons into two groups -neurons with long axons (Imp L) and neurons with short axons (Imp Sh)- using the k-means algorithm. 54% of the neurons were assigned to Imp Sh and 46% to Imp L, consistent with the percentage reported by Medioni et al. [4]. Figure 6 shows the main axon length histograms for each group, Imp divided into Imp L and Imp Sh. We decided to keep this division of the Imp group (Imp L and Imp Sh) through all the analysis in this paper, in the attempt to detect other morphological differences between the two subgroups.

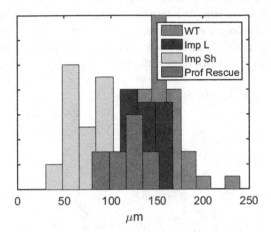

Fig. 6. Main axon length distributions for each biological sample.

To assess which groups present significant differences regarding the main axon length, non-parametric Kruskal Wallis tests were carried out between all the possible pairs of groups (Table 1). We chose this test for the sake of consistency, as it can be applied to analyze all the features independently of each model. For p values inferior to 5%, we consider the null hypothesis that both distributions are the same can be rejected. Thus, the only pair not presenting a significant difference is WT vs. Imp L. It is interesting to highlight also that Prof Rescue distribution lies in between the distributions for Imp L and Sh and is more similar to Imp L, meaning it does not present extremely short axons.

Table 1. p values from the non-parametric Kruskal Wallis test comparing the main axon length between the studied groups.

	Imp L	Imp Sh	Prof Rescue
WT	0.1219	5.0098E–12	0.000144
Imp L		3.2627E–09	0.0013
Imp Sh		·	2.48E–06

The likelihood of a given neuron n of length l_n to belong to a given group X is defined by the Normal probability density function

$$L_l(l_n | n \in X) = P(l_n | n \in X) = \frac{1}{\sigma^X_{m.a} \sqrt{2\pi}} e^{\frac{-(l_n - \mu^X_{m.a})^2}{2\sigma^{X2}_{m.a}}}. \tag{1}$$

3.2 Main Axon Morphology

To define the shape model, we considered as random variable the unit vector $\vec{x}_t$ that accounts for the shift of the axon tip between $t - 1$ and t. Because we used the 6-connectivity and backwards moves are not allowed, each $\vec{x}_{t+1}$ (shift of the axon tip between t and $t + 1$) can take five different values, as shown in Fig. 7. Assuming the main axon development follows a second order Markov property, we have [15, 16]

$$P\big(\vec{x}_{t+1} | \vec{x}_i \, i \le t\big) = P\big(\vec{x}_{t+1} | \vec{x}_t, \vec{x}_{t-1}\big). \tag{2}$$

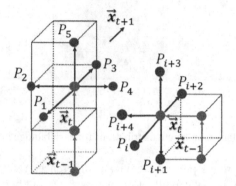

Fig. 7. Two examples of three vector configurations on a 3D 6-connected path. Each future direction has a probability of occurrence conditioned by the present and past directions and is numbered from 1 to 150 (30 possible configurations for past plus present and 5 possible future configurations for each of them).

The shape model is then completely defined by the conditional probabilities $P\big(\vec{x}_{t+1} | \vec{x}_t, \vec{x}_{t-1}\big)$. There are 30 possible combinations of the two unit vectors $[\vec{x}_t, \vec{x}_{t-1}]$ and each of these combinations has five possible future jumps $\vec{x}_{t+1}$, giving a total of 150 possible transitions in $t + 1$, each of them with probability P_i (conditionally to $[\vec{x}_t, \vec{x}_{t-1}]$). The order of the Markov chain was chosen in order to combine a discriminative efficiency between similarly shaped axons and a reasonable combinatorial to robustly estimate the conditional probabilities.

Figure 7 presents two basic configurations of a pair of unit vectors $[\vec{x}_t, \vec{x}_{t-1}]$ and their corresponding five possible $\vec{x}_{t+1}$. The one on the left depicts one of the six possible cases where the vectors $\vec{x}_t$ and $\vec{x}_{t-1}$ are aligned. The second configuration exemplifies the 24 cases where the vectors $\vec{x}_t$ and $\vec{x}_{t-1}$ are not aligned.

We estimate the conditional probabilities from data using the empirical estimator (3), where $\#_n$ accounts for the number of times the nth configuration of three unit vectors $[\vec{x}_{t+1}, \vec{x}_t, \vec{x}_{t-1}]$ appears.

$$P_{5s+j} = \frac{\#_{5s+j}}{\sum_{k=1}^{5} \#_{5s+k}}, \quad \begin{array}{l} j = 1, \ldots, 5 \\ s = 0, \ldots, 29 \end{array} \tag{3}$$

We performed the Kruskal Wallis non-parametric test between populations for each $P_i, 1 \leq i \leq 150$. Table 2 shows the number of parameters P_i that present a p value inferior to 5% between each pair of populations. In addition, regarding the possible two past unit vectors $[\vec{x}_t, \vec{x}_{t-1}]$, the results of the estimation show that all the groups share the six most frequent configurations, representing together between 65 and 76% of the total.

Table 2. Number of parameters with $p < 0.05$ for the non-parametric Kruskal Wallis test.

	Imp L	Imp Sh	Prof Rescue
WT	12	22	28
Imp L		16	19
Imp Sh			14

The computation of the Markov chain likelihood appears to lack of robustness to compare populations. This can be explained by the limited length of the axons in pixels (~1500) and the combinatorial of the problem (150 conditional probabilities). Indeed, some of the three vector configurations, even though with non-zero probability, may not appear in the learning sample. When this is the case, if the axon to classify does present at least one time this configuration the likelihood becomes zero. This means that the likelihood is extremely sensible to fluctuations in the presence of low probable events, which is statistically inevitable with the size of our data. To overcome this inconvenience and add robustness to the likelihood analysis, an original approach was applied (4). We assume the 30 probability distributions $P(\vec{x}_{t+1} | \vec{x}_t, \vec{x}_{t-1})$ as independent, and defined a multinomial Bernoulli distribution [17] for the variable $\vec{x}_{t+1}$ for each given $[\vec{x}_t, \vec{x}_{t-1}]$.

For each neuron n, the likelihood of each group X according to the shape model of X, $P_{i,X}$, and the frequencies of appearance of three unit vectors corresponding to n, $\#_i$, is then defined as follows

$$
L_{sh}(\#_1 - \#_{150} | n \in X) = P(\#_1 - \#_{150} | n \in X)
$$
$$
= \prod_{s=1}^{30} \left(\binom{N_s}{\#_{k+1}} P_{k+1}^{\#_{k+1}} \binom{N_s - \#_{k+1}}{\#_{k+2}} P_{k+2}^{\#_{k+2}} \binom{N_s - \#_{k+1} - \#_{k+2}}{\#_{k+3}} \right.
$$
$$
\left. P_{k+3}^{\#_{k+3}} \binom{N_s - \#_{k+1} - \#_{k+2} - \#_{k+3}}{\#_{k+4}} P_{k+4}^{\#_{k+4}} P_{k+5}^{\#_{k+5}} \right), \tag{4}
$$
$$
k = 5(s - 1), \quad N_s = \sum_{j=1}^{5} \#_{5(s-1)+j}.
$$

3.3 Branch Density

We propose a model that describes the branching point distribution independently of the axon length and that is based on the biological process of interstitial branch formation during development. This process can be described in three simple steps (Fig. 8): A. the main axon grows following particular external and internal guiding cues. B. When the growth cone senses external guiding cues indicating the formation of an interstitial branch, the main axon decreases its growing speed until it totally stops while it accumulates molecular material in its tip. C. After some time the main axon continues growing following its particular cues, leaving the accumulated material in a specific zone of its shaft. The left material has been organized into an independent growing tip and starts elongating an interstitial branch towards its particular target [18].

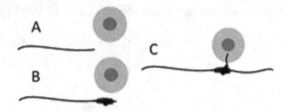

Fig. 8. Interstitial branch formation during axonal development described schematically in three main steps, adapted from Szebenyi et al. [18]. A. the main axon grows. B. When the growth cone senses external guiding cues indicating to branch, the growth speed is decreased until it stops. C. After some time the main axon continues growing, leaving accumulated material in a specific zone of its shaft. The left material has been organized into an independent growing tip and starts elongating an interstitial branch towards its particular target.

Modelling this process becomes initially unreachable as none of two main actors (growing rate, guiding cues presence) can be measured from the adult stage static images available as data. Regarding these limitations, we propose a model to mimic this dynamic process from our static data. We focused our study on the behavior of the axon growing rate v, starting with a certain initial speed v_o and evolving until $v = 0$, when a new branch point appears.

We can measure the number k of pixels between every two successive branching points along the main axon of a segmented neuron. Then, we suppose that each of this pixels represents a differential progress in the axonal growth where, during development, the axon had a certain growing rate v. Our model assumes random decreases in speed which we call Δv, with a probability of occurrence p. When a certain number of decreases Δv occurs, the speed v equals zero thus the growing tip stops, allowing the material needed to form a branch to accumulate. After some time the process starts again, with initial speed v_o.

Because at each one of the k pixels a decrease in v may or not happen, we describe the problem using a Bernoulli probability distribution [19] where each success means the occurrence of a decrease in speed Δv. We consider that the growing rate goes to zero after $A + 1$ steps of speed decreasing. The probability to reach $v = 0$ after k steps is given by an accumulation process and is written as follows:

$$P(k) = \binom{k-1}{A} p^{A+1}(1-p)^{k-A-1}. \tag{5}$$

Equation (5) gives the probability of having A successes in $k - 1$ trials and a success in the kth trial. This means that the axon tip decreases its speed A times before stopping completely (which happens in $A + 1$), or equivalently that the length between two branching points is k (Fig. 9). Thus, our accumulation Bernoulli-based, time-mimicking branching point distribution model has two parameters, A and p, to be estimated from data. Knowing all the distances k_i between successive branching points for every axon in each group, we can calculate their mean and variance μ_k and σ_k^2. From

$$\mu_k = \sum_i k_i \binom{k_i - 1}{A} p^{A+1} (1 - p)^{k_i - A - 1} \tag{6}$$

and

$$\sigma_k^2 = \sum_i k_i^2 \binom{k_i - 1}{A} p^{A+1} (1 - p)^{k_i - A - 1} - \mu_k^2 \tag{7}$$

it can be shown that $\mu_k(A, p)$ and $\sigma_k^2(A, p)$ have the simple forms

$$\mu_k(A, p) = \frac{A}{p}; \sigma_{k^0}^2(A, p) = \frac{(1 - p)A}{p^2} \tag{8}$$

allowing to easily estimate A and p from data (proof described in Appendix). Once A and p are estimated, A needs to be rounded as it has to be an integer. Then p can be recalculated knowing the value of A as

$$p = \frac{\sqrt{A(\mu_k + \sigma_k^2)}}{\mu_k + \sigma_k^2}. \tag{9}$$

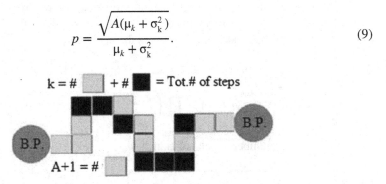

Fig. 9. 2D 4-connected path showing an axonal trajectory between two branching points (3D not shown for simplicity). Light colored pixels depict Δv and occur with a probability p, decreasing the growing rate. When the number of light colored pixels equals $A + 1$, $v = 0$ and a new branching point appears. (Color figure online)

The number $A + 1$ of needed Δv in order to form a branch, and p their probability to happen will define each axonal group regarding their branch density.

Tables 3 and 4 present the resulting values of the parameters for each group and the p values from the non-parametric Kruskal Wallis test comparing all the distances between two consecutive branches k_i among neuron groups, respectively.

Table 3. p values from the non-parametric Kruskal Wallis test comparing the distances in pixels between consecutive branches between the studied groups.

	Imp L	Imp Sh	Prof Rescue
WT	0.9398	4.20E–03	0.5704
Imp L		2.16E–02	0.6478
Imp Sh			1.32E–02

Table 4. Value of the parameters A and p describing the branching point distribution.

	A	p	p for $A = 1$
WT	1.2	0.0087	0.0078
Imp L	1.0	0.0068	0.0067
Imp Sh	0.9	0.008	0.0084
Prof Rescue	1.2	0.0074	0.0068

While each group has the same value of A, Imp Sh presents the highest value of p. This means that Δv occurrence is more probable, thus it takes less time to reach $v = 0$, and consequently it is the most branched group. This difference is significant ($p < 0.05$) between Imp Sh and every other group.

To calculate the likelihood of each neuron n to belong to the group X regarding this model, we use the Binomial probability density function considering the distances between each pair of branches $k_{n,m}$ independent between them, obtaining

$$
L_{bp}\left(k_{n,m}|n \in X\right) = P\left(k_n = \left\{k_{n,1}, \ldots, k_{n,M}\right\}|n \in X\right)
$$
$$
= \prod_{m=1}^{M} P\left(k_{n,m}|n \in X\right) = \prod_{m=1}^{M} \binom{k_{n,m} - 1}{A_X} p_X^{A_X+1} p^{k_{n,m}-1-A_X} \tag{10}
$$

where M is the total number of pairs of branches.

3.4   Branch Length Distribution

To study the branch length distribution within the neuron groups, we established four length categories (μm); L_1:(0, 1], L_2:(1, 5], L_3:(5, 10] and L_4:(10, ∞) following Tessier and Broadie study [20]. The length was measured in the same way as described for the main axon, and branches of all levels were taken into account. The probability distribution modelling the relative amount of branches within these length categories and for each group was considered as Gaussian. For each group of axons X we calculated the mean and standard deviation $\left(\mu_{bi}^X, \sigma_{bi}^X\right)$, $1 \leq i \leq 4$ of the relative number of branches corresponding to each length category per axon $b_1 - b_4$ (i.e. number of branches in each length category normalized by the total number of branches, per axon).

To know between which groups and for which length category the differences displayed in Table 5 are significant, we performed the Kruskal Wallis non-parametric test for the four length groups. Significant results ($p < 0.05$) are only present in L_2 and L_4 categories. The p values are shown in Tables 6 and 7.

Table 5. Branch length distribution by length and neuron group (%).

	L_1	L_2	L_3	L_4
WT	10.6	49.2	11.7	28.5
Imp L	8.4	66.5	10.1	15
Imp Sh	19.8	48.2	14.5	17.5
Prof Rescue	19.5	48.3	10.2	22

Table 6. p values from the non-parametric Kruskal Wallis test comparing the branch length distribution in L_2 between the studied groups.

L_2	Imp L	Imp Sh	Prof Rescue
WT	8.92E–05	0.9392	0.7884
Imp L		9.04E–04	0.0014
Imp Sh			0.9134

Table 7. p values from the non-parametric Kruskal Wallis test comparing the branch length distribution in L_4 between the studied groups.

L_4	Imp L	Imp Sh	Prof Rescue
WT	3.45E–04	1.29E–04	0.1822
Imp L		0.7383	0.1238
Imp Sh			0.1387

Imp L presents significantly more branches in L_2 than any other group while WT has a bigger proportion of L_4 branches than Imp L and Sh, but not Prof Rescue. For further analysis, we took only the discriminant categories L_2 and L_4.

To calculate the likelihood of each neuron n with each group X regarding the branch length distribution in L_2 and L_4 -$b_{n,2}$ and $b_{n,4}$- we considered a bivariate Gaussian distribution with mean $\vec{\mu}_b^X = \left(\mu_{b2}^X, \mu_{b4}^X \right)$ and Σ^X the covariance matrix between b_2^n and b_4^n.

$$L_{bl}\left(\vec{b}_n | n \in X\right) = P\left(\vec{b}_n | n \in X\right) = \frac{1}{2\pi\sqrt{|\Sigma^X|}} e^{-\frac{1}{2}\left(\vec{b}_n - \vec{\mu}_b^X\right)^T \Sigma^{X-1}\left(\vec{b}_n - \vec{\mu}_b^X\right)}, \tag{11}$$

where $|\Sigma^X|$ is the determinant of the covariance matrix Σ^X.

4 Likelihood Analysis

For a neuron n, we calculated the value of each feature and then compute the likelihood for each group of neurons X, ($X \in \{WT, Imp, Prof Rescue\}$). The neuron n is then classified in the group that maximizes the global likelihood. All the classifications present in this work were done using the leave one out technique, which consists in classifying an element of the sample that has been removed from the database to perform the learning stage (i.e. the estimation of the models parameters). This maximum likelihood classification provides some assessment about the discriminative properties of the proposed models but is also used to analyze the mixture of feature values between the populations.

Considering our four features to be independent from each other, the global likelihood is given as follows

$$L\left(\left\{l_n, \#_{1,n} - \#_{150,n}, k_n, \vec{b}_n\right\} | n \in X\right)$$
$$= L_l\left(l_n | n \in X\right) L_{sh}\left(\#_{1,n} - \#_{150,n} | n \in X\right) L_{bp}\left(k_n | n \in X\right) L_{bl}\left(\vec{b}_n | n \in X\right), \tag{12}$$

and the maximum likelihood estimation results

$$n \in X_o \leftrightarrow X_o = \text{argmax}_X L\left(\left\{l_n, \#_{1,n} - \#_{150,n}, k_n, \vec{b}_n\right\} | n \in X\right), \tag{13}$$

$$X = \{WT, Imp\ L, Imp\ Sh, Prof\ Rescue\}.$$

Equation (13) allows to classify each neuron by resemblance to each group considering the four morphological features (main axon length and shape, branch length distribution and branch point distribution) and their mathematical models. Table 8 presents the results of the global resemblance analysis.

Table 8. Global likelihood analysis considering the four features.

Predicted (%)		WT	Imp L	Imp Sh
Actual class	WT	82.6	17.4	0
	Imp L	54.5	45.5	0
	Imp Sh	19.2	3.9	76.9
	Prof Rescue	40	26.7	33.3

These results suggest a relevant global difference between neurons belonging to Imp L and Imp Sh, as well as between WT and Imp Sh; while between WT and Imp L this difference is weaker. More than half of Imp L neurons are likely to be WT while for Imp Sh this proportion is less than 20%. Some WT axons are classified as Imp L but none as Imp Sh. Interestingly, the percentage of Prof Rescue neurons likely to be WT lies in between those percentages for Imp Sh and Imp L. This result points in the direction of a partial rescue of the *imp* neuron morphology.

To understand how each morphological feature contributes to the results in Table 8, we carried out the maximum likelihood analysis regarding each of them separately. For the main axon length, as expected from Fig. 6, WT neurons are shared between WT and Imp L categories; and Imp L is correspondingly mixed with WT. Imp Sh is completely separated from the rest of the groups (Table 9). Regarding Prof Rescue, our results agree with those in Medioni et al. [4] about the main axon length being partially rescued by *profilin* overexpression. Our study pointed out that 54% of *imp* mutant neurons present a conserved main axon length while 46% are significantly shorter than WT (Sect. 3.1). Here we show that Prof Rescue neurons are distributed by a 67% (Imp L + WT) vs. 33% (Imp Sh), moving the tendency towards a WT phenotype.

Table 9. Likelihood analysis according to the main axon length feature.

L		Predicted (%)		
		WT	Imp L	Imp Sh
Actual class	WT	39.1	54.4	6.5
	Imp L	22.7	77.3	0
	Imp Sh	0	0	100
	Prof Rescue	6.7	60	33.3

According to the main axon shape in Table 10, WT and Imp L look again similar and, interestingly, Imp Sh looks more similar to WT than to Imp L. Prof Rescue behavior is opposite to that of Imp Sh.

Table 10. Likelihood analysis according to the main axon shape feature.

SH		Predicted (%)		
		WT	Imp L	Imp Sh
Actual class	WT	54.3	43.5	2.2
	Imp L	50	50	0
	Imp Sh	61.5	38.5	0
	Prof Rescue	40	60	0

Table 11 presents the likelihood analysis results regarding the branch point density. It can be noticed that every group is mainly classified as Imp Sh, which our previous analysis revealed as the most branched group. The reason for this behavior relies on the nature of the model. Even though the means of the distances between branches are different between the biological groups, axons frequently display one or more pairs of branches which are close. Because for close branches the likelihood is maximum for Imp Sh, with a significant difference from the other groups, the presence of near branches automatically classifies a neuron as Imp Sh. Nevertheless, the branch density coherence is respected for each group as the resemblance with Imp Sh is maximum for the most branched group (itself) and is followed in the correct order: WT first, followed by Imp L and Prof Rescue.

Finally, according to the branch length distribution (Table 12) WT, Imp L and Imp Sh show a higher resemblance to their own groups, suggesting a significant difference between them regarding this feature. Prof Rescue has a slight preference for Imp Sh is understandable as both have the same proportion of branches in L_2 which is the most abundant group of branches. Nevertheless, its resemblance to WT regarding this feature is notoriously higher than those for Imp L and Imp Sh. This results reveal that *profilin* overexpression partly rescues branch length distribution –i.e. it presents a bigger

Table 11. Likelihood analysis according to the branching point feature.

BP		Predicted (%)		
		WT	Imp L	Imp Sh
Actual class	WT	0	13	87
	Imp L	13.6	18.2	68.2
	Imp Sh	7.7	11.5	80.8
	Prof Rescue	6.7	26.7	66.7

proportion of long branches- in addition to the main axon length. Thus Profilin may be involved in the branching process as well.

Table 12. Likelihood analysis according to the branch length distribution feature.

BL		Predicted (%)		
		WT	Imp L	Imp Sh
Actual class	WT	60.9	23.9	15.2
	Imp L	18.2	72.7	9.1
	Imp Sh	15.4	30.8	53.8
	Prof Rescue	33.3	20	46.7

In order to better understand the morphological changes induced by Profilin over-expression in *imp* mutant axons, we performed the global maximum likelihood analysis considering *imp* mutants altogether (i.e. Imp Sh + Imp L), and the possible classification groups either altogether (Table 13) or split between Imp L and Imp Sh (Table 14).

Table 13. Global likelihood analysis considering the four features. Imp englobes Imp Sh and Imp L.

		Predicted (%)	
		WT	Imp
Actual class	WT	80.4	19.6
	Imp	37.5	62.5
	Prof Rescue	60	40

From the analysis in Table 13, we can highlight that while only 37.5% of *imp* mutants present a WT phenotype, 60% of Profilin rescue neurons exhibit this behavior. Moreover, it is interesting to analyze how Prof Rescue is classified regarding Imp L and Imp Sh (Table 14). The percentage of neurons classified as Imp Sh decreases compared to *imp* mutants from 42 to 33% while the tendency for Imp L and WT is increased in Prof Rescue.

Table 14. Global likelihood analysis considering the four features. Imp is splitted between L and Sh for possible classification groups.

		Predicted (%)		
		WT	Imp L	Imp Sh
Actual class	WT	82.6	17.4	0
	Imp	35.5	23	41.5
	Prof Rescue	40	26.7	33.3

Finally a brief comparison can be done regarding our classification results with those in Mottini et al. [21], who also analysed wild type as well as *imp* mutant gamma neurons. The authors report an 80.4 and 91.7% of accurate classifications for WT and *imp* mutants respectively with the ESA curve distance method and 85 and 79.2% with RTED. It is relevant to highlight that the goal in their work was to discriminate between populations, thus they considered exclusively highly discriminative parameters. On the contrary, our results –80.4 and 62.5% for WT and Imp respectively- show and value not only the differences but also the existing similarities between phenotypes, considering all the relevant morphological features (including those that may be known as not discriminative). Our work also allows to correlate the conclusions with biological parameters. In addition, our sample size doubles the one used in the cited work.

5   Discussion

5.1   Axon Growing Rate and Branch Formation

The value of $A = 1$ indicates that the axon tip diminishes its growing speed only two times before stopping to create a branch, instead of doing it gradually. The first time may be related to the moment when it senses the external guiding cues. Then it continues growing more slowly, which may facilitate other cues detection, until it finally stops, consequence of the second and last speed lost. When this happens, branching material could be accumulated and after some time an interstitial branch is created. An increased value of p may indicate a higher sensibility to external cues as well as an increased concentration of internal cues triggering branching. Another interpretation can be that axons with a defective growing rate (i.e. slower speed, or high p) are more susceptible to stop independently from external cues, and therefore to branch more.

All the groups present the same value of A indicating that this two-step behavior may be conserved and therefore independent from Imp. Regarding p, Imp Sh is significantly more branched than the rest of the groups, including Imp L, even though they have the same genotype. We suggest a correlation between the size of the main axon and the branch density for *imp* mutants. More interestingly, Prof Rescue axons present the same value of p than Imp L. This suggests that the phenotype presenting high branch density may be rescued by *profilin* overexpression.

5.2 Wild Type Neurons are Mostly Differentiated by Their Branch Length Distribution

The global maximum likelihood analysis results in more than 80% of WT axons to be correctly classified (Tables 8, 13 and 14). Nevertheless, when looking at each particular feature it becomes evident that WT shares most of those with Imp L. Regarding the main axon length (Table 9), 54% of WT neurons are likely to be Imp L and 43% for the main axon shape (Table 10). The analysis following the branching point density results in 13% of WT neurons likely to be Imp L, while no WT neuron was correctly classified. This results are validated by the p values for main axon length and branch length distribution that do not show significant differences between WT and Imp L. We encounter a similar situation regarding the shape model, as between Imp L and WT the amount of significantly different parameters is the minimum of all the group pairs and it is only 12 in 150.

The maximum likelihood analysis taking only the branch length distribution into account is the sole to correctly classify WT axons (Table 12). While WT and Imp L present both 80% of branches in L_2 and L_4 altogether (Table 5), the difference between them is that WT shows statistically more branches in L_4 while Imp L in L_2. We can relate our results to those of Tessier and Broadie [20] and Medioni et al. [4]. The first publication reports that a loss of L_2 branches by a late pruning process occurs in wild type neurons and not in dFMRP mutants (dFMRP is also a *profilin* regulator) and the second one concludes a defective development of long branches (L_4) in *imp* mutants.

The maximal percentage of correct classification for WT considering the features separately is 60% for the branch length distribution (Table 12), followed by 54, 39 and even 0% corresponding to main axon shape, length and branching point distribution (Tables 9, 10 and 11). Interestingly, the global classification mixing the four features improves these percentages to 80% (Tables 8, 13 and 14). This suggests that WT neurons are well defined and different from *imp* mutants but it is necessary to consider all the morphological features together for a correct classification. This also highlights the advantages of our method as it goes beyond a simple statistical analysis, allowing to mix different features as well as to consider each neuron independently.

5.3 *Imp* Knockdown Presents Two Different Phenotypes

Medioni et al. (2014) [4] reported that *imp* mutants could either present a conserved main axon length or an aberrant one, with a ~50% of occurrence each. Following these results we applied the k-means automatic algorithm to separate the Imp population in Imp L and Imp Sh, and obtained a 46 vs. 54% of incidence correspondingly. This bimodal behavior can also be seen in the length distribution (Fig. 6). Surprisingly, we have found other relevant morphological differences between this two groups that have not been yet reported in the bibliography. The main one is the branching point distribution, as Imp Sh is significantly more densely branched than Imp L (Tables 3 and 4). Also, the percentage of branches ranging from 1 to 5 μm, while aberrant in Imp L, is conserved in Imp Sh (which shows no differences from WT (Table 6)).

Regarding the global likelihood analysis (Table 8), while less than 20% of Imp Sh neurons can be considered to have a WT phenotype, 55% of Imp L do, allowing to conclude that Imp L presents a generally more conserved phenotype. Globally, we conclude that the penetrance of the *imp* phenotype is ~63%, following our global likelihood analysis (Tables 13 and 14).

These results are consistent with an essential role of Imp in main axon elongation and branch formation as well as branch elongation during remodelling. Nevertheless, the phenotypical variability within *imp* mutants (i.e. from globally aberrant to completely WT-like neurons) indicates the existence of other –maybe Imp independent- important actors with the capability of controlling these processes and neutralize Imp absence; or that the knockdown of the gene is not 100% efficient.

5.4   *Profilin* Overexpression Partially Rescues the Main Axon Length as Well as the Branch Length Distribution

The global likelihood analysis (Table 13) considering Imp altogether shows that Profilin decreases the percentage of *imp* mutant phenotype from 63 to 40%.

Regarding the main axon length, while aberrant neurons represent 54% of the Imp population, they represent only 33% in Prof Rescue (in Prof Rescue 67% of neurons present a conserved length (WT + Imp L) and only 33% do not). Following the branch length distribution resemblance analysis, 33% of Prof Rescue neurons are classified as WT and represent the second maximum percentage after WT itself (only 18 and 15% correspond to Imp L and Sh, respectively). Looking at the *p* values between branch length categories (Tables 6 and 7), we can conclude that Profilin rescues the late pruning of small branches [20] showing a conserved percentage of L_2 branches and also allows to develop more long branches. Even though the percentage of branches in L_4 is slightly smaller for Prof Rescue than WT (Table 5), this difference is not significant in the statistical tests, suggesting a conserved percentage of long branches in Prof Rescue which is not seen in Imp Sh nor in Imp L.

Finally, regarding the global likelihood analysis considering Imp L and Imp Sh separately (Table 14), we conclude that Profilin rescue diminishes the general morphological aberration, as it moves the tendency towards WT and Imp L phenotypes and lowers the percentage of neurons with an Imp Sh phenotype.

This study suggests that Profilin is also involved in branch formation and elongation, in addition to main axon elongation during remodelling. Nevertheless, because the phenotypical rescue is not complete, we can conclude either that its regulation by Imp is still an essential step in remodelling or that other Imp targets are also essential in these processes; or –most probably– both simultaneously.

6   Conclusions

In this work we proposed a framework to compare neuron groups based on their morphology. Our procedure consists in applying probabilistic models to describe the behavior of selected morphological features (i.e. main axon length and shape as well as branch length and density), the estimation from data of the associated parameters and a resemblance analysis combined with statistical tests. We applied this framework to understand the effects of *imp* knockdown –as well as its rescue by Profilin- in Drosophila gamma adult neuron morphology. The similarities and differences we are able to highlight between wild type and mutant neurons allow to better understand the role of Imp and Profilin during axonal remodelling, particularly on axon elongation and branch formation.

We propose that this method consisting in feature selection, model application and likelihood analysis could be applied to any case of study between species where similarities are as important as differences. We can also conclude that the study of individuals is relevant and more enriching than just population analysis driven by ordinary statistics. Finally, we highlight the importance of combining different features to achieve a global result.

Acknowledgements. This work was supported by the French Government (National Research Agency, ANR) through the « Investments for the Future » LABEX SIGNALIFE: program reference # ANR-11-LABX-0028-01.

All the authors are within Morpheme (a joint team between Inria CRI-SAM, I3S and IBV).

Appendix

From Eq. 6 we can express $\mu_k(A + 1, p)$ as

$$\mu_k(A + 1, p) = \sum_k k \binom{k - 1}{A + 1} p^{A+2}(1 - p)^{k-A-2}. \tag{14}$$

Using

$$\binom{k - 1}{A} + \binom{k - 1}{A + 1} = \binom{k}{A + 1}, \tag{15}$$

Equation 14 can be rewritten as

$$\mu_k(A + 1, p) = \sum_k k \binom{k}{A + 1} p^{A+2}(1 - p)^{k-A-2}$$
$$- \sum_k k \binom{k - 1}{A} p^{A+2}(1 - p)^{k-A-2}. \tag{16}$$

Taking out $\frac{p}{1 - p}$ as a common factor from the second sum in Eq. (16), we obtain

$$\sum_k k \binom{k - 1}{A} p^{A+2}(1 - p)^{k-A-2}$$
$$= \frac{p}{1 - p} \sum_k k \binom{k - 1}{A} p^{A+1}(1 - p)^{k-A-1} = \frac{p}{1 - p} \mu_k(A, p) \tag{17}$$

Similarly, the first sum in (16) can be worked out to obtain

$$\mu_k(A+1,p) = \frac{1}{1-p}\mu_k(A+1,p) - \frac{1}{1-p}. \tag{18}$$

From (17) and (18) we finally obtain

$$\mu_k(A,p) = \frac{A}{p}. \tag{19}$$

From Eqs. (7) and (19) we can express

$$\sigma_k^2(A,p) = \sum_k k^2 \binom{k-1}{A} p^{A+1}(1-p)^{k-A-1} - \frac{1}{p^2}A^2, \tag{20}$$

from where

$$\sigma_k^2(A+1,p) = \sum_k k^2 \binom{k-1}{A+1} p^{A+2}(1-p)^{k-A-2} - \frac{1}{p^2}(A+1)^2. \tag{21}$$

Equation (21) can be rewritten using Eq. (15) to get $\sigma_k^2(A+1,p)$:

$$
\begin{aligned}
\sigma_{k_0}^2(A+1,p) = &\sum_k k^2 \binom{k}{A+1} p^{A+2}(1-p)^{k-A-2} \\
&- \sum_k k^2 \binom{k-1}{A} p^{A+2}(1-p)^{k-A-2} - \frac{1}{p^2}(A+1)^2,
\end{aligned}
\tag{22}
$$

Working out Eq. (22) similarly to Eq. (16) we finally get

$$\sigma_k^2(A,p) = \frac{(1-p)A}{p^2}. \tag{23}$$

References

1. Xie, Z., Huang, C., Ci, B., Wang, L., Zhong, Y.: Requirement of the combination of mushroom body γ lobe and α/β lobes for the retrieval of both aversive and appetitive early memories in Drosophila. Learn. Mem. **20**(9), 474–481 (2013)
2. Redt-Clouet, C., et al.: Mushroom body neuronal remodelling is necessary for short-term but not for long-term courtship memory in Drosophila. Eur. J. Neurosci. **35**(11), 1684–1691 (2012)
3. Williams, D.W., Truman, J.W.: Remodeling dendrites during insect metamorphosis. J. Neurobiol. **64**(1), 24–33 (2005)
4. Medioni, C., Ramialison, M., Ephrussi, A., Besse, F.: Imp promotes axonal remodeling by regulating profilin mRNA during brain development. Current Biol. **24**(7), 793–800 (2014)
5. Luo, L.: Actin cytoskeleton regulation in neuronal morphogenesis and structural plasticity. Ann. Rev. Cell Dev. Biol. **18**(1), 601–635 (2002)
6. Schlüter, K., Jockusch, B.M., Rothkegel, M.: Profilins as regulators of actin dynamics. Biochimica et Biophysica Acta (BBA)-Mol. Cell Res. **1359**(2), 97–109 (1997)

7. Verheyen, E.M., Cooley, L.: Profilin mutations disrupt multiple actin-dependent processes during Drosophila development. Development **120**(4), 717–728 (1994)
8. Kong, J.H., Fish, D.R., Rockhill, R.L., Masland, R.H.: Diversity of ganglion cells in the mouse retina: unsupervised morphological classification and its limits. J. Comp. Neurol. **489**(3), 293–310 (2005)
9. Guerra, L., McGarry, L.M., Robles, V., Bielza, C., Larranaga, P., Yuste, R.: Comparison between supervised and unsupervised classifications of neuronal cell types: a case study. Dev. Neurobiol. **71**(1), 71–82 (2011)
10. López-Cruz, P.L., Larrañaga, P., DeFelipe, J., Bielza, C.: Bayesian network modeling of the consensus between experts: an application to neuron classification. Int. J. Approx. Reason. **55**(1), 3–22 (2014)
11. Mottini, A., Descombes, X., Besse, F.: From curves to trees: a tree-like shapes distance using the elastic shape analysis framework. Neuroinformatics **13**, 175–191 (2014)
12. Wu, J.S., Luo, L.: A protocol for mosaic analysis with a repressible cell marker (MARCM) in Drosophila. Nature Protoc. **1**(6), 2583–2589 (2006)
13. Schindelin, J., et al.: Fiji: an open-source platform for biological-image analysis. Nature Meth. **9**(7), 676–682 (2012)
14. Myatt, D.R., Hadlington, T., Ascoli, G.A., Nasuto, S.J.: Neuromantic–from semi-manual to semi-automatic reconstruction of neuron morphology. Front. Neuroinform. **6**, 4 (2012)
15. Mottini, A., Descombes, X., Besse, F., Pechersky, E.: Discrete stochastic model for the generation of axonal trees. In: EMBS, pp. 6814–6817 (2014)
16. Kemeny, J.G., Snell, J.L.: Finite Markov Chains, vol. 356, 1st edn. Princeton, van Nostrand (1960)
17. Keller, M.T., Trotter, W.T.: Applied Combinatorics. Georgia, Atlanta (2015)
18. Szebenyi, G., Callaway, J.L., Dent, E.W., Kalil, K.: Interstitial branches develop from active regions of the axon demarcated by the primary growth cone during pausing behaviours. J. Neurosci. **18**(19), 7930–7940 (1998)
19. Forbes, C., Evans, M., Hastings, N., Peacock, B.: Statistical Distributions, 4th edn. Wiley, Hoboken (2011)
20. Tessier, C.R., Broadie, K.: Drosophila fragile X mental retardation protein developmentally regulates activity-dependent axon pruning. Development **135**(8), 1547–1557 (2008)
21. Mottini, A., Descombes, X., Besse, F.: Tree-like shapes distance using the elastic shape analysis framework. In: British Machine Vision Conference (2013)

Sensitivity Analysis of Granularity Levels in Complex Biological Networks

Sean West and Hesham Ali[✉]

College of Information Science and Technology,
University of Nebraska at Omaha, Omaha, USA
{scwest,hali}@unomaha.edu

Abstract. The influx of biomedical measurement technologies continues to define a rapidly changing and growing landscape, multi-modal and uncertain in nature. The focus of the biomedical research community shifted from pure data generation to the development of methodologies for data analytics. Although many researchers continue to focus on approaches developed for analyzing single types of biological data, recent attempts have been made to utilize the availability of multiple heterogeneous data sets that contain various types of data and try to establish tools for data fusion and analysis in many bioinformatics applications. At the heart of this initiative is the attempt to consolidate the domain knowledge and experimental data sources in order to enhance our understanding of highly-specific conditions dependent on sensory data containing inherent error. This challenge refers to granularity: the specificity or mereology of alternate information sources may impact the final data fusion. In an earlier work, we employed data integration methods to analyze biological data obtained from protein interaction networks and gene expression data. We conducted a study to show that potential problems can arise from integrating or fusing data obtained at different granularity levels and highlight the importance of developing advanced data fusing techniques to integrate various types of biological data for analytical purposes. In this work, we explore the impact of granularity from a more formulized approach and show that granularity levels significantly impact the quality of knowledge extracted from the heterogeneous data sets. Further, we extend our previous results to study the relationship between granularity and knowledge extraction across multiple diseases, examining generalizability and estimating the utility of a similar methodology to reflect the impact of granularity levels.

Keywords: Data integration · Knowledge extraction · Gene expression data · Protein-protein interaction · Co-regulation · Correlation networks · Clusters

1 Introduction

The histories of granularity and data fusion are relatively young, but they have broken free in the last century of unbridled scientific excitement. A rapidly expanding wave of data through the rise of biomedical instrumentation encouraged previously impossible methodologies which rely on heterogeneous data integration or fusion. Along with data fusion,

© Springer International Publishing AG 2017
A. Fred and H. Gamboa (Eds.): BIOSTEC 2016, CCIS 690, pp. 167–188, 2017.
DOI: 10.1007/978-3-319-54717-6_10

several uncertainties were inaugurated into the realm of popular, scientific attention, interrupting a 300 year monopoly on uncertainty by probability. In a benchmark paper, Jerry Hobbs used indistinguishable sets to define granularity (Hobbs 1985).

The history of data fusion is even more recent. The Joint Directors of Labs designated a Data Fusion Lexicon in 1991 to delineate its various features (White 1991). For a decade, data fusion was rarely discussed without a militant theme. However, it spread to every domain as data processing has become multi-modal.

Everyone pictures a different understanding when they hear the words "data fusion". This variance in internal definitions has led many researchers to create clever ways of identifying its attributes. Ultimately in their purest form, these data fusions definitions are about making a decision. It is here, through the novel knowledge extraction component, where it differentiates with data integration. The impact of an uncertainty term, such as granularity, on uniquely data fusion should be measured through the effect on information and knowledge extraction.

Bioinformatics has seen a push towards data aggregation and integration (Halevy et al. 2006), and encountered a series of challenges, highlighted by rapidly changing bioinformatics data standards (Brazma 2009). These challenges are exacerbated when considering data fusion, whose entire nature is dependent on sensitivities to data source variance.

To illustrate these challenges, consider the instance of microarray data. Even for a technology where the use is standardized across experiments (Brazma 2009), many researchers have identified high false positive and false negative rates. One data fusion methodology is to put microarray in the context of domain knowledge as a component of network creation (Zhang and Horvath 2005), through enrichment (Bindea et al. 2009), or examining expression differences within the protein-protein interaction (PPI) network (Obayashi and Kinoshita 2009). Previously, we expanded a similar methodology, except we incorporated domain knowledge directly into the network topology rather than through enrichment. Our assumption in doing so was one which many other researchers assume as well: *The promise of data fusion is that more data means more accurate results*. Unfortunately, this was not the case, as we show in this study. Data fusion between domain knowledge and experimental data emphasizes the impact of granularity on knowledge extraction. Rather than depend on the promise of data fusion, we should instead ask: *Where along the scale from domain knowledge to experimental data does optimal knowledge extraction exist?*

For this question, granularity can be defined through mereology (Bittner and Smith 2003) or indiscernibility (Hobbs 1995). These two granularity dimensions were originally specified as abstraction, shifts in specificity, and aggregation, shifts in part-whole relations (McCalla et al. 1992). Later the aggregation dimension was adapted to into granularity parthood, molecules in a cell, and determinate parthood, functioning members of the cell (Bittner et al. 2004). These scales have not seen a lot of change in recent publications, and the term granularity is usually attributed to specificity. However, increased differentiation of granularity scales have been specified for ontology purposes (Rector et al. 2006; Vogt et al. 2012). In this study we use the terms *abstraction* and *aggregation* in reference to their associated dimensions of granularity, instead of their traditional data processing definitions (West et al. 2016).

With these two dimensions, we can focus the key question into two smaller research questions:

1. How does the data fusion of low and high abstraction sources effect information extraction?
2. What is the relationship between aggregation and information extraction?

To approach these research questions we present a study with three stages.

1. We combine low and high abstraction data sources, defining information extraction success as disease-specific outcomes.
2. We formulize aggregation granularity and identify its relationship with disease-specific information extraction scores.
3. We extend the study to multiple diseases, and examine the consistency of granularity as an uncertainty term.

We use microarray data as our focus of low abstraction, high granularity data, since each series usually represents just a few experimental conditions across a limited number of tissues. The variability of cellular function within these tissues necessitates that microarray data is not the epitome of high granularity data, rather it exists at a granularity level where differences between cellular conditions can be extracted. To combat high false-positive and false-negative rates, microarray is often enriched through low-granular domain knowledge.

Protein-protein interaction databases may contain high abstraction, low granularity data. Some recent PPI databases are cell-specific or even molecule specific (Veres et al. 2014; Liu et al. 2011). Additionally, many integrated PPI databases, such as the Search Tool for the Retrieval of INteracting Genes/Protein (STRING), compile a list of potential relationships, not taking unique cellular conditions into account. STRING scores the interaction between proteins across a set of data sources in a union-like fashion (Franceschini et al. 2013). Here, we use PPI data sources that are non-integrated and not condition specific, in order to bias the data towards a low granularity. These non-integrated PPI databases use manual curation methods to extract PPI information from scientific literature. So, even non-integrated PPI data sources are examples of multi-modal systems. Yet, since the manual curation methodologies employed to create PPI databases may innately increase the granularity of the data, the diversity in the curation methods may lead to the lower-abstraction levels. In this work, we examine the structural and biological attributes of several popular PPI databases in order to characterize their unique contributions towards data integration. We further examine their pathway enrichment of each database to determine any specificity or unique bias towards similar groupings of biological functionality, which would indicate increased levels of granularity.

Although the differences may be explicit between cellular conditions from the expression data and since PPI data comes from high abstraction data sources, integrating microarray data with PPI data that is not tissue or cellular condition specific does not model the true protein-protein interaction network. Therefore, the consequences of alternate expression and PPI network structure changes may not depict true biological reality. If the variability in granularity levels between PPI databases and microarray data biases the data away from high-granularity, potentially questionable biological information will be extracted after the data fusion implementation. To test this critical point, we can fuse the PPI and the microarray data and compare the information extraction between the original experimental microarray data and the fused datasets. In this study, we test to see the effect of fusing low abstraction, microarray data with high abstraction, PPI data on extraction of Type II Diabetes specific pathways.

Granularity along the *aggregation dimension* requires a more formulized definition, which is specified within the methods section. This formulization includes three suggestions for aggregation definition through Rough Set Theory, the number of attributes defining the fused network, the number of indistinguishable sets given a relevant set of attributes, and the number of original data sources.

We use these three definitions of aggregation to test the relationship between information extraction and aggregation. This relationship represents the bias of granularity away from the biological reality. The consistency of this bias can be determined by examining the relationship between information extraction and aggregation across multiple diseases and within each disease.

So, in summary, we have four hypotheses:

H1: The different curation methods of non-integrated PPI databases do not offer unique bias towards specific biological functionality.
H2: Fusion of low-abstraction and high-abstraction data will decrease experimental-specific information extraction.
H3: There exists a definition of aggregation such that a relationship between granularity and information extraction can be seen.
H4: Definitions of aggregation are consistent within a disease and across multiple diseases.

Section 2 describes the methods for this study, including the formulizations for the definition of granularity in the aggregation dimension. Section 3 depicts the results. Section 4 discusses the outcome of the study and its impact on the hypotheses. The paper concludes with Sect. 5.

2   Methods

Throughout the study, protein-protein interaction data and microarray data are modeled as networks, where the nodes represent the biological elements and the edges connect elements that are related by interaction or high correlation. In the first part, we attempt to find unique biological themes or functionalities associated with the PPI databases queried in order to answer hypothesis H1. We use structural similarity between the PPI networks and the quality of the clusters obtained from the networks using standard pathway, disease, and ontology enrichments. In this manner, we identify the biological functionality associated with each protein-protein interaction network and enriched clusters are mapped to human pathway hierarchies to search for significant patterns.

In the second part, we address hypothesis H2. To test the hypothesized relationship between abstraction and data extraction, we use a case study with a Type II Diabetes microarray series. We create the integrated network using PPI and microarray data. We then enrich obtained network clusters to identify network-specific biological functions. We choose a list of 24 diabetes associated pathways or diseases curated from Reactome, the Online Mendelian Inheritance in Man (OMIM), and the Kyoto Encyclopedia of Genes and Genomes (KEGG). We assess the enrichments of these pathways across the original and fused networks, and discuss the potential information loss that may occur due to the lack of consistent granularity levels.

To validate the relationship between aggregation granularity and knowledge extraction, hypothesis H3, we formulize granularity using three different approaches and add additional sources to expand on the number of discrete granularity levels that can be formed. Extraction scores based on enrichment results are assigned to each network and correlation is measured.

We can further test hypothesis H3 by identifying the robustness of this methodology across multiple diseases. By incorporating several microarray series for both pancreatic cancer and Alzheimer's disease, a more stable look at the relationship between aggregation granularity and information extraction can be specified. In addition, an increase in information extraction score is expected for data sets which incorporate disease-specific signals. However, the strength of the relationship between information extraction and granularity, which represents the bias of granularity, ought not to be dependent on the disease condition. So, we can test to see if granularity bias is dependent on data structure rather than biological influences by comparing the strengths across each disease data set, and their control inverses.

2.1 Protein Protein Interaction Databases

The following protein-protein interaction databases were selected for this study since they reflect variability associated with experiments used to obtain them, they do not have a high-degree of integration among each other, and they were initially, seemingly sources of low granularity. The databases used were Database of Interacting Proteins (DIP) (Salwinski et al. 2004), BioGRID (Chatr-aryamontri et al. 2013), Human Protein Reference Database (HPRD) (Prasad et al. 2009), IntAct (Karrien et al. 2011), and Molecular INTeraction database (MINT) (Ceol et al. 2009) (Fig. 1).

DIP focuses on extracting experimental knowledge from publications and stores binary interactions between proteins, clarifying source and evidence. BioGRID is a database of protein and genetic interactions that are extracted from manually annotated publications, by a team of PhD curators. Text mining is used to rank relevant publications where interactions are manually extracted and added to BioGRID. HPRD uses laboratory submitted data through a tool called BioBuilder which helps researchers interact with the database and submit experimental information. In this way, HPRD has protein-protein interactions that are post-translational modification, disease, and tissue specific. It also has an overarching binary PPI source. IntAct takes an open-source approach, with all data and repository code available to the public. The stored interactions are publicly curated from literature but also have a design to allow for direct researcher annotation. Rules on curation are specified on the EBI website and interactions are reviewed by a second curator. MINT is highly similar to IntAct, using the same infrastructure and curation rules. The difference is the set of MINT curators.

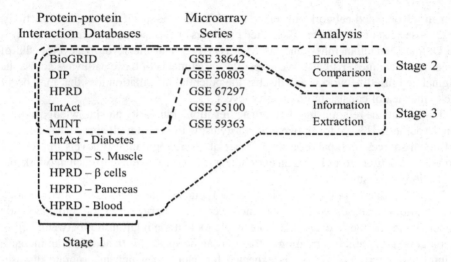

Fig. 1. Data sources of this study. Stage one corresponds to hypothesis 1 and uses only the PPI databases. Stage 2 uses a selection from the PPI databases and conducts an enrichment comparison. Stage 3 uses all the data sources in an information extraction test to understand its interplay with aggregation

For the third (aggregation) component of the study, we expand on those PPI databases with curation conducive to tissue or disease specificity. We add a diabetes subnetwork for IntAct. HPRD has many tissue specific curations, but we use only the HPRD subnetworks with attribute overlap for the microarray series used in the third part. So, we include skeletal muscle, β-cell, pancreas, and blood HPRD PPI networks.

2.2   Network Creation

Although the protein-protein interaction databases contained evidence codes which may affect edge weights through confidence variance, the granting of specific edge weights was not implemented. This alleviated the necessity of consolidating the PPI edge weights with the microarray edge weights. Instead, edges exist where evidence supports an edge. Further, all types of experiments, including high-throughput evidences, were included if they were present in the original PPI data source. This may introduce a technology bias beyond what is incorporated into the research bias. However, correction of a technology bias may introduce unknown sensitivities. So, networks created were binary and non-directional. Protein-protein interaction networks were derived from the overarching sets of database information, such that tissue specific information was included without its specificity. Only complete proteins which correspond to at least one Ensembl gene Id were utilized. We attempt to highlight the issues of removing granularity from domain knowledge sources. Yet for validating the concept of the interaction between aggregation and information extraction, we use PPI networks with higher levels of granularity as mentioned above.

Microarray data was initially downloaded from the Gene Expression Omnibus series, GSE 38642. This series was chosen since it is human, has a large set of biological replicates, demonstrates a disease with a long list of well-characterized pathways, Type II Diabetes, and obtains expression through a relevant and specific tissue, pancreatic islets. Additionally for the third part of the study, we included series GSE 30803, a treatment based study on healthy β-cells, GSE 67297, a study on cold acclimation effects of diabetic adipose tissue, GSE 55100, a blood tissue study of diabetes, and GSE 59363, which uses skeletal muscle tissue in healthy and diabetic samples with exercise stages. These additional microarray series were chosen as they have at least a moderate number of biological replicates, and overlapping values across "tissue", "disease state", "treatment", and "technology" attributes.

The raw expression files were downloaded, and robust multi-array (RMA) normalized. Pearson correlation was implemented to find expression relationships. The microarray networks then took two different paths, those filtered through false-discovery rate p-value correction and those with hard thresholds at 0.8 power and a 0.05 p-value. Base mapping of probes to Ensembl gene Ids was completed through the Biomart API. Ensembl gene Ids which correspond to multiple probes were assigned edge weights which matched the highest correlation scoring probe for each individual interaction. The strong influence of some protein domains (e.g. probes which correspond to multiple transcripts) reduce the accuracy of the correlations which use the probe's expression values. The conjugated expression values are representative of multiple transcripts. Depending on the abundance distribution of these transcripts either correlations may be assigned to the wrong protein, or more likely, the correlations will favor random correlation values, which are more likely to be insignificant and negligible. These multi-transcript probes are considered negligible in this study as they make up only a small percentage of the total probes.

With the lack of PPI exact interaction strength values, the integrated networks created were the union of the PPI matrices and the microarray matrices. Union is a surprisingly common kernel function when integrating and fusing biological data sources. We use it here as an example of an integration-based data fusion approach.

2.3 Identification of Unique Contributions from Protein-Protein Interaction Data and Type II Diabetes Case Study

PPI networks were clustered with the Speed and Performance In Clustering (SPICi) algorithm, a fast and biologically driven clustering approach (Jiang and Singh 2010). The standard parameters produced ideally sized clusters for enrichment. An in-house tool for enrichment which downloads source groups and group information for Reactome, OMIM, and KEGG datasets. It uses the multivariate hypergeometric function to find overly expressed source groups within network clusters. Then, it uses the Benjamini-Hochberg-Yekutieli false discovery rate p-value correction to address multiple hypothesis testing and dealing with the lack of independence for enrichment terms on a single cluster.

Unique contributions were determined by finding those enrichments for a PPI source that were not identified in any other PPI source. For visualization, unique Reactome enrichments were mapped to the Reactome pathway hierarchy and grouped by pathway similarity. Further, structural differences between PPI networks were uncovered at the node, edge, and cluster levels.

The microarray networks were fused with the PPI networks in a union fashion so that there were control microarray, diabetes microarray, PPI, control fused, and diabetes fused networks. These networks were filtered as to only include only those biological elements present in the microarray sets. Then they were clustered and enriched using SPICi and the in-house enrichment tool. Diabetes pathways were manually determined for Reactome, OMIM, and KEGG. Enrichments of these pathways were examined across the networks to identify biological differences between control and diabetes networks.

2.4   Validation of Relationship Between Aggregation and Information Extraction

A more formulized definition of granularity is required to characterize the relationship between granularity and information extraction. So far, we use the dimension of abstraction. This allows only for direct comparisons between objects or networks along the same scale. An extended discrete comparison scale is needed along the dimension of aggregation.

Shortly after the initial introduction of rough sets into uncertainty theory, Hobbs began to distinguish granularity as a significantly contributing factor towards uncertainty (Hobbs 1985). This received formulization (Greer and McCalla 1989) and then developed into concepts of discrete granularity scales (Hobbs 1995). We use Hobbs scales of granularity with the concept of minimum rough sets to define levels of granularity from our *universe of objects* (i.e. our set of original data sources). Granularity over multiple universes in rough sets is currently used in decision support and management science (Słowiński et al. 2014; Sun and Ma 2015). We use it here as a formulized approach to measuring granularity.

Given a universe, U, consisting of a set P of predicates over a number of objects in O. R is the relevant subset of predicates from P. So, we can define objects x and y as indistinguishable if they meet:

$$\forall(x, y)\, x \sim y \cong (\forall p \in R)(p(x) \cong p(y)) \tag{1}$$

Two objects are indistinguishable if their values for every relevant predicate are equal. Expanding on this, given a set of predicates (or attributes), we can separate the complete list of objects into sets of indistinguishable elements or equivalence sets.

In the first definition of aggregation, we define granularity by the number of attributes used to create the equivalence sets. In the second definition, each of these equivalence sets have membership at a discrete granularity level defined by the number of these indistinguishable sets. So, given a set of attributes we can determine the granularity level as well as group data sources for network fusion. In the third definition of aggregation, we can define granularity by the number of data sources in the equivalence set and separate these sets according to granularity.

We created three universes of objects as our complete set of data sources, using "source", "tissue", "disease state", "treatment", "technology", "aggregation method", and "species" as attributes. The first universe, the *overarching universe,* contained all data sources. The second two, *diabetes* and *control* universes, representatively used diabetic or non-diabetic data sources. In doing so we can see the effect of experimental condition on information extraction. We used each combination of every length of these attributes, defining a set of fused networks and a defined granularity level. We can use the enrichment and granularity level to characterize the relationship between the two, given our defined universe.

To score information extraction, we use a similar method as above, measuring the enrichment of diabetes related terms from Reactome, KEGG, and OMIM. We define the information extraction score in two ways. First, we use the proportion of relevant enrichment terms found over the total number of relevant enrichment terms. To standardize enrichment term impact on the score, we also use an information extraction score which weights the contribution of an enriched term by the probability of finding the term, as defined by the calculated probability of finding the term across all used networks produced by the universe.

In this equation, N is the set of enrichment terms in a given network, T is the complete set of diabetic enrichment terms, and $P(t)$ is the probability of finding term t in any network.

$$weighted score = \frac{\sum_{t \in N} 1 - P(t)}{\sum_{t \in T} 1 - P(t)} \qquad (2)$$

Then we find the correlation between discrete levels of granularity and information extraction scores, utilizing equivalent set derived fused networks as each point. In total, we obtain six correlation and p-value pairs for each aggregation definition from the three universes and the two scoring techniques. These correlations are applied to each of the definitions for aggregation granularity: the number of attributes, the number of equivalence sets, and the number of contributing data sources. High correlations indicate linear covariance across the information extraction score and the alternate granularity levels from the associated networks.

2.5   Consistency of the Relationship Between Granularity and Information Extraction Score

Measuring the consistency of the relationship between granularity and information extraction score can use two references. First, we can examine the consistency across granularities for separate diseases. Second, examining it across the each universe. The first tells us the overarching impact of each type of granularity in a generalized format. The second indicates if granularity influence is higher in one disease or another.

The original list of data sources was extended to include microarray data from pancreatic cancer and Alzheimer's disease studies. The microarray series added were GSE 28735, a comparison between pancreatic cancer tumors and nearby non-tumor tissue, GSE 41372, a pancreatic cancer study including a miRNA analysis, GSE 49515, which included blood mononuclear cells of patients with pancreatic cancer, GSE 36980, three brain regions in Alzheimer's disease patients, GSE 48350, which determined expression across different areas of the brain, and finally, GSE 53890, a microarray series without disease patients that compared transcription as patients age. Further, all micro-array networks were hard thresholded at 0.0001 since p-value adjustment for the larger networks failed due to p-value distribution and size. This threshold value was chosen since it was similar to each of the p-value adjustment thresholds from the previous phases.

Along with the additional data sources, new universes were defined. In total there were seven universes: all data sets, Alzheimer's disease positive data sources only, its inverse, diabetes disease positive data sources only, its inverse, pancreatic cancer positive data sources only, and its inverse. Each of these universes underwent similar analysis as done in the previous section. All combinations of all attributes were used to create relevant attribute sets.

These sets were each used to determine the equivalence sets. The data sources in each equivalence set were fused to create a network, which was clustered and an information extraction score was determined. Further, along with the diabetes-specific score, information extraction scores were determined using term enrichment lists for Alzheimer's disease (21 terms) and pancreatic cancer (10 terms). Both of these information extraction scores were calculated using the proportion and weighted measures above. In all for this stage, there are 7 universe and 6 information extraction scores.

3   Results

3.1   Protein-Protein Interaction Databases Show Low Structural Similarity

Structural differences between networks were examined at the node, edge, and cluster levels. Figure 2 shows the number of biological elements found in each database, i.e. proteins for the PPI databases, and transcripts for the microarray series. The overlap percent of these node sets were calculated by dividing the intersection of the two sets by their union. As would be expected the larger databases have low overlap percent with the smaller databases since their potential overlap is small. Table 1 shows this overlap between the PPI networks and the 0.8 power threshold control microarray network. The larger, more inclusive networks tend to have higher similarity, but DIP and MINT, even with a close number of nodes, had a low overlap. the intersection of the two sets by their union. As would be expected the larger databases have low overlap percent with the smaller databases since their potential overlap is small. Table 1 shows this overlap between the PPI networks and the 0.8 power threshold control microarray network. The larger, more inclusive networks tend to have higher similarity, but DIP and MINT, even with a close number of nodes, had a low overlap.

More so than the overlap between biological elements, the interactions derived from each data source showed almost no overlap. The number of edges in each network seemingly enhanced the distance in size between data sources. Figure 2 shows the number of interactions in each network; Table 2 shows the overlap, calculated by taking the intersection over the union of two interaction sets.

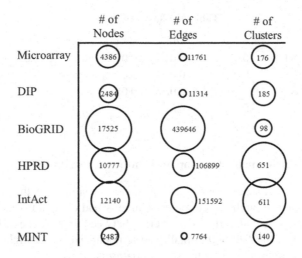

Fig. 2. Data source sizes – the relative sizes for the data sources as the number of nodes, number of edges, and number of clusters are displayed as numbers and represented as the area of their corresponding circles in order to show relative size.

Table 1. Node overlap.

	Microarray	DIP	Biogrid	HPRD	IntAct
MINT	0.08	0.22	0.14	0.20	0.20
IntAct	0.18	0.21	0.65	0.50	
HPRD	0.16	0.23	0.56		
Biogrid	0.19	0.16			
DIP	0.09				

Table 2. Edge overlap

	Microarray	DIP	Biogrid	HPRD	IntAct
MINT	0.00	0.02	0.00	0.02	0.04
IntAct	0.00	0.03	0.02	0.06	
HPRD	0.00	0.04	0.03		
Biogrid	0.00	0.02			
DIP	0.00				

The overlap of clusters from each network was determined. Only clusters of size five or higher were used and two clusters needed a 70% member overlap as determined by the smallest cluster to be determined the same. Figure 4 shows the number of clusters in each network; Table 3 shows the overlap between these clusters. The BioGRID network had a high density and clustered into small, yet huge clusters. Once again, the structural overlap of these networks is negligible.

Table 3. Cluster overlap.

	Microarray	DIP	Biogrid	HPRD	IntAct
MINT	0.00	0.01	0.01	0.01	0.02
IntAct	0.00	0.01	0.08	0.06	
HPRD	0.00	0.02	0.06		
Biogrid	0.00	0.01			
DIP	0.00				

3.2 Reactome Enrichment: Overlap and Unique Contributions

Although the structural aspects of the networks had low similarity, the biological properties as determined through Reactome enrichments had comparatively high overlap. Table 4 shows the potential enrichment overlap between data sources. Here, the intersection is divided by the size of the smaller enrichment size to highlight the low availability for unique contributions by the smaller datasets.

Table 4. Potential enrichment overlap.

	Microarray	DIP	Biogrid	HPRD	IntAct
MINT	0.30	0.52	0.60	0.56	0.57
IntAct	0.74	0.43	0.40	0.40	
HPRD	0.71	0.49	0.46		
Biogrid	0.72	0.49			
DIP	0.54				

For visualization, we examine only the Reactome enrichments of each PPI network. The unique enrichments for individual networks would delineate any source as specific towards an individual biological domain. Figure 3 shows the unique Reactome pathway enrichments for each individual PPI source in the setting of the entire set of human Reactome pathways. Although the structural differences between networks are small, unique pathway enrichments only make up a small proportion of the total potential pathway space. Initially, this indicates that any bias that does exist towards a specific biological condition or theme, is weak. However, the impact of these biases can only be examined by segmenting the Reactome hierarchy into biologically relevant clusters.

The manual grouping of these pathways into groups of similar function, as shown in Fig. 4, does not distinguish confident unique themes of biological extraction. Yet a few sets of unique contributions show significant grouping within the pathway hierarchy. HPRD highlights gamma carboxylation; IntAct highlights GAG protein metabolism; and BioGRID highlights ER to Golgi transport and single-nucleotide replacement. However, these tendencies are insufficient to specify particular granularity for individual PPI data sources. Rather, they each demonstrate generic pathway enrichment across the human pathway hierarchy (Fig. 5).

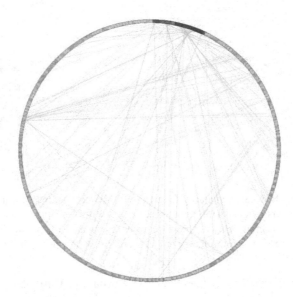

Fig. 3. Full hierarchy of Reactome human pathways. The entire list of human Reactome pathways and their hierarchy was graphed so that nodes represent pathways and edges represent the child-parent relationships in the Reactome hierarchy. The unique Reactome enrichments for each PPI network are highlighted in color: HPRD (green), IntAct (gold), BioGRID (blue), DIP (pink), and MINT (red). These unique enrichments represent a small proportion of the total human pathway hierarchy. (Color figure online)

Fig. 4. Organized hierarchy of Reactome human pathways - Fig. 3 is restructured to group pathways from the same branches of the Reactome hierarchy together. The unique pathway enrichments show low grouping tendency, and the PPI sources do not demonstrate biological specificity. Enrichments for each PPI network are highlighted in color: HPRD (green), IntAct (gold), BioGRID (blue), DIP (pink), and MINT (red). (Color figure online)

		Granularity: # of Attributes	Granularity: # of Eq. Sets	Granularity: # of Sources
Union Universe	proportion	-0.153, 0.120	-0.306, 0.009	0.492, 0.000
	weighted	-0.146, 0.140	-0.259, 0.027	0.467, 0.000
Diabetes Universe	proportion	-0.266, 0.244	-0.381, 0.456	0.718, 0.006
	weighted	-0.338, 0.134	-0.806, 0.053	0.774, 0.002
Control Universe	proportion	-0.011, 0.945	-0.257, 0.248	0.450, 0.016
	weighted	-0.034, 0.827	-0.223, 0.319	0.457, 0.014

Fig. 5. Relationship between granularity (aggregation) definitions and information extraction scores – for each universe and each information extraction score type, the correlation and p-value between score type and granularity is shown (correlation, pvalue). Only significant scores are highlighted and the darker the coloring, the higher the correlation value. (Color figure online)

3.3 Data Fusion with PPI Sources Drowns Microarray Conditional Differences

We compare the differences between control and diabetic conditions for the microarray networks, the PPI networks, and the fusion networks. For the microarray networks, the false-discovery rate adjustment diminished the interactions and enrichment so that there are only three differences between the control and diabetic conditions. The power and p-value threshold networks showed ten pathway differences between the control and diabetic conditions out of 24 total diabetes related pathways. After fusion, the network structure for the FDR adjusted and the thresholded networks were enriched for nearly every single available pathway. Table 5 shows these enrichments, only showing 0.05 p-value adjustment.

The ten differences from the microarray networks are not seen in the fusion networks. Instead, every enrichment available was found enriched in at least one cluster of both the control and diseased fusion networks. In one case towards the bottom of Table 5, the diabetes microarray was not enriched for term "maturity: onset of diabetes in the young". None of the PPI networks contained clusters enriched for this term, yet the fused network for diabetes did contain this it. So, the topological changes of networks, when fused, encouraged an expansion of available enrichment terms. This may be caused by an increase in available clusters due to the additional edges, or a shift in relative edge densities led to choosing of clusters which may have been dormant in the original PPI or diabetes microarray networks.

3.4 Granularity and Information Extraction

The currently defined universes of predicates and objects has seven relevant attributes; however, we remove the "source" attribute as it does not produce equivalence sets greater than one, leaving 60 total sets of attributes. A total of 57 equivalence sets were determined across 7 discrete granularity levels for the first definition of aggregation granularity, 24 discrete granularity levels for the second, and 24 for the third. We found

Table 5. Pathway enrichment network spectrum.

	Micro Array Control 0.05 p	Micro Array T2D 0.05 p	DIP	Bio GRID	HPRD	IntAct	MINT	Fusion Control 0.05p	Fusion T2D 0.05p
Metabolism of lipids and lipoproteins		■	■	■	■	■	■	■	■
PERK regulates gene expression				■	■	■		■	■
Protein processing in endoplasmic reticulum		■	■	■	■	■		■	■
Adrenaline, noradrenaline inhibits insulin secretion			■	■		■		■	■
Calcitonin-like ligand receptors			■	■				■	■
Glucagon-like Peptide-1 (GLP1) regulates insulin secretion			■	■	■			■	■
Signaling by Leptin			■	■	■	■	■	■	■
Notch signaling pathway		■	■	■	■	■		■	■
Wnt signaling pathway	■	■	■	■	■	■	■	■	■
TGF-beta signaling pathway			■	■	■	■	■	■	■
Hormone-sensitive lipase (HSL)-mediated triacylglycerol hydrolysis			■	■		■	■	■	■
PPAR signaling pathway			■	■	■	■	■	■	■
Cell cycle		■	■	■	■	■		■	■
p53 signaling pathway		■	■	■	■	■	■	■	■
Advanced glycosylation end-product receptor signaling			■	■		■		■	■
Regulation of insulin secretion		■	■	■	■	■		■	■
Unfolded Protein Response (UPR)			■	■	■	■	■	■	■
Type II diabetes mellitus		■	■	■	■	■		■	■
Diabetes Mellitus, noninsulin-dependent; NIDDM	■		■	■	■	■		■	■
Pancreatic secretion			■	■	■	■		■	■
Maturity onset diabetes of the young	■							■	■
mTOR signaling			■	■	■	■	■	■	■
Insulin secretion	■	■	■	■	■	■		■	■

the correlations between the information extraction scores based on enrichment. Of these, we found no significant correlations between granularity and information extraction when using the number of attributes determining equivalence sets. When using the number of equivalence sets, half of the correlations were significant. Finally, when considering the number of sources as the scale of granularity, each of the correlations was significant. The highest correlation was found when considering the number of equivalence sets as the granularity definition in the diabetes universe. However, the relative increase in correlation strength was only 0.030.

3.5   Relationship Between Granularity Level and Information Extraction Across Multiple Diseases

The relationship between granularity and information extraction was stronger using the third way to define granularity across each disease. This is shown in Fig. 6, where the information extraction scores for the three aggregation definitions are displayed for each disease and each score type. There are three exceptions to this consistency, however. The Diabetes Control Universe had low information extraction scores for the third granularity definition for pancreatic cancer and diabetes. While these were larger for the other two granularity definitions, the mean difference was not statistically significant. Also, The Alzheimer's Control Universe had a poor pancreatic cancer information extraction score, but the score was even lower for the first and second granularity types. So, this significant interaction between granularity (as number of data sources) and information extraction exists across multiple diseases. This suggests that the impact of granularity may exist independent of disease condition. Further, defining aggregation as the number of equivalence sets did see some strong relationships, particularly for the Pancreatic Cancer Universe and the Alzheimer's Universe.

While the mean information extraction score is higher on average for the third definition of granularity, the variance of this relationship is large across each universe. Therefore, we cannot speculate as to the robustness of using this third granularity definition as an outlet for removing the bias of granularity.

4   Discussion

The cascading importance of high confidence necessary in medical and biological data puts an emphasis on reproducibility, stability, and sensitivity of analytical workflows. Due to the complexity of biological data analysis, results have a high sensitivity to small changes in parameters, producing results that are unstable and difficult to reproduce. This concept is accentuated and expanded in data fusion, granularity being just a single potential for sensitivity. The data fusion approach presented here demonstrates the sensitivity of these biological data towards the various levels of granularity along two different dimensions.

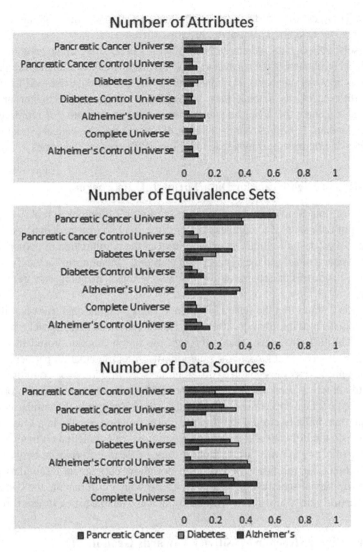

Fig. 6. For the three types of granularity (1. Number of Attributes, 2. Number of Equivalence Sets, 3. Number of Data Source), disease-specific, weighted information extraction scores were calculated for each of the 7 universes. Bars in green are the correlations between granularity and pancreatic cancer information extraction; orange is between granularity and diabetes information extraction; and blue is between granularity and Alzheimer's information extraction. (Color figure online)

4.1   Protein-Protein Interaction Database Differences and Sensitivity

The structural differences between the protein-protein interaction networks are byproduct of slight differences in curation methods. Although each PPI database retrieves their information from the same population of publications, the knowledge presented in the databases is different. These curation differences which lead to vast node and edge network differences, also lead to different structural or clustering differences.

Conversely, although the structure of the networks is sensitive to the curation methods, the biological enrichment of the networks is not. The biological enrichment had a relatively large overlap between databases. Had the creation of the PPI databases led to significant biological differences, as a total, they would have represented a higher granularity. However, as it stands, the small unique pathway enrichments are not representative of highly granular data. To address our hypothesis, *H1*, the unique biological extractions of the PPI networks do not represent any apparent biological themes or conditions.

4.2   Flooding of High Granular Data Through Data Fusion

The promise of data fusion is a more accurate depiction of biological reality through the combination of data sources to compensate for individual source inadequacies. Yet a key component to data fusion is within its own definition: the combination of two or more data elements to create a novel and meaningful data element. "Meaningful" is vitally important. A good data fusion result captures a biologically relevant meaning and treats the data accordingly.

Hypothesis 2 states that the fusion of low and high granular data sources will remove experimentally derived information. The union function utilized here, innately, favors a low granularity. The PPI databases, created through the union function, result in "potential" networks. These networks illustrate the potential of protein interaction partners and structure, but may not be specific enough to differentiate specific biological conditions or pathways, and ultimately as seen in Table 5 the larger PPI databases capture the majority of this selection of Type 2 Diabetes enrichments. So, the union between high granularity data, and low granular data initially created through the union function, results in a low granularity data set. As with the PPI networks covering the potential of interaction partners, this fusion creates a potential of pathways list, making it impossible to differentiate between experimental conditions. I.e. the fusion networks do not show enrichment dissimilarity. In this case, the experimental specific differences are those which differentiate the control tissue from the diabetic tissue. These differences are flooded and unable to be extracted after fusion.

4.3   Information Extraction Sensitivity Towards Granularity

Our third hypothesis, *a granularity scale exists in which aggregation is associated with information extraction*, was not supported by each of the defined scales of aggregation. The first scale of aggregation, number of attributes, did not demonstrate any relationship with information extraction. When aggregation was defined as the number of equivalence sets, half of the correlations were significant. Explicitly, we find that the diabetes universe and weighted score using this definition of granularity had the highest correlation out of all the tested conditions. The lack of significant correlation in the control universe indicates that these diabetes data sources are the reason that significant correlation was found in the overarching universe. We note that, biologically, diabetic data sources are likely to have increased diabetic information extraction; however, the relationship between granularity and information extraction is not innately evident. Defining aggregation by the number of equivalence sets is only satisfactory under certain data source combinations and a weighted information extraction score.

The final definition of granularity, as the number of data sources used in the fused network, had a complete suite of significant correlations. Yet the diabetic set of data sources carried a higher bias in generating a strong relationship between granularity and information extraction. This result defends the proposition that more available information innately present in a data fusion indicates a higher potential for information extraction after the fusion has taken place.

By supporting the third hypothesis, we suggest that sensitivity to granularity contributes to the confidence in the limits of a data fusion function. Yet the field of data fusion is diverse and we are not certain that granularity is important for all data fusion functions, specifically those which may correct for abstraction or aggregation among sensor technologies. Further, granularity is a high-level uncertainty term and can be defined in various ways beyond the two dimensions suggested in this study. An intelligent data fusion must consider the biology, the technology, and the sensitivity of the function to initial parameters, including granularity, in order to obtain confidence in its results.

4.4   Generalizability Across Biomedical Data Sources

When conducting multivariate data analysis, univariate normality does not guarantee multivariate normality. In the same way, the sensitivities of individual biomedical data sources, including the sensitivity to granularity, must be examined in a multi-modal perspective. We can only speculate to the sensitivity of biomedical data sources not included in this study, but we suggest that while granularity may not be an issue in an individual data source, data fusion approaches should check for sensitivity to granularity.

Further, per the final portion of the results section, the consistency of the impact an individual granularity definition is suspect (at least those used in this study). We suspect that other ways of defining granularity exist where consistency can be obtained, and therefore, used as a means to correct data for granularity bias. A solution to the granularity bias will leverage impactful granularity definitions to filter data towards a biological reality.

5   Conclusion

As the technology associated with biomedical research continues to advance, larger and more diverse data sources are becoming available to researchers. Each data source has its own attributes that influence the way its data can be used or integrated with other data. As a result, there is a growing need for sophisticated ways to effectively integrate different types of biological data and improve the outcome of using data mining algorithms. In this study, we proposed several tests for characterizing granularity within the integration of protein-protein interaction and gene expression data using the network model through testing four hypothesis.

The PPI databases that we examined in this study each were informed through a manual curation process. While each source contain relics of bias, their unique community contributions are not biased in a patterned way. So, their combined fusion (H2) with Microarray data did not demonstrate significant topological bias towards a single biological condition. Yet, this union between high-granularity and low-granularity data did not uncover experiment-specific differences in the Microarray data.

The third hypothesis, which guesses at a relationship between granularity and information extraction was only partially validated for these data. Only defining granularity as the number of data sources demonstrated a consistent correlation, which themselves were not higher than 0.8. Perhaps a more biologically driven granularity definition will demonstrate a stronger relationship. Any attempt to overcome the granularity bias should seek first to define the impacting granularity scale.

While the mean differences between information extraction scores between diseases were not significantly different, proving hypothesis 4 requires further analysis, the variances between information extraction scores is high enough to suggest further approaches which can stabilize these scores are necessary.

This study serves as a case study to highlight the need to study data integration methods further in the domain of biomedical informatics and explore different ways to characterize the impact of uncertainty variables throughout alternate data integration methodologies. These characterizations must also include topological information regarding substructure changes in order to further classify the relationships among elements in biological networks. Yet according to our results, the consistency of this granularity bias is suspect, especially within a single disease. While a consistent bias would be easier to remove, the inconsistencies highlight the complexity of biomedical data. We suspect that as we investigate granularity biases further, a solution to granularity bias in data fusion methodologies may need to be adapted with each unique fusion.

The underlying principle here is that each network represents a form of an expert system, the more relevant data incorporated in the network, the more knowledgeable the network becomes. Yet the dependency of the extraction of this knowledge is dependent on data source variables (including granularity) which impact the topology. In turn, proper handling of these variables would allow the researchers to extract more biologically relevant signals while limiting the impact of noise that will always be associated with raw biological data. Ultimately, the attainment of the more useful biological networks, dependent on the type and environment of a network or biological replicate, is contingent on the ability to successfully integrate data types through characterization of their sensitivities.

References

Bindea, G., Mlecnik, B., Hackl, H., Charoentong, P., Tosolini, M., Kirilovsky, A., Fridman, W., Pages, F., Trajanoski, Z., Galon, J.: ClueGO: a Cytoscape plug-in to decipher functionally grouped gene ontology and pathway annotation networks. Bioinformatics **25**(8), 1091–1093 (2009)

Bittner, T., Smith, B.: A theory of granular partitions. Found. Geogr. Inf. Sci. **7**, 124–125 (2003)

Bittner, T., Donnelly, M., Smith, B.: Individuals, universals, collections: on the foundational relations of ontology. In: Proceedings of the Third Conference on Formal Ontology in Information Systems, pp. 37–48, November 2004

Brazma, A.: Minimum information about a microarray experiment (MIAME)–successes, failures, challenges. Sci. World J. **9**, 420–423 (2009)

Ceol, A., Aryamontri, A. C., Licata, L., Peluso, D., Briganti, L., Perfetto, L., Castagnoli, L., Cesareni, G.: MINT, the molecular interaction database: 2009 update. Nucleic Acids Res. (2009). doi:10.1093/nar/gkp983

Chatr-aryamontri, A., Breitkreutz, B.J., Heinicke, S., Boucher, L., Winter, A., Stark, C., Nixon, J., Ramage, L., Tyers, M.: The BioGRID interaction database: 2013 update. Nucleic Acids Res. **41**(D1), D816–D823 (2013)

Franceschini, A., Szklarczyk, D., Frankild, S., Kuhn, M., Simonovic, M., Roth, A., Santos, A., Tsafou, K., Kuhn, M., Bork, P., Jensen, L.J., von Mering, C.: STRING v9. 1: protein-protein interaction networks, with increased coverage and integration. Nucleic Acids Res. 41(D1), D808-D815 (2013)

Greer, J.E., McCalla, G.I.: A computational framework for granularity and its application to educational diagnosis. In: IJCAI, pp. 477–482, August 1989

Halevy, A., Rajaraman, A., Ordille, J.: Data integration: the teenage years. In: Proceedings of the 32nd international Conference on Very Large Data Bases, pp. 9–16. VLDB Endowment, September 2006

Hobbs, J.R.: Granularity. In Proceedings of the Ninth International Joint Conference on Artificial Intelligence (1985)

Hobbs, J.R.: Sketch of an ontology underlying the way we talk about the world. Int. J. Hum. Comput. Stud. **43**(5), 819–830 (1995)

Jiang, P., Singh, M.: SPICi: a fast clustering algorithm for large biological networks. Bioinformatics **26**(8), 1105–1111 (2010)

Kerrien, S., Aranda, B., Breuza, L., Bridge, A., Broackes-Carter, F., Chen, C., Duesbury, M., Dumousseau, M., Feuermann, M., Hinz, U., Jandrasits, C., Jimenez, R.C., Khadake, J., Mahadevan, U., Masson, P., Pedruzzi, I., Pfeiffenberger, E., Porras, P., Raghunath, A., Roechert, B., Orchard, S., Hermjakob, H.: The IntAct molecular interaction database in 2012. Nucleic Acids Res. (2011). doi:10.1093/nar/gkr1088

Liu, Z., Cao, J., Gao, X., Zhou, Y., Wen, L., Yang, X., Xuebiao, Y., Ren, J., Xue, Y.: CPLA 1.0: an integrated database of protein lysine acetylation. Nucleic Acids Res. **39**(suppl. 1), D1029–D1034 (2011)

McCalla, G., Greer, J., Barrie, B., Pospisil, P.: Granularity hierarchies. Comput. Math Appl. **23**(2), 363–375 (1992)

Obayashi, T., Kinoshita, K.: Rank of correlation coefficient as a comparable measure for biological significance of gene coexpression. DNA Res. **16**(5), 249–260 (2009)

Prasad, T.K., Goel, R., Kandasamy, K., Keerthikumar, S., Kumar, S., Mathivanan, S., Telikicherla, D., Raju, R., Shafreen, B., Venugopal, A., Balakrishnan, L., Marimuthu, A., Banerjee, S., Somanathan, D.S., Sebastian, A., Rani, S., Ray, S., Harrys Kishore, C.J., Kanth, S., Ahmed, M., Kashyap, M.K., Mohmood, R., Ramachandra, Y.L., Krishna, V., Rahiman, B.A., Mohan, S., Ranganathan, P., Ramabadran, S., Chaerkady, R., Pandey, A., Pandey, A.: Human protein reference database—2009 update. Nucleic Acids Res. **37**(suppl. 1), D767–D772 (2009)

Rector, A., Rogers, J., Bittner, T.: Granularity, scale and collectivity: when size does and does not matter. J. Biomed. Inform. **39**(3), 333–349 (2006)

Salwinski, L., Miller, C.S., Smith, A.J., Pettit, F.K., Bowie, J.U., Eisenberg, D.: The database of interacting proteins: 2004 update. Nucleic Acids Res. **32**(suppl. 1), D449–D451 (2004)

Słowiński, R., Greco, S., Matarazzo, B.: Rough-set-based decision support. In: Burke, E.K., Kendall, G. (eds.) Search Methodologies, pp. 557–609. Springer, New York (2014)

Sun, B., Ma, W.: Multigranulation rough set theory over two universes. J. Intell. Fuzzy Syst. Appl. Eng. Technol. **28**(3), 1251–1269 (2015)

Veres, D.V., Gyurkó, D.M., Thaler, B., Szalay, K. Z., Fazekas, D., Korcsmáros, T., Csermely, P.: ComPPI: a cellular compartment-specific database for protein–protein interaction network analysis. Nucleic Acids Res. (2014). doi:10.1093/nar/gku1007

Vogt, L., Grobe, P., Quast, B., Bartolomaeus, T.: Accommodating ontologies to biological reality–top-level categories of cumulative-constitutively organized material entities. PLoS ONE 7(1), e30004 (2012)

West, S., Ali, H.: On the impact of granularity in extracting knowledge from bioinformatics data. In: The 7th International Conference on Bioinformatics Models, Methods, and Algorithms (Bioinformatics 2016), Rome, Italy, 22–25 February 2016

White, F.E.: Data fusion lexicon. Joint Directors of Labs Washington DC(1991)

Zhang, B., Horvath, S.: A general framework for weighted gene co-expression network analysis. Stat. Appl. Genet. Mol. Biol. 4(1), 1128 (2005)

Bio-inspired Systems and Signal Processing

Optimal Lead Selection for Evaluation Ventricular Premature Beats Using Machine Learning Approach

Pedro David Arini[1,2] and Drago Torkar[3(✉)]

[1] Facultad de Ingeniería, Instituto de Ingeniería Biomédica,
Universidad de Buenos Aires, Paseo Colón 850, Buenos Aires, Argentina
[2] Instituto Argentino de Matemática, 'Alberto P. Calderón', CONICET,
Saavedra 15, Buenos Aires, Argentina
pedro.arini@conicet.gov.ar
[3] Jožef Stefan Institute, Jamova cesta 39, 1000 Ljubljana, Slovenia
drago.torkar@ijs.si

Abstract. Repolarization heterogeneity (RH) has been shown to increase with ventricular premature beats (VPBs). Moreover, several differences between left ventricle (Lv) and right ventricle (Rv), such as fibrillation threshold and anatomic properties have been presented. Nevertheless, few results exist regarding the influence of the origin site of VPBs on modulation of ventricular RH, as well as the optimal electrode location to assess the origin of VPBs. We studied electrocardiographic indices as a function of the coupling interval and the site of VPBs, in an isolated rabbit heart preparation (n = 18) using ECG multi-lead (5 rows × 8 columns) system. In both ventricles, results have shown significant increases in ventricular depolarization duration. Also, we have observed that when the VPBs were applied to the Lv, a significant decrease of the total repolarization duration was detected, while in the Rv premature stimulation we have not found significant changes of total repolarization duration. Also, we compared twenty machine learning classification techniques with the aim to find the optimal electrode placement (row4–column4 to Lv stimulation and row5–column3 to Rv stimulation) and interpret the site of origin of VPBs. It was observed that the Random Forest classifier has shown the best performance among all the techniques studied. Finally, we found differences in the overall duration of repolarization associated to transmural RH.

Keywords: ECG · Ventricular transmural dispersion · ECG multi-lead · Learning machine · Ventricular premature beats

1 Introduction

Repolarization heterogeneity (RH) is a measure of inhomogeneous recovery of excitability during the ventricular repolarization phase. This phenomenon is mainly attributable to differences in the action potential duration (APD) in

© Springer International Publishing AG 2017
A. Fred and H. Gamboa (Eds.): BIOSTEC 2016, CCIS 690, pp. 191–204, 2017.
DOI: 10.1007/978-3-319-54717-6_11

different myocardium areas. The APDs differs not only between myocites of different ventricular layers [1] but also between posterior and anterior endocardial layers, apex and base [2], and left and right ventricles [3]. Clinical and experimental studies have shown RH constitute a substrate for malignant ventricular arrhythmias [4,5]. In this way, changes in repolarization heterogeneity values that are higher than normal have been linked with an increased risk of developing reentrant arrhythmias [6].

Some authors have shown that alterations in ventricular repolarization dispersion (VRD) are correlated with changes in the total repolarization duration (T_{RD}) or T-wave width [7]. Our study has also shown that T-wave widening can result from a differential shortening or lengthening of the APD in both apex-base and transmural [8]. Moreover, the T-wave peak-to-end (T_{PE}) interval has been suggested as a marker of transmural repolarization dispersion [9,10], consequently the interval between the J-point and the T-wave peak position has been considered as the full repolarization of epicardium or early repolarization duration (E_{RD}). The translation of these concepts to the standard ECG is not straightforward, making it difficult the interpretation of the relationship between T-wave peak-to-end and transmural dispersion in a clinical population [10].

In this regard, several investigations showed that ventricular premature beats (VPBs) produce a significantly increased of the repolarization heterogeneity and that these changes were markedly associated with an increase in the induction of ventricular arrhythmias [11]. Also, ventricular vulnerability, as evaluated by the ventricular fibrillation threshold technique, was shown different when studied at the left ventricle (Lv) or at the right ventricle (Rv). The left ventricular epicardium presented higher fibrillation threshold when compared with left ventricular endocardium or both epicardium and endocardium of right ventricular, respectively [12].

Because a different ventricular fibrillation threshold and differences in the anatomic properties may exist between both ventricles, we hypothesized that there also would be differences in the E_{RD}, T_{PE} and T_{RD} values depending on the site where VPBs were elicited.

The aims of this work were to: (1) Determine the optimal lead to detect changes of repolarization heterogeneity and depolarization using the best machine learning classifier system obtained from one group of twenty classifiers. (2) Evaluate electrocardiogram (ECG) indices associated to repolarization heterogeneity and depolarization depending on the site of VPBs.

2   Materials and Methods

2.1   Isolated Heart Rabbit Preparation

This study conformed to the *Guide for the Care and Use of Laboratory Animals* published by the US National Institutes of Health (NIH Publication No. 85-23, revised 1996). To obtain isolated Langendoff-perfused rabbit hearts, male New Zealand white rabbits of 2.8–3.8 Kg (n = 18) were heparinized (500 U/Kg IV) and anesthetized by the intramuscular injection of a combination of lidocaine

(5 mg/Kg) and ketamine (35 mg/Kg). The rabbits were euthanized by cervical dislocation. The chest was opened via a median sternotomy and, immediately, the heart was removed with scissors and immersed in cold Tyrode's solution. After the remaining connective tissue, lungs, and pericardium were removed, the heart was placed in a vertical Langendorff device through cannulation of the aorta. Time from chest opening to cannulation of the aorta oscillated between 2 to 3 min. The heart was retrogradely perfused through the aorta with Tyrode's solution and immersed in a tank filled with the same solution [13]. The temperature of both solutions were maintained at $38° \pm 0.5 °C$ and bubbled with O_2 using a flow of 700–900 ml/h at a pressure of 70 mmHg. To regulate the flow rate of the aortic perfusion, a variable speed roller pump (Extracorporeal, 2102 Infusion Pump) was used. Care was taken to fix the hearts in the same position by alignment of the left anterior descending coronary artery (LAD) with the electrode matrix reference system on the tank (see Fig. 1).

The composition of Tyrode's solution was (in mM): 140 NaCl, 5 KCl, 1 $MgCl_2$, 0.33 NaH_2PO_4, 5 Hepes, 11.1 glucose and 2 $CaCl_2$. The pH was adjusted to 7.4 using NaOH. The sinus node was destroyed by applying radiofrequency energy through a customized device.

The artificial pacemaker was a rectangular pulse that had a 2 ms duration and twice the diastolic threshold stimuli amplitude. In the ventricular premature beats (VPBs) experimental protocol, the bipolar pacing electrodes made of teflon-coated stainless-steel wires were positioned in the middle of the base of each ventricle, below the auricle appendage (see Fig. 1). To ensure stability in the preparation, the heart activity was monitored for 30 min to determine that the heart was arrhythmia-free, stable in amplitude, and had no manifest ischemia. We used an In Vitro rabbit heart model because it provides advantages such as a high level of experimental reproducibility, has a greater throughput compared to complicated in vivo models, provides a better evaluation over a range of concentrations and different combinations of drugs to be tested. In addition, it can be manipulated to mimic clinical conditions, such as hypokalemia and bradycardia that support these comments. Also, it has been well established that with VPBs [14] a significant increase in ventricular repolarization heterogeneity (VRH) is induced.

2.2   Experimental Model and Protocol

The experimental model consisted of the In Vitro system, which used a multiple recording system to obtain the beat-to-beat electrical activity of isolated rabbit heart. The VPBs protocol used a circular tank (diameter = 7 cm, height = 7 cm) that had 40silver-silver chloride electrodes (diameter = 2 mm) distributed homogeneously within an array of 5 rows and 8 columns (see Fig. 1). The distance between electrodes was 10 mm and the angular distance was 45°. The dimensions of the tank simulated a rabbit's thorax. Four additional electrodes were allocated in an "Einthoven-like" configuration (see Fig. 1). Two of them were positioned on the base of the tank and the other two were on the upper left and right side

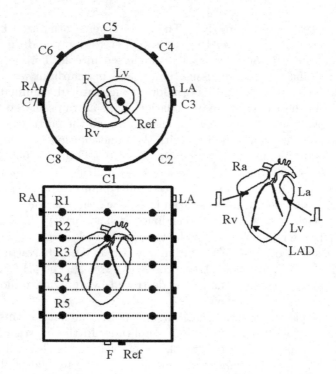

Fig. 1. ECG multilead system: 40-electrodes configuration for electrocardiographic recording in the VPBs protocol. Schematic view showing the superior and frontal 5 × 8 matrix electrodes, as well as the standard lead foot (F), left arm (LA), right arm (RA), reference (Ref), column (C) and row (R). Also are shown the positions of both the stimulating electrodes located in the base of the Lv and Rv, below the atrial appendages

of the tank wall and served as arm electrodes. The four electrodes were designed to build the electrical reference by configuring the Wilson Central Terminal.

In this study VRD was modified by VPBs. Due to heterogeneous distribution of APD lengthening induced by potassium-channel blocking drugs [15] or the heterogeneous shortening of APD caused by the heterogeneous distribution of restitution kinetics [14], a real increase in VRH phenomena can be obtained. Besides, it can be noted that we measured the increase in VRH, not dispersion as an absolute value, so our gold standard was the same heart in the control condition in each experiment.

In the VPBs protocol, the heart was stimulated from the Rv or Lv at basal frequency during a train (S1) of 49 beats. Single premature stimuli (S2, beat number 50th) were applied after pulse trains, at two different S1-S2 coupling intervals: 200 ms and Effective Refractory Period (E_{rp}) plus 5 ms.

In each case, E_{rp} was estimated prior to the VPBs application. To estimate E_{rp}, premature coupling intervals (distance from the last beat to the premature stimulation time) were diminished step by step at 5 ms until period refractoriness was reached. We used the average of 48th and 49th beats from each S1 as Control. The premature beat (50th) was elicited to generate changes of VRH paced either at Rv or Lv.

In the VPBs protocol, the hearts were paced from the Rv ($n = 9$) or the Lv ($n = 9$) by stimuli trains at a basic cycle length of 430 ms for control condition. Then, single premature stimuli were applied after a pulse train at a frequency equal to $E_{rp} + 5$ ms (167 ± 7.2 ms for Rv stimulation (RVS) and 168 ± 11.5 ms for Lv stimulation (LVS); p value = NS). The hearts were paced using an artificial pacemaker (DTU 101, Bloom Associates Ltd. Reading, PA, USA).

2.3   Acquisition of ECGs Signals

ECG data were recorded using instrumentation amplifiers that had a gain factor of 1000 and a bandwidth of 0.05–300 Hz. The signals were digitalized at a sampling rate $f_s = 1$ KHz and 12-bit resolution using a digital acquisition board (Lab PC+, National Instruments, Austin, TX, USA). When necessary, a bandstop filter was used to remove 50-Hz. The baseline movement was compensated using a cubic spline algorithm. All of the data were acquired and monitored using customized software made in C++.

2.4   Construction of Data Matrix

The ECGs from the first row of leads were recorded simultaneously, and the same procedure was applied sequentially to the remaining rows. The ith beat was selected in the ECG recordings of each rth row, $r = 1, \ldots, 5$, obtaining the i_r^{th} beat. After selecting and segmenting the i_r^{th} beat from each row, a signal, $x_{c,r}(n)$, $n = 0, \ldots, N - 1$, was determined for each channel characterized by the (c, r) pair, where c is the column in the electrode matrix ($c = 1, \ldots, 8$) and r is the row, being $M = 5 \times 8$ the number of register electrodes in each experimental protocol, respectively. Expressing that signal as a vector, $\mathbf{x}_{c,r}$, we obtain

$$\mathbf{x}_{c,r} = [x_{c,r}(0), \ldots, x_{c,r}(N - 1)]^T \tag{1}$$

The five i_r^{th} selected beats were aligned using the QRS complex maximum upstroke slope. The beats extend a time window composed of N samples corresponding to 400 ms, and include the repolarization phase. For each experimental condition (Control, 200 ms and $E_{rp} + 5$ ms), recordings were obtained from 40 ECG leads for the experimental protocol. Expressing in matrix notation the selected segmented signals, $\mathbf{X}$ ($M \times N$), we obtain

$$\mathbf{X} = [\mathbf{x}_{1,1}, \ldots, \mathbf{x}_{L,1}, \ldots, \mathbf{x}_{1,5}, \ldots, \mathbf{x}_{L,5}]^T \tag{2}$$

From $\mathbf{X}$, the ECG-derived parameters were measured. A matrix $\mathbf{X}$ characterize each experimental condition.

2.5 ECG Indices

The QRS fiducial points (QRS_{ON} and QRS_{END}) and T-wave location (T_{END}, T_{PEAK}) were obtained from the ECG delineation system based on the Wavelet Transform [16]. Also, ECG indices have been computed to describe the characteristics of VRD on the electrocardiographic multilead system. For each ith beat, we have computed as:

(1) Ventricular depolarization index: the Q_{RS} interval measured in milliseconds from the onset of the Q wave to the offset of the S wave, has been calculated as;

$$Q_{RS_i} = QRS_{END_i} - QRS_{ON_i} \qquad (3)$$

(2) Total ventricular repolarization duration index (measured in milliseconds): the T_{RD} quantifying the total ventricular repolarization time, has been computed as;

$$T_{RD_i} = T_{END_i} - QRS_{END_i} \qquad (4)$$

(3) Early repolarization duration index (measured in milliseconds): the E_{RD} which several authors have linked to the full repolarization of epicardium, has been calculated as;

$$E_{RD_i} = T_{PEAK_i} - QRS_{END_i} \qquad (5)$$

(4) T-wave peak-to-end interval index (measured in milliseconds): the T_{PE} associated to transmural ventricular repolarization [9], has been computed as;

$$T_{PE_i} = T_{END_i} - T_{PEAK_i} \qquad (6)$$

2.6 ECG Recording Stability

To quantify the stability of ECG recordings we have measured the coefficient of variation (C_V) parameter [8] for each feature, the Q_{RS}, T_{RD}, E_{RD} and T_{PE}. These variables were repeatedly measured at each electrode of the multi-lead ECG recording system every 20 min during an hour.

The C_V is defined as

$$C_V = \sqrt{\sigma^2} \times 100\% \qquad (7)$$

and the σ^2 (variance within) is estimated as

$$\sigma^2 = \sum_{i=1}^{k} \sum_{j=1}^{n_i} \frac{(y_{ij} - \overline{y}_i)^2}{n - k} \qquad (8)$$

For each variable evaluated (Q_{RS}, T_{RD}, E_{RD} and T_{PE}) we assume that there are k groups of measurements with n_i measurements in the ith group. The jth measurements in the ith group will be denoted by y_{ij} and $n = \sum_{i=1}^{k} n_i$. The term $(y_{ij} - \overline{y}_i)$ represents the deviation of an individual measurement from the group mean for that measurement and is a clue of within group variability.

2.7   Classification

There were 3 data classes: $E_{rp} + 5$ ms, 200 ms coupling interval and Control data. Among all available data we used 840 samples from 9 rabbit hearts (93.33 samples/heart) for LVS and 732 samples also from 9 rabbit hearts (81.33 samples/heart) for RVS. The data was perfectly balanced with 1/3 (280 samples) $E_{rp} + 5$ ms, 1/3 (280 samples) 200 ms, and 1/3 (280 samples) control samples of LVS data, and 1/3 (244 samples) $E_{rp} + 5$ ms, 1/3 (244 samples) 200 ms and 1/3 (244 samples) control samples of RVS data. The LVS and RVS data were processed separately. Each dataset was divided into the training set (60%) and the test set (40%). The classification results were compared on the test set which was not used during classifier training.

2.8   Multiclass Classification

We used a *one-versus-rest* approach to multiclass classification problem. We trained a single classifier per each class thus turning a multiclass classification into a series of binary classifications. Each stimulation type data (LVS and RVS) were divided into 3 sets, one per each class, resulting in 6 datasets all together. Each set contained all the data samples of particular class and the same number of samples of the other two classes picked with random sampling resulting in 50%–25%–25% data distribution. For example, the LVS $E_{rp} + 5$ ms class dataset contained 280 samples of $E_{rp} + 5$ ms stimulated from the left ventricle (labelled as E_{rp}), 140 LVS samples of 200 ms coupling interval (labelled as not_E_{rp}) and 140 LVS control samples (labelled also as not_E_{rp}). The dataset thus contained 280 E_{rp} samples and 280 not_E_{rp} samples. The classifiers were evaluated on the basis how well they perform binary classification on all 6 data sets.

2.9   Classifiers

We utilized most of the popular advanced machine learning classification techniques and we compared their performances on ECG indices based on intervals duration. In this sense, we tested 20 classifiers of different types using Weka software [17]. We have chosen 3 classifiers from category Bayes, 5 from Functions, 3 from Lazy, 4 from Trees, 2 from Rules, 2 from Meta and 1 from category Miscellaneous. For all classifiers, we used the default parameter settings. The two meta classifiers in addition to their own parameters take one *base* classifier. In our case the Adaboost M1 method uses REPTree and The Bagging method uses the Decision stump classifier (this is the Weka software default setting). Beside these two, we have chosen several modern supervised classification approaches, like Support Vector Machine, Naive Bayes, Artificial Neural Network (Multi-Layer Perceptron and radial basis functions), Sequential minimal optimization, K-Nearest Neighbour, J48, Classification and Regression Trees, Locally Weighted Learning, and others (see Table 1). Each classifier was applied 10 times to each data set and the average performance on test set of these 10 runs was calculated. Then using these averages we calculated for each classifier the average of standard performance

Table 1. Classifier comparison with average values achieved across all data sets. The best performances are bolded. Bagging with REP Tree (Bagging with Reduced-Error Pruning Tree), Classification and R Trees (Classification and Regression Trees), Ada Boost M1 with DS (Ada Boost M1 with Decision Stump, *Acc.* (accuracy), *Sens.* (sensitivity), *Spec.* (specificity), *AUC* (area under curve), *F1 Score*, *T.Perf.* (Total performance).

Algorithm	Acc.	Sens.	Spec.	AUC	F1 Score	T.Perf.
Bayes Net	0.7471	0.7764	0.7179	0.8048	0.7552	0.76
Naive Bayes	0.7368	0.7773	0.6963	0.8210	0.7472	0.76
Naive Bayes Multinomial	0.6681	0.6819	0.6545	0.7419	0.6701	0.68
Logistic Regression	0.7079	0.7245	0.6913	0.7815	0.7129	0.72
ANN Multilayer Perceptron	0.7656	**0.8024**	0.7288	0.8413	0.7754	0.78
K-Nearest Neighbour	0.7491	0.7616	0.7367	0.7495	0.7519	0.75
K*	0.7613	0.7904	0.7323	0.8338	0.7687	0.78
Random Forest	**0.7841**	0.8019	**0.7664**	**0.8621**	**0.7886**	**0.80**
Classification and R Trees	0.7643	0.7747	0.7539	0.7970	0.7673	0.77
J48	0.7571	0.7934	0.7210	0.7986	0.7666	0.77
Decision Table	0.7510	0.7987	0.7035	0.8054	0.7634	0.76
Locally Weighted Optimization	0.7337	0.7916	0.6760	0.8244	0.7464	0.75
Extra Tree	0.7270	0.7309	0.7230	0.7268	0.7273	0.73
Sequential Minimal Optimization	0.7139	0.7571	0.6708	0.7139	0.7282	0.72
CHIRP	0.7710	0.7912	0.7509	0.7711	0.7771	0.77
One Rule	0.7220	0.7534	0.6906	0.7220	0.7323	0.72
Radial Basis Function Classifier	0.7514	0.7724	0.7304	0.8328	0.7570	0.77
Bagging with REP Tree	0.7773	0.7989	0.7558	0.8561	0.7832	0.79
Ada Boost M1 with DS	0.7521	0.7673	0.7381	0.8126	0.7558	0.76
Support Vector Machines	0.6203	0.5228	0.7177	0.6202	0.5425	0.60

measures for binary classification: accuracy (see Eq. 9), sensitivity (see Eq. 10), specificity (see Eq. 11), F1 score (see Eq. 12), and area under ROC (AUC) across all 6 data sets, as we can observed in Table 1. The first four are based on confusion matrix with 4 standard quantities: true positive (TP), true negative (TN), false positive (FP), false negative (FN), and the last one is based on Receiver Operating Characteristics (ROC) curve. The performance measures were computed as following:

$$Acc = \frac{(TP + TN)}{(TP + TN + FP + FN)} \tag{9}$$

$$Sens = \frac{TP}{(TP + FN)} \tag{10}$$

$$Spec = \frac{TN}{(TN + FP)} \tag{11}$$

$$F1 = 2\frac{(Prec \cdot Sens)}{(Prec + Sens)} \tag{12}$$

where

$$Prec = \frac{TP}{(TP + FP)} \tag{13}$$

2.10   Statistical Analysis

In order to quantify the discrepancy between the parameters' distribution and the Gaussian distribution, we have analyzed the normality of these values using the D'Agostino-Pearson test. It has been observed that the underlying variables distribution was Gaussian. Data were expressed as mean value ± standard deviation (SD). Comparison between ECG indices was performed by means of unpaired Student t-test for normally distribution variables. Significance was considered at a value of $p < 0.05$.

3   Results

We have evaluated the C_V for each ECG index. The C_V was <2% for Q_{RS}, <2% for E_{RD}, <3% for T_{PE} and <3% for T_{RD}. So, we have verified that the estimated variables have not shown significant statistical differences over the one hour In Vitro experiment.

Table 2. Mean values ± standard deviation of the experiments (n = 9 during LVS and n = 9 during RVS) showing all heterogeneity ECG indices measured in control, 200 ms and in E_{rp} + 5 ms. The values were computed in the optimal ECG leads. Significant differences against control (C) are marked as *($p < 0.05$), †($p < 0.005$) and ‡($p < 0.0005$). Non-significant differences were marked as NS.

Indice	Control	200 ms	E_{rp} + 5 ms	VPBs	Lead
Q_{RS}	74.0 ± 4.7	82.5 ± 6.9†	93.2 ± 18.1†	Lv	r4c4
	74.9 ± 8.7	81.4 ± 10.7NS	101.7 ± 19.3‡	Rv	r5c3
E_{RD}	127.1 ± 7.2	100.6 ± 11.9‡	87.3 ± 19.7‡	Lv	r4c4
	117.5 ± 22.9	98.6 ± 13.4*	81.6 ± 13.9‡	Rv	r5c3
T_{PE}	51.9 ± 7.6	62.8 ± 10.0*	62.0 ± 14.2 NS	Lv	r4c4
	52.2 ± 9.4	65.5 ± 12.0†	75.2 ± 15.8‡	Rv	r5c3
T_{RD}	178.6 ± 14.4	163.9 ± 13.3*	152.0 ± 17.7†	Lv	r4c4
	169.7 ± 20.7	164.89 ± 13.9NS	158.8 ± 19.9NS	Rv	r5c3

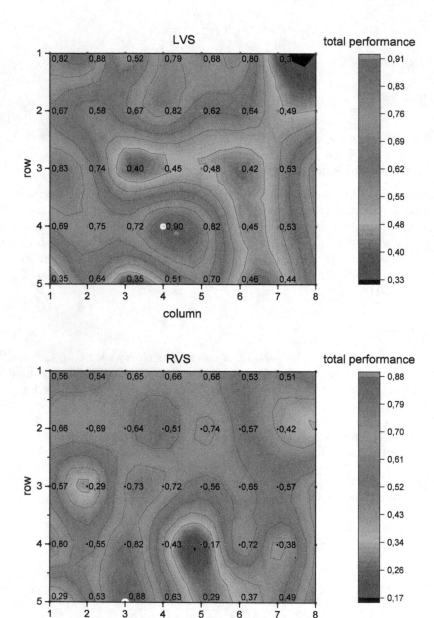

Fig. 2. Contour maps of Total Performance distributed over the 5 × 8 matrix electrodes. Top panel: LVS optimal lead, located in row #4 and column #4 (white circle). Bottom panel: RVS optimal lead located in row #5 and column #3 (white circle).

Since it is a known fact that we can not define which classifier is the best for certain data sets in advance our approach was to test several algorithms from different groups. We have used most of the popular advanced algorithms and compared their performances on our data. We have compared classifiers of different types using Weka software [17], such as 3 classifiers from Bayes, 5 from Functions, 3 from Lazy, 4 from Trees, 2 from Rules, 2 from Meta and 1 from Miscellaneous. We used the default parameter settings. The two meta classifiers in addition to their own parameters take one *base* classifier. The aim of this was to find the best classifier for our 6 data sets. We have observed that Random Forest classifier achieved best average performance in all categories except in sensitivity where it was second best. Thus it has the best average Total Performance (*T.Perf.* in Table 1) and we can proclaim it for the most suitable classifier for our data sets. These aforementioned comparisons are presented in Table 1.

The classification results have shown that there is one optimal lead during LVS, which was located in row #4 and column #4 and so called as r4c4. Moreover, we have located during RVS, the optimal lead in row #5 and column #3, denominated as r5c3.

Mean and standard deviation of Q_{RS}, T_{RD}, E_{RD} and T_{PE} indices in Control condition and VPBs are presented in Table 2. It can be observed that these results were computed in the optimal electrocardiographic lead obtained. On the other hand, with the aim to visualize the spatial position of the optimal lead in the 5×8 matrix electrodes, we have produced colored contour maps, as we can see in Fig. 2.

4   Discussion

A total of 40 unipolar ECG leads, in 18 isolated rabbit heart preparation, were studied. VPBs was introduced at different coupling intervals. The longest S1-S2 was equal to the basic cycle length (430 ms), the intermediate was equal to 200 ms, and the shortest S1-S2 was equal to refractoriness plus 5 ms ($\sim$165.5 ms). Also, in order to find the optimal lead to detect RH changes during VPBs, we compared twenty machine learning classification techniques and we evaluated several ECG indices.

We have observed that when the premature stimuli were applied in the Lv, the repolarization heterogeneity changes were detected preferentially in the ECG lead located in row4–column4. Whilst, when the premature stimuli were applied in the Rv the repolarization heterogeneity changes were detected in the ECG lead located in row5–column3. It is important to highlight that optimal leads during LVS is located opposite of left ventricle. On the contrary the optimal lead in RVS is not located exactly opposite of the right ventricle. These positions can be observed in Fig. 1 and note that all hearts were fixed in the same position, as we have described in Sect. 2.1.

Otherwise, we have observed, in both Lv and Rv premature stimulation, a statistical significant increase of Q_{RS} index duration (see Table 2). We have concluded that depolarization cardiac phase did not seem to be different when the stimuli were elicited from either the Lv or Rv, because both ventricles exhibited a similar response to the VPBs.

Moreover, it can be observed that VPBs in Lv produced a significantly decrease of T_{RD}, while in the RVS this index did not change. We can explain this phenomena analyzing T_{PE} and E_{RD} (see Table 2). Regarding the T_{PE} index, associated to transmural ventricular repolarization dispersion, we have only found statistically significant increases during 200 ms premature stimuli, while $E_{rp} + 5$ ms increase its value but not significantly. Conversely, significantly changes in T_{PE}, for both kind of VPBs, have been observed in RVS.

Random Forest classifier is an ensemble learning technique introduced by [18]. It generates many random classification trees and aggregates their results using the bagging technique where successive trees do not depend on earlier trees. Each is constructed independently using a bootstrap sample of the data set. In addition to this, each node is split using the best among a subset of randomly chosen predictors. This method is robust to over fitting to training data [19]. The used attributes are all continuous numerical data (ECG indices) from similar intervals. The classes are perfectly balanced. It seems that random forest performs very well on such data since it outperformed even newer methods. It is reasonable to expect good performance when using other or additional ECG indices which was already tested [20].

Moreover, the mechanism responsible for different response by premature stimulation depending on the site of pacing is not clearly explainable only with the present results. There are anatomic differences between ventricles, such as the 3D structure or the anisotropic properties linked with dissimilar wall thickness and cardiac fibers orientation. We have concluded that all of these parameters might contribute to the different results obtained between both ventricles. Finally, the present results have shown that changes of VRH during premature stimuli can be very well captured by means of machine learning classification methods in a multi-lead ECG system.

5   Study Limitations

No attempt was made to measure VRD on the epicardial surface or endocardial muscle layers, we have limited our analysis to ECG signals obtained from recording electrodes embedded in the tank wall. The conclusions were made using only ECG interval durations, leaving for future work other characteristics such as areas, amplitudes and slopes.

6   Conclusions

During VPBs we have observed significant decreases in early repolarization duration for both ventricles, while in the Lv and Rv we have measured increases of transmural dispersion. Moreover, we have found optimal lead to detect VPBs, when the premature ventricular stimuli were elicited from left or right ventricles. We conclude that Random Forest classifier achieved the best Total Performance and we have shown it is the most suitable classifier for ECG interval duration during VPBs.

Acronyms

APD	Action potential duration
E_{RD}	Early repolarization duration
ECG	Electrocardiogram
LAD	Left anterior descending artery
Lv	Left ventricle
LVS	Left ventricle stimulation
Rv	Right ventricle
RVS	Right ventricle stimulation
RH	Repolarization heterogeneity
T_{RD}	Total repolarization duration
T_{PE}	T-wave peak-to-end duration
VPBs	Ventricular Premature Beats
VRD	Ventricular Repolarization Dispersion
VRH	Ventricular Repolarization Heterogeneity

References

1. Yan, G., Jack, M.: Electrocardiographic T wave: a symbol of transmural dispersion of repolarization in the ventricles. J. Cardiovasc. Electrophysiol. **14**, 639–640 (2003)
2. Noble, D., Cohen, I.: The interpretation of the T wave of the electrocardiogram. Cardiovasc. Res. **12**, 13–27 (1978)
3. Di Diego, J.M., Sun, Z.Q., Antzelevitch, C.: Ito and action potential notch are smaller in left vs. right canine ventricular epicardium. A. J. Physiol. **271**, H548 (1996)
4. Surawicz, B.: Ventricular fibrillation and dispersion of repolarization. J. Cardiovasc. Electrophysiol. **8**, 1009–1012 (1997)
5. Kuo, C.S., Munakata, K., Reddy, P., Surawicz, B.: Characteristics and possible mechanism of ventricular arrhythmia dependent on the dispersion of action potential. Circulation **67**, 1356–1367 (1983)
6. Shimizu, W., Antzelevitch, C.: Cellular basis for the ECG features of the LQT1 form of the long QT syndrome: effects of β adrenergic agonist and antagonist and sodium channel blockers on transmural dispersion of repolarization and torsades de pointes. Circulation **98**, 2314–2322 (1998)
7. Fuller, M.S., Sándor, G., Punske, B., Taccardi, B., MacLeod, R.S., Ershler, P.R., Green, L.S., Lux, R.L: Estimates of repolarization and its dispersion from electrocardiographic measurements: direct epicardial assesment in the canine heart. J. Electrocardiol. **33**, 171–180 (2000)
8. Arini, P.D., Bertrán, G.C., Valverde, E.R., Laguna, P.: T-wave width as an index for quantification of ventricular repolarization dispersion: evaluation in an isolated rabbit heart model. Biomed. Signal Proc. Control **3**, 67–77 (2008)
9. Antzelevitch, Ch., Viskin, S., Shimizu, W., Yan, G., Kowey, P., Zhang, L., Sicouri, S., Diego, J., Burashnikov, A.: Does Tpeak-Tend provide an index of transmural dispersion of repolarization? J. Mol. Biol. **4**(8), 1114–1119 (2007)

10. Smetana, P., Schmidt, A., Zabel, M., Hnatkova, K., Franz, M., Huber, K., Malik, M.: Assessment of repolarization heterogeneity for prediction of mortality in cardiovascular disease: peak to the end of the T wave interval and nondipolar repolarization components. J. Electrocardiol. **44**, 301–308 (2011)
11. Yuan, S., Blomström-Lundqvist, C., Pherson, C., Wohlfart, B., Olsson, S.B.: Dispersion of ventricular repolarization following double and triple programmed stimulation: a clinical study using the monophasic action potential recording technique. Eur. Heart J. **17**, 1080–1091 (1996)
12. Horowitz, L.N., Spear, J.F., Moore, E.N.: Relation of endocardial and epicardial ventricular fibrillation thresholds of the right and left ventricles. Am. J. Cardiol. **48**, 698–701 (1981)
13. Zabel, M., Portnoy, S., Franz, M.R.: Electrocardiographic indexes of dispersion of ventricular repolarization: an isolated heart validation study. J. Am. Coll. Cardiol. **25**, 746–752 (1995)
14. Laurita, K.R., Girouard, S.D., Fadi, G.A., Rosenbaum, D.S.: Modulated dispersion explains changes in arrhythmia vulnerability during premature stimulation of the heart. Circulation **98**, 2774–2780 (1998)
15. Spear, J., Moore, E.: Modulation of arrhytmias by isoproterenol in a rabbit heart model of d-Sotalol induced long QT intervals. American J. Physiol. **279**, H15–H25 (2000)
16. Mendieta, J.G.: Algoritmo para el delineado de señales ECG en un modelo animal empleando técnicas avanzadas de procesamiento de señales. Facultad de Ingeniería de la Universidad de Buenos Aires (2012)
17. Hall, M., Frank, E., Holmes, G., Pfahringer, B., Reutemann, P., Witten, I.H.: The WEKA data mining software: an update. SIGKDD Explor. **11**(1), 10–18 (2009)
18. Breiman, L.: Random forests. Mach. Learn. **45**(1), 5–32 (2001)
19. Liaw, A., Wiener, M.: Classification and regression by random forest. R News **2/3**, 18–22 (2002)
20. Emanet, N.: ECG beat classification by using discrete wavelet transform and random forest algorithm. In: Fifth International Conference on Soft Computing, ICSCCW 2009 (2009)

Detailed Estimation of Cognitive Workload with Reference to a Modern Working Environment

Timm Hörmann[✉], Marc Hesse, Peter Christ, Michael Adams,
Christian Menßen, and Ulrich Rückert

Cognitronics and Sensor Systems Group, CITEC, Bielefeld University,
Universitätsstr. 21-23, 33615 Bielefeld, Germany
thoerman@techfak.uni-bielefeld.de
http://www.ks.cit-ec.uni-bielefeld.de

Abstract. In modern industry, employees are confronted with ever more complex working tasks. As a consequence, cognitive workload of the employees rises. This makes automatic estimation of cognitive workload a key subject of research. Such an estimate would enable adaptive Human-Machine Interaction that could be used to fit the employees' workload accordingly to their needs. In this work, a tablet interaction study is presented that is designed to induce cognitive workload. Supervised machine learning methods are used to estimate the induced cognitive workload based on features taken from heart rate, electrodermal activity and user interaction (touch input). Ground truth data is obtained from the subjects' self-reported cognitive workload. Inter-subject accuracy of the best learner is 74.1% for the detailed 5-class problem and 96.0% for the simplified binary problem.

Keywords: Cognitive workload · Stress · Heart rate · Electrodermal activity · Tablet computer · Human-machine interaction

1 Introduction

Today's modern working environments are increasingly challenging for the employees. As one example the concept of "Industry 4.0" sketches the design of new flexible working environments, in which employees are constantly confronted with new requirements. This implies manufacturing processes with very small lot sizes, which will result in a higher diversity of working processes. It forces the employees to be highly flexible and to adapt rapidly to changing work tasks. For instance, the employees will have to memorize and apply new knowledge more often [3]. Therefore, *adaptive* Human-Machine Interaction (HMI) becomes more important. Especially by means of the implementation and utilization of adaptive assistant systems, which shall be used to guide an employee through a new and unfamiliar task [26]. With the ability to balance the cognitive workload (CW) of a specific task, the ergonomic design of working tasks could be

© Springer International Publishing AG 2017
A. Fred and H. Gamboa (Eds.): BIOSTEC 2016, CCIS 690, pp. 205–223, 2017.
DOI: 10.1007/978-3-319-54717-6_12

improved. This makes the estimation of CW a key factor towards human centric design, concerning the development of *adaptive* HMI and assistant systems [21, 26].

To fulfill the requirement of adaptability, an assistant system needs to *know* the human user's cognitive capacity. Therefore, in order to adjust correspondingly to the user's needs, it is important to precisely model the user's perceived CW. The goal is to balance the complexity of a given task. This is because, on the one hand, if the user is not sufficiently assisted, it might lead to mistakes. But on the other hand, if the user feels unchallenged, it might decrease his attention [30]. Because both lead to frustration, the estimation of CW has to be detailed and therefore as fine-grained as possible. To prevent such situations, an adaptive assistant system, as an example, could increase or decrease the amount of supporting information provided or the general working speed correspondingly.

We present a tablet computer interaction study, during which different levels of CW are induced. The proposed experiment abstracts and emulates typical tasks employees have to fulfill in modern working environments. In total, 15 subjects participated in the experiment. To estimate the CW, we evaluated the heart rate, the heart rate variability, the electrodermal activity and the tablet computer's touch features (duration and pressure). A sparse feature subset was identified and tested by comparing the accuracy of multiple machine learners.

The work is structured as follows: In Sect. 1 we introduce the theoretical background of CW and summarize related work. An overview of the used hardware and the conducted experiment is given in Sect. 2. Furthermore, the applied machine learning methods are described. In Sect. 3 the results of our feature selection and classification are shown. Subsequently, in Sect. 4 follows a discussion of the results. Finally, we summarize our work in Sect. 5 and give prospect on our future work.

1.1   Background

Up to now, there is no universal definition of mental or cognitive workload. The term mental workload was summarized by [5] as: "the capabilities and effort of the operators [or users] in the context of [a] specific [situation]". Hence, CW is not a univariate, but a "multifaceted" [5] entity. A comprehensible definition states CW to be: "an all-encompassing term that includes any variable reflecting the amount or difficulty of one's work" [4]. We will follow that definition within this work.

The measurement of CW is as divergent as its definition. CW can either be measured subjectively (self-reported) via performance measures (e.g. error rate or time-on-task) or by utilizing psycho-physiological measures [5]. In addition to psycho-physiological responses, behavioral changes (i.e. facial expressions or smart phone usage) and contextual information (i.e. location and ambiance) are used as predictors as well [1].

Psycho-physiological measures are based on the physiological responses of the human body, resulting from a psychological strain (e.g. cognitive workload). These physiological responses are controlled by the autonomic nervous system, which consists of the sympathetic and parasympathetic nervous system. Both systems regulate bodily functions accordingly to environmental conditions (e.g. increase alertness in challenging situations). Well known and most used signals to quantify these bodily functions and therefore to estimate CW are based on the heart rate (HR) or heart rate variability[1] (HRV) [13], as well as on the electrodermal activity (EDA; or galvanic skin response - GSR) [12]. Additionally, other physiological signals are used, namely, blood pressure, skin temperature, respiration rate or the electroencephalogram. A comprehensive overview of the methods used for the estimation of stress is given by [1].

In this work we aim for a model that utilizes readily accessible psychophysiological measures (HR, HRV, EDA) in order to estimate CW. Subjective measure (self-report) is used as ground truth to set up the estimation.

1.2   Related Work

The possibility of estimating psychological strain has frequently been presented in recent research. Physiological strain is thereby often referred to as mental or cognitive workload or more generally as stress[2]. A survey of recent work on the estimation of stress can be found in the methodological review given by [1]. Selected examples related to our work are presented in the following:

The effectiveness of heart rate monitors in detecting mental stress was demonstrated by [7]. They highlighted the importance of an unobtrusive design to obtain high user acceptance rates. With their approach they were able to distinguish between stressed and non-stressed mental states with an accuracy of 69%.

Within the work of [28] and [6] it was shown that the combination of the heart rate and additional predictors (e.g. respiration rate and EDA) improves the estimation's accuracy (79% and 81%).

Most recent work also emphasizes the problem of detecting CW (or stress) by considering physical activity as an additional predictor. [14] used physical activity information in order to prevent it from becoming a confounding factor. Their approach resulted in an estimation with an accuracy of up to 92.4%.

Additionally, [24] focused on short term signal processing, which enables the detection of short term stress events. They presented remarkable results with a classification accuracy of up to 95%. Also, fine-grained estimation of stress has been addressed in the work of [11]. They estimated the perceived stress level of drivers in three distinct gradations with an accuracy up to 97%. A similar work was presented by [16], which reported an accuracy of 96% for a five-class problem.

Yet, a combined contemplation of a detailed estimation of stress or CW based on short-term signals has not been individually addressed. In this work we focus

---

[1] Variation of the time interval between successive heartbeats. Also known as RR-interval.

[2] In some applications, e.g. the automotive industry, related parameters like arousal or fatigue are considered.

on both, short-term signal processing of multiple parameters and fine-grained estimation of CW. Both are mandatory requirements in order to implement CW estimation into adaptive assistant technology.

2   Methods

In the following, we give an overview of the used sensory equipment (Sect. 2.1). Detailed explanation of the conducted experiment (Sect. 2.2) and the definition of ground truth (Sect. 2.3) is provided afterwards. Hereinafter, we outline mandatory signal processing steps (Sect. 2.4) and refer the feature selection (Sect. 2.5). Finally, the machine learning methods used within this work (Sect. 2.6) are presented.

2.1   Hardware

The hardware setup is based on the Google Nexus 10 tablet computer[3], which has sufficient computing power for the desired task and allows an easy integration of the external sensors.

The EDA was captured by using the Mindfield eSense Skin Response system[4], which is a portable solution designed for tablet computers and smartphones. Its microphone jack is connected to the tablet computer and the two finger (hook and loop) electrodes are placed around the subject's index- and middle finger.

The Mindfield system was compared to a Brainproducts EDA sensor connected to an appertaining QuickAmp Amplifier[5] as a reference system (Fig. 1). Although both systems produced different outputs in terms of absolute value, their signals showed close agreement (Pearson's $r > 0.8$). Therefore, we used the mobile and inexpensive Mindfield system.

The heart rate was captured by two redundant systems. Firstly, we used an ECG based Polar H6 heart rate sensor[6], which is attached to a chest strap. Secondly, the photoplethysmogram (PPG) based Mio Alpha watch[7] was used, which is worn around the wrist. Both heart rate sensors communicate wirelessly with the tablet computer via Bluetooth Low Energy. Measurement readings from both devices were comparable (mean deviation 3.85%). However, we noted that the Mio Alpha smooths the measured values. For this reason, we only use data obtained from the Polar sensor in the following.

2.2   Experiment

We conducted an experiment to induce varying levels of CW during the interaction with a tablet computer. In total, 15 subjects volunteered to participate in

---

[3] GT-P8110; Google Inc., Samsung Electronics.
[4] Mindfield Biosystems Ltd., http://www.mindfield.de.
[5] Brain Products GmbH, http://www.brainproducts.com.
[6] Polar Electro Oy, http://www.polar.com.
[7] Physical Enterprises Inc. (Mio Global), http://www.mioglobal.com.

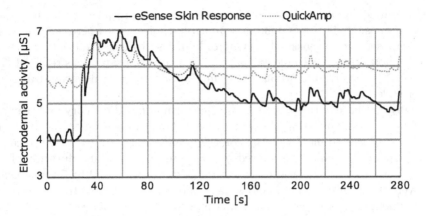

Fig. 1. Comparison of the portable Mindfield eSense Skin Response sensor against the Brainproducts EDA sensor.

the experiment. Subjects were mainly male students (14 male, 1 female, mean age 25.9 ± 2.1). All subjects were aware about the design of the experiment and gave their informed consent.

The total experiment lasted approximately 20 to 25 min for each participant and was repeated after a short break. During the break, the sensors were reapplied to increase robustness in terms of repeatability concerning the various sensors' attachment. Each pass of the experiment was divided into five phases:

1. Relaxation video (2 min)
2. Memorize items (3 to 4 min)
3. Stroop test (3 to 4 min)
4. Recall items (4 to 5 min)
5. Memory and reaction test (3 to 4 min)

The experiment started with a resting phase in which a relaxation video was presented to the subject (phase 1). This was done in order to prevent possible effects resulting from the excitement of the ongoing experiment.

Afterwards, a memory test was initiated (phase 2). During this phase, 12 items of learning content were provided to the subject. The learning content consisted of demographic and economic data of the United States (first pass) and the Czech Republic (second pass). For each item, the time to memorize the provided information was limited to 10 s.

Before the memorized content had to be recalled by the subject (phase 4) a Stroop test [23] was carried out (phase 3). During the Stroop test the user had to touch the button with the color that is identical to the color of a shown text on the screen. The background color, the number of possible solutions (buttons) and the available time to answer was altered randomly. Hence, the Stroop test challenged the user with varying intensity levels. Overall, the subject was asked to reply to 90 Stroop items during 6 repetitions (15 items each). A short break preceded every repetition.

Afterwards, in phase 4 the subject was asked to recall the learning content from phase 2. This was done in a multiple-choice way, whereas 7 questions were composed into 3 blocks of varying difficulties. To increase the CW for the multiple-choice test in each block, the available time to answer was reduced (7 s, 6 s and 5 s, respectively). Additionally, in the last block, only invalid answers were provided.

Finally, in phase 5, the subject had to perform a mixed memory and reaction test. For this test, colored circles were consecutively drawn on the screen. The subject's task was to memorize the color sequence and to immediately recall it afterwards. The difficulty was altered by changing the count and duration of the circles shown (3 to 7 circle were shown for a duration of 700 to 500 ms each). Moreover, the number of used colors was changed randomly (3 to 7). To recall the color sequence, a checker board was presented to the subject (Fig. 2). The checker board was sparsely filled with colored circles (randomly distributed). The subject was asked to recall the color sequence, which was shown beforehand, by touching the corresponding circles.

The proposed experiment covers typical tasks workers are faced with in an abstract way. The abstraction focuses on the tasks to memorize and recall various working steps, e.g. while assembling a work piece or wiring a cable harness at the production line (mixed reaction and recall test, phase 5). The worker has to recall a new working process under time pressure. Another example is performing and following a diagnostic sequence. In this case, the worker has to memorize facts and later on recall and compare the results (memory test: phase 2 and 4).

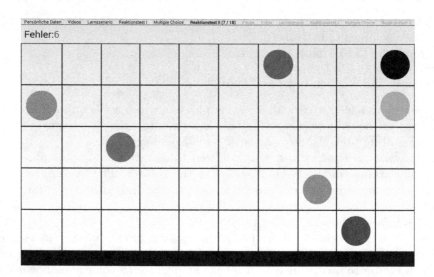

Fig. 2. Checker board used to recall the color sequences in phase 5 of our tablet based experiment. (Color figure online)

2.3 Ground Truth

During the experiment, we simulated short-term stress events with varying intensities. Each event was assigned with an estimated or demanded CW by the experimenter. The annotation scale reached from 1 to 5. Yet, it is unclear if the subjects' perceived CW corresponds to the demanded CW. Therefore, in order to obtain ground truth data, all participants were asked to self-report their perceived CW on a scale from 1 to 5. The self-report was enquired directly after a specific task was finished. Thereby, during each pass of the experiment, the subject was asked 17 times to give self-report of the perceived CW. This self-report was then assigned as ground truth (target label) for the previously performed task.

2.4 Preprocessing and Feature Extraction

The utilized Polar H6 provides the heart rate and the RR-interval for each recognized heartbeat. For this reason, the data stream is recorded in non-uniform time intervals. To enable a common frequency based analysis the data is re-sampled to 4 Hz as suggested by [22]. For the transformation into the frequency domain Welch's method in combination with a Hamming window is used. Prior to the feature extraction, the RR-interval is normalized and detrended as demonstrated by [25]. Furthermore, heart rate for each subject is min-max normalized to increase inter-subject comparability.

The EDA is captured with a sample rate of 10 Hz. In order to remove outliers, we applied a low pass filter with a cut-off frequency of 0.5 Hz. The raw EDA signal is decomposed into the skin conductance level (SCL) and skin conductance response (SCR), as described by [6]. Their method is based on the approach from [25], which was also used to detrend the RR-interval beforehand.

Statistical data (minimum, maximum, mean, standard deviation) is calculated from HR, HRV, EDA, SCR and SCL. In addition, amplitude, duration, area and frequency of the EDA and SCR signals are computed. Furthermore, commonly known features based on heart rate variability are used [17]. As the experiment was carried out using a tablet computer, we additionally recorded mean pressure, mean duration and total count of touch events on the touchscreen during the experiment. These features shall be used to reflect behavioral changes of the users. A comprehensive overview of all extracted features is given in Sect. 3.2.

Because the extracted features are not all commensurate, min-max scaling (1) or z-transformation (2) is used (depending on the classifier used, Sect. 3.3).

$$\text{Min-Max}(X) = \frac{X - min(X)}{max(X) - min(X)} \tag{1}$$

$$\text{Z-Trans.}(X) = \frac{X - \bar{X}}{\sigma(X)} \tag{2}$$

$$\text{with} \quad \sigma(X) = \sqrt{\frac{1}{n-1}\sum_{i=1}^{n}(x_i - \bar{X})^2} \quad \text{and} \quad \bar{X} = \frac{1}{n}\sum_{i=1}^{n}x_i.$$

2.5   Feature Selection

To identify the optimal window size and overlap, we derive multiple feature subsets based on the corresponding sensory element (HR, EDA and Touch). Then, we empirically explore the performance of the models for each combination of subset, window size and overlap. For this purpose, we refer to the mean accuracy from stratified 10-fold cross-validated Decision-Trees. Afterwards, we reduce the feature space to avoid redundancies. Therefore, all features are ranked by their information gain, utilizing Weka 3 data mining software [29].

2.6   Classification

With a comparative analysis we want to identify the potential of the derived feature set for the fine-grained and short-term estimation of CW. Therefore, we train multiple fine-grained supervised classification models with the optimal feature set and window size that was evaluated beforehand (Sect. 3.2). We compare various well-known classifiers, using the correspondent MATLAB Toolbox [18] implementations. Evaluated methods are: Naïve Bayes, Decision-Tree, k-Nearest Neighbor and Support Vector Machine. Additionally, we set up a Gaussian Process Regression model utilizing the GMPML MATLAB Toolbox [20].

The Naïve Bayes classifier provides a generative model of the feature space. It is used to estimate the probability distribution of the feature space given a specific class label. Thereby, the estimate is based on the (naïve) assumption that, given a certain class label, the corresponding predictors are conditionally independent to each other [27].

The Decision-Tree classifier follows the divide-and-conquer approach, meaning that multiple decision rules are created and arranged in a tree like structure. Thus, Decision-Trees allow non-parametric modeling, which at the same time, however, can lead to over-fitting [9].

The k-Nearest Neighbor classifier belongs to the group of lazy or instance-based learners. The classification is based on querying the similarity (or distance) of a new observation to the known observations from the training set. Typically, a Euclidean distance measure is used. Each new observation is then classified by majority vote in respect to its k nearest neighbors [15].

The Support Vector Machine is a kernel-based discriminative classifier. Utilizing the kernel trick [2] the Support Vector Machine constructs a hyperplane that allows non-linear separation of the feature space. Often polynomial, Gaussian or radial-basis functions are used as kernel functions. In order to enable multi-class classification with Support Vector Machine, we make use of the MATLAB Error-Correcting Output Codes implementation [8].

Lastly, we train a Gaussian Processes Regression (also known as Kriging), which is a non-parametric kernel-based model. In the Gaussian Process Regression, the observations of the training set are seen as random samples from a multivariate Gaussian distribution. The estimation is based on a Gaussian process, which is defined by a mean and a covariance function. To attain class labels, we round the output values of the regression [2].

To compare the models' performances, we refer to accuracy (3), sensitivity (true positive rate, 4), specificity (true negative rate, 5) and precision (positive predictive value, 6). To prevent overfitting and to assure validity of the classifier, we make use of stratified 10-fold cross-validation.

$$\text{Accuracy} = \frac{TP + TN}{TP + FP + TN + FN} \tag{3}$$

$$\text{Sensitivity} = \frac{TP}{TP + FN} \tag{4}$$

$$\text{Specificity} = \frac{TN}{TN + FP} \tag{5}$$

$$\text{Precision} = \frac{TP}{TP + FP} \tag{6}$$

TP - True Positive   FP - False Positive
TN - True Negative   FN - False Negative

3   Results

In this section, we present findings from the experiment (Sect. 3.1) and reveal the selected feature subset (Sect. 3.2). Finally, we compare results of the trained classifiers (Sect. 3.3).

3.1   Experiment

Firstly, we verify that the subjects were adequately challenged during the experiment. Therefore, we compare the self-reported CW (ground truth) for the different phases of the experiment. Additionally, we examine the repeatability of the experiment by comparing the first and the second pass of the experiment (255 data points for each pass).

We found similar mean and variance concerning the self-reported CW levels during the different experimental phases (Fig. 3). The null hypothesis that the self-reported CWs are equal between the first and the second pass could not be rejected with One-Way ANOVA ($p = 0.42$). Thus, we conclude there was no significant difference in the perceived stress level during both passes of the experiment.

In direct comparison, the different phases of the experiment mostly coincided with the subjects' self-reported CW level (Fig. 3). However, during phase 4 (recall test) and 5 (memory and reaction test) no significant difference between the self-reported CW levels were found (1-way ANOVA, $p = 0.40$). The same applies for the comparison of the CW levels during the 2nd (memorize test) and 3rd (Stroop test) phase of the experiment (1-way ANOVA, $p = 0.75$). We conclude that the subjects were equally challenged during both tasks.

In all, 14 subjects (93%) self-reported 4 or 5 different CW levels during the experiment. Only one subject reported CW with just three different levels (1 to 3). Hence, the self-reported CW with 5 distinct levels can be used for the automatic estimation.

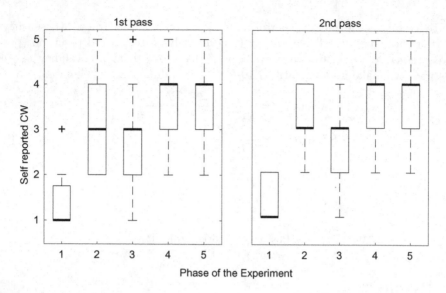

Fig. 3. Box-plot of the self-reported CW levels, grouped by the different phases of the experiment (Sect. 2.2). Results for the 1st (left) and 2nd (right) pass of the experiment are plotted individually. The bold bars and crosses denote median value or outliers, respectively.

3.2 Feature Selection

We extracted a total of 49 features (Table 1) from the different sensor elements (HR, EDA, touch). Detailed information can be found in Table 2.

To determine an appropriate window size for the feature extraction, we defined multiple feature subsets. For every subset we extracted features and varied the length of the time window from 10 to 60 s in 5 s steps. Additionally, the time window overlap was altered. To generate overlapping windows the signal window is shifted by 25%, 50%, 75% or 100% (no overlap) of the length of the time window. To pre-estimate the usability and to determine the optimal window size and overlap, we evaluated the accuracy of 10-fold cross-validated Decision-Trees for each of the 308 possible feature sets[8] (Table 3).

For all tested combinations, the optimal accuracy for each feature subset was found with 75% overlap. With regard to the window length, the results were not equally consistent. Except for the heart rate feature set (Pearson's r = 0.9503, $p < 0.05$), we found no significant trend or correlation between the classifier's performance and the length of the time windows. We conclude that there is no all-encompassing optimal window size or overlap, but each subset has its own optimum (Fig. 4).

Following the objective to set up a short-term estimation of CW, the window size needs to be as short as possible. On the other hand, we need to keep

---

[8] For each of the 7 subsets all combinations of window sizes (11) and overlaps (4) are evaluated.

Table 1. Overview of all extracted features.

Source	Feature
HR	mean, standard deviation, min., max.
HRV	mean, standard deviation, min., max., pRR50, RMSSD, SD1, SD2, SD1/2, skew, kurtosis, VLF, LF, nLF, nHF, LF/HF
EDA, SCR	mean, standard deviation, min., max., peak count, peak prominence, max. peak prominence, mean peak prominence, median peak prominence, peak duration, peak area
SCL	mean, standard deviation, min., max.
Touch	mean duration, mean pressure, count

a minimal length in order to obtain reliable features, e.g. from the heart rate sensor. We found that the models based on a 40 s window performed well concerning the heart rate features as well as the EDA features (Fig. 4). Hence, for further analysis, we chose a window size of 40 s with an overlap of 75%. With this compromise, we fit with the classification accuracies and keep the window size short at the same time. Nevertheless, due to the overlap, we obtain a new estimate every 10 s. The chosen window length is by 20 s smaller compared to related work from [14] or [24].

Next, we select an optimal feature subset. From the first test, we found maximum accuracy by using the full feature set. However, to reduce interdependencies and redundancies within the full feature set we want to identify the most valuable features and deduce a sparse feature subset. Therefore, we ranked all features by their information gain (Table 4).

Taking information gain into account, we found EDA features to be most important. Although the models based on the touch features showed the worst performance beforehand (Decision-Tree, Table 3), they were ranked second most important after the EDA features. With further analysis, we have to assume that this result is due to spurious relationship within the experimental design. We must note that the expected count of touch events was not evenly distributed among the different phases of the experiment or the expected CW levels. Furthermore, there was no control setting for touch pressure or duration, between the touch intensive and challenging phases (phase 3 and 5) and those phases that required only few or no touch inputs (phase 1, 2, 4). For this reason, we withdraw touch features from further analysis.

In order to reduce the overall complexity of the feature space, the maximum prominence peak feature (derived from the EDA signal) was also withdrawn. The resulting sparse feature set contains the 9 most valuable features (regarding information gain), which include 5 features based on EDA and 4 heart rate based features (Table 4).

Table 2. Selected methods used for the feature extraction.

Signal	Category	Function	Definition
Heart rate (HR)	time	average	$\mu = \frac{1}{n}\sum_{i=1}^{n} x_i$
		standard deviation	$\sigma = \sqrt{\frac{1}{n-1}\sum_{i=1}^{n}(x_i - \mu)^2}$
Heart rate variability (HRV)	time	average	$1/n\sum_{i=1}^{n} a_i$
	statistical	skew	$1/n\sum_{i=1}^{n}(x_i - \mu)^3$
		kurtosis	$1/n\sum_{i=1}^{n}(x_i - \mu)^4$
	heart rate variability	NN50	$\sum_{i=1}^{n-1}(x_i - x_{i+1} > .05)$
		RMSSD	$\sqrt{1/n\sum_{i=1}^{n}(x_i - x_{i+1})^2}$
		SDSD	$\sigma((x_1 - x_2)\dots(x_{n-1} - x_n))$
		SD1	$\sqrt{.5 \cdot SDSD^2}$
		SD2	$\sqrt{(2 \cdot SDSD^2) - (.5 \cdot \sigma^2(x))}$
		SD12	$SD1/SD2$
	spectral	VLF	energy 0.00 to 0.04 Hz
		LF	energy 0.04 to 0.15 Hz
		HF	energy 0.15 to 0.40 Hz
		nLF	normalized energy $(LF/LF + HF)$
		nHF	normalized energy $(HF/LF + HF)$
		LF/HF	LF/HF
Electro-dermal-activity (EDA, SCR, SCL)	geometric (peak)	count	number of peaks
		prominence	distance between to successive peaks
		width	distance between the two minimums surrounding a peak
		area	integral between the two minimums surrounding a peak

3.3 Estimation Performance

To evaluate the quality of the selected feature subset, we tested multiple classifiers and compared their accuracy (Table 5).

Lowest accuracy resulted from Naïve Bayes classifier (45.09 ± 2.08%). We tested normal distributions as well as multiple kernel smoothing density

Table 3. Best classification accuracy for each feature subset in respect to window size and overlap. Results from 10-fold cross-validated Decision-Trees.

Subset	Window size	Overlap	Accuracy
FULL	35 s	75%	62.22%
HR	60 s	75%	51.54%
EDA	30 s	75%	50.50%
TOUCH	45 s	75%	46.01%
HR & EDA	25 s	75%	60.16%
HR & TOUCH	50 s	75%	58.21%
TOUCH & EDA	35 s	75%	55.89%

Table 4. Average information gain (IG) with standard deviation for the top 12 ranked features. Selected features for the sparse feature subset are printed bold.

Feature	IG
Minimum EDA	0.486 ± 0.006
Average SCL	0.451 ± 0.004
Average EDA	0.451 ± 0.004
Maximum EDA	0.416 ± 0.003
Average touch duration	0.361 ± 0.004
Average touch pressure	0.333 ± 0.003
Minimum heart rate	0.323 ± 0.020
Maximum heart rate	0.228 ± 0.012
Average heart rate	0.199 ± 0.004
Standard deviation SCR	0.151 ± 0.003
Average RR	0.116 ± 0.003
Maximum GSR peak prominence	0.098 ± 0.002

estimates for the probability density. Regardless of the configuration, no perceptibly difference in the accuracy could be found. One explanation for the low accuracy is the lack of independence concerning the feature set. However, a thorough investigation of the cause is not part of this work.

For the Decision-Tree based classifier an average accuracy of $60.13 \pm 4.05\%$ was achieved (Fig. 5(a)). In order to avoid over-fitting, we chose a limit of 100 splits per tree. Maximum average sensitivity is found on level 1 ($80.49 \pm 7.56\%$). Yet, the mean sensitivity considering levels 2, 3 and 4 reached only $56.75 \pm 9.45\%$. Thus, the misclassifications (or inaccuracy) mainly resulted from the confusion on the CW levels 2, 3 and 4. Comparable results are found with the classifier's specificity.

The usage of k-Nearest Neighbor resulted in an enhanced accuracy and overall sensitivity. Again, the highest sensitivity is found with CW 1 ($82.14 \pm 5.35\%$).

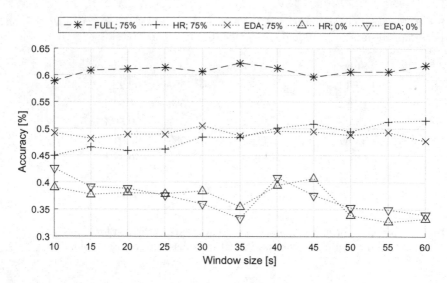

Fig. 4. Mean accuracy from 10-fold cross-validated Decision-Trees trained on the full dataset, the heart rate subset, and the EDA subset. Features were extracted on time windows with length of 10 to 60 s in 5 s steps. For the sake of clarity only 75% and 0% overlap are depicted.

Compared to the Decision-Tree based classifier, the critical confusion on self-reported CW levels 2, 3 and 4 is reduced (sensitivity: $65.38 \pm 6.75\%$). However, we noticed a continuous drop of the accuracy with a growing neighborhood. Concerning Euclidean distance measure, best results were found with $k = 1$, which could suggest an over-fitted model. For instance if the neighborhood is set to $k = 10$, accuracy declines to $58.96 \pm 2.05\%$.

Using Support Vector Machine we were able to further reduce confusion in the mid-levels (sensitivity: $69.10 \pm 7.74\%$) and therefore increase the overall accuracy to $71.00 \pm 3.36\%$. Best results were archived with radial-basis kernel, although usage of Gaussian or polynomial kernel did only slightly affect the model's performance.

In consideration of the observed confusion in the mid-levels of the CW estimation, we infer both the target values (self-reported CW) and the predictors (EDA, HR) to be noisy. Taking the assumption of noisy predictors and target values into account, we chose Gaussian Process Regression as an additional learner for the comparison. Gaussian Process Regression is well known to act as a linear smoother and therefore generally provide good performance in noisy settings [19]. Indeed, the Gaussian Process based classification outperformed the other methods with an accuracy of 74.05% (Fig. 5(b)). Additionally, the mean sensitivity concerning estimated CW level 2, 3 and 4 was enhanced ($72.95 \pm 8.10\%$).

Still, not all uncertainties are covered by the Gaussian Process. This can easily be seen by shrinking the classification task to a binary problem. In this

Table 5. Comparison of different classifiers based on the sparse feature subset in descending order of 5-class accuracy. All tests are 10-fold cross validated. Standard deviation during the cross correlation is given with the mean accuracy.

Classifier	Normalization	Accuracy 5-class	Accuracy 2-class
Gaussian Process	min-max	74.05 ± 3.11%	96.03 ± 147%
Support Vector Machine	z	71.00 ± 3.36%	91.43 ± 151%
k-Nearest Neighbor	z	67.90 ± 398%	92.10 ± 2.34%
Decision-Tree	min-max	60.13 ± 4.05%	90.85 ± 1.76%
Naïve Bayes	min-max	45.09 ± 2.08%	84.37 ± 3.20%

case, self-reported CW level 1 is interpreted as no CW. All remaining levels are taken as present CW. By reducing the machine learning task to this binary problem, the average accuracy for the Gaussian Process reaches up to 96.03 ± 1.47%. For the binary classification task, the ranking of the models' performances (accuracy) of the other tested classifiers remains mainly unchanged. In contrast to the fine-grained tasks, the Naïve Bayes classifier also provided an acceptable classification rate.

4   Discussion

Within this work, we successfully demonstrated a fine-grained estimation of CW. By focusing on a fine-grained estimation based on short-term signals, we extended the complexity of the classification task. Additionally, we reached the accuracy of today's state of the art publications for the binary classification task. In comparison, the fine-grained classification resulted in a lower overall accuracy. This was explained by a low sensitivity regarding mid-level CW levels. This observed variation in the self-reported CW levels is partly explained due to the subjective perception of CW. In future work the usage of more detailed self-reports (e.g. based on NASA-Task Load Index [10]) could overcome this issue. Additionally, performance measures like error rate or time-on-task could further clarify the level of subjectively perceived CW. Nevertheless, regarding the Gaussian Process model, misclassification rarely exceeded more than one class (or level). Therefore, despite the lower overall accuracy, the fine-grained estimation should be favorable, because it facilitates a detailed specification of the perceived CW.

Although Gaussian Process showed best accuracy, Support Vector Machine yielded comparable accuracy. As Support Vector Machines are more widespread and computationally efficient implementations are commonly available, they might be used preferentially.

Ranking of the extracted features revealed EDA features to contain maximum information content, directly followed by the heart rate features. As emphasized by [24] care has to be taken if heart rate is chosen as a predictor, because it

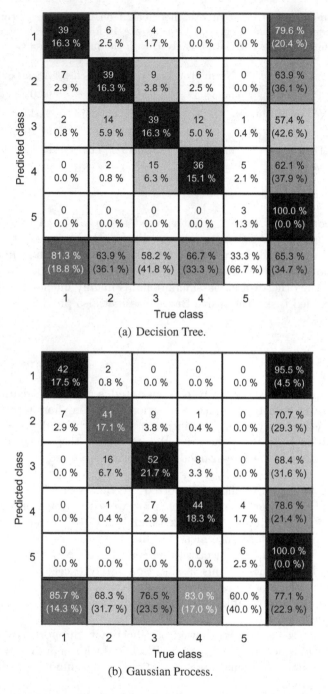

(a) Decision Tree.

(b) Gaussian Process.

Fig. 5. Confusion matrix for the Decision Tree (a) and Gaussian Process (b) based CW level estimation. Each plot shows the results for the classifier with the best accuracy during 10-fold cross validation. Last row contains true positive rate or sensitivity and false negative rate (bracketed). Last column contains true negative rate or specificity and false positive rate (bracketed).

is possibly influenced by means of physical activity. However, during our experiment the subjects were monitored by the experimenter, thus we can exclude physical activity as confounding factor. Nevertheless, the observed confounding influence of the touch features has to be considered in future tablet computer based experiments.

Yet, we found that even a narrow short-term feature subset is sufficient to precisely estimate a person's cognitive workload. This is a mandatory requirement in order to set up an adaptive assistant system, which is capable of balancing a given task's complexity accordingly to the user's cognitive capacity.

5   Summary and Conclusion

We were able to achieve a fine-grained estimation of cognitive workload (stress), which exceeds the complexity of the ordinary binary classification task. Additionally, short-term features were utilized. To reproduce a realistic setting, modern working environments were simulated in the presented experimental setup. The subjects self-reported their perceived CW directly after each task. After the preprocessing we were able to extract a total of 49 features. The most significant features and their ideal window size and overlap were determined with an initial estimate based on 10-fold cross-validated Decision-Trees. The identified sparse feature subset contains 9 features, which include 5 features based on EDA and 4 heart rate based features. The feature subset was then evaluated by comparing the accuracy of multiple well established machine learning methods.

In conclusion, we achieved a classification accuracy of 96.03% for the binary CW estimation task and an accuracy of 74.05% for the fine-grained estimation model. This is likely to enable the development of more advanced assistance technology that can precisely adjust to the user's requirement in modern working environments.

In future work we plan to integrate the utilized sensors into a wearable and hands-free system. This will allow field studies in real working environments including skilled manual work. Additionally, the usage of more detailed self-reports is planned. Furthermore, we want to investigate how our fine-grained estimation of CW can be used to adapt the complexity of a task to the user's needs.

Acknowledgments. This research was supported by the DFG CoE 277: Cognitive Interaction Technology (CITEC), the German Federal Ministry of Education and Research (BMBF) within the Leading-Edge Cluster "Intelligent Technical Systems OstWestfalenLippe" (it's OWL), managed by the Project Management Agency Karlsruhe (PTKA), the BMBF project ALUBAR, and the PhD program "Design of Flexible Work Environments - Human-Centric Use of Cyber-Physical Systems in Industry 4.0" supported by the North Rhine-Westphalian funding scheme "Fortschrittskolleg". The authors are responsible for the contents of this publication.

The authors would like to thank Mindfield for providing the API for their eSense Skin Response system.

References

1. Alberdi, A., Aztiria, A., Basarab, A.: Towards an automatic early stress recognition system for office environments based on multimodal measurements: a review. J. Biomed. Inform. **59**, 49–75 (2016). doi:10.1016/j.jbi.2015.11.007
2. Bishop, C.M.: Pattern recognition and machine learning (information science and statistics). In: Kernel Methods, pp. 291–323. Springer, New York (2006)
3. Botthof, A., Hartmann, E.: Zukunft der Arbeit in Industrie 4.0 - Neue Perspektiven und offene Fragen. In: Botthof, A., Hartmann, E.A. (eds.) Zukunft der Arbeit in Industrie 4.0, pp. 161–163. Springer, Heidelberg (2015). doi:10.1007/978-3-662-45915-7_15
4. Bowling, N.A., Kirkendall, C.: Workload: a review of causes, consequences, and potential interventions. In: Houdmont, J., Leka, S., Sinclair, R.R. (eds.) Contemporary Occupational Health Psychology, vol. 2, pp. 221–238. Wiley, Chichester (2012). doi:10.1002/9781119942849.ch13
5. Cain, B.: A review of the mental workload literature. In: RTO-TR-HFM-121-Part-II. NATO Science and Technology Organization (2007)
6. Choi, J., Ahmed, B., Gutierrez-Osuna, R.: Development and evaluation of an ambulatory stress monitor based on wearable sensors. IEEE Trans. Inf. Technol. Biomed. **16**(2), 279–286 (2012). doi:10.1109/TITB.2011.2169804
7. Choi, J., Gutierrez-Osuna, R.: Using heart rate monitors to detect mental stress. In: Sixth International Workshop on Wearable and Implantable Body Sensor Networks, pp. 219–223, June 2009. doi:10.1109/BSN.2009.13
8. Dietterich, T.G., Bakiri, G.: Solving multiclass learning problems via error-correcting output codes. J. Artif. Intell. Res. **2**, 263–286 (1995)
9. Fürnkranz, J.: Decision tree. In: Sammut, C., Webb, G. (eds.) Encyclopedia of Machine Learning, pp. 263–267. Springer New York (2010). doi:10.1007/978-0-387-30164-8_324
10. Hart, S.G., Staveland, L.E.: Development of NASA-TLX (task load index): results of empirical and theoretical research. In: Human Mental Workload, vol. 52, pp. 139–183. North-Holland (1988). doi:10.1016/S0166-4115(08)62386-9
11. Healey, J., Picard, R.: Detecting stress during real-world driving tasks using physiological sensors. IEEE TITS **6**, 156–166 (2005). doi:10.1109/TITS.2005.848368
12. Isshiki, H., Yamamoto, Y.: Instrument for monitoring arousal level using electrodermal activity. In: Proceedings of IEEE International Conference on Instrumentation and Measurement Technology, pp. 975–978. IEEE (1994). doi:10.1109/IMTC.1994.351943
13. Jorna, P.G.: Spectral analysis of heart rate and psychological state: a review of its validity as a workload index. Biol. Psychol. **34**(2–3), 237–257 (1992). doi:10.1016/0301-0511(92)90017-O
14. Karthikeyan, P., Murugappan, M., Yaacob, S.: Detection of human stress using short-term ECG and HRV signals. J. Mech. Med. Biol. **13**(02), 1350038 (2013). doi:10.1142/S0219519413500383
15. Keogh, E.: Nearest neighbor. In: Sammut, C., Webb, G. (eds.) Encyclopedia of Machine Learning, pp. 714–715. Springer, New York (2010). doi:10.1007/978-0-387-30164-8_204
16. Li, X., Chen, Z., Liang, Q., Yang, Y.: Analysis of mental stress recognition and rating based on Hidden Markov Model. J. Comput. Inf. Syst. **10**(18), 7911–7919 (2014). doi:10.12733/jcis11559

17. Malik, M.: Heart rate variability. Ann. Noninvasive Electrocardiol. **1**(2), 151–181 (1996). doi:10.1111/j.1542-474X.1996.tb00275.x
18. MATLAB: Version 8.6.0 (R2015b). The MathWorks Inc., Natick, Massachusetts (2015)
19. Quadrianto, N., Kersting, K., Xu, Z.: Gaussian process. In: Sammut, C., Webb, G. (eds.) Encyclopedia of Machine Learning, pp. 428–439. Springer, New York (2010). doi:10.1007/978-0-387-30164-8_324
20. Rasmussen, C.E., Nickisch, H.: Gaussian processes for machine learning (GPML) toolbox. J. Mach. Learn. Res. **11**, 3011–3015 (2010)
21. Rouse, W., Edwards, S., Hammer, J.M.: Modeling the dynamics of mental workload and human performance in complex systems. IEEE Trans. Syst. Man Cybern. **23**, 1662–1671 (1993). doi:10.1109/21.257761
22. Singh, D., Vinod, K., Saxena, S.: Sampling frequency of the RR interval time series for spectral analysis of heart rate variability. J. Med. Eng. Technol. **28**(6), 263–272 (2004). doi:10.1080/03091900410001662350
23. Stroop, J.R.: Studies of interference in serial verbal reactions. J. Exp. Psychol. **18**(6), 643–662 (1935). doi:10.1037/h0054651
24. Sun, F.-T., Kuo, C., Cheng, H.-T., Buthpitiya, S., Collins, P., Griss, M.: Activity-aware mental stress detection using physiological sensors. In: Gris, M., Yang, G. (eds.) MobiCASE 2010. LNICSSITE, vol. 76, pp. 211–230. Springer, Heidelberg (2012). doi:10.1007/978-3-642-29336-8_12
25. Tarvainen, M., Ranta-aho, P., Karjalainen, P.: An advanced detrending method with application to HRV analysis. IEEE Trans. Biomed. Eng. **49**(2), 172–175 (2002). doi:10.1109/10.979357
26. Wallhoff, F., Ablassmeier, M., Bannat, A., Buchta, S., Rauschert, A., Rigoll, G., Wiesbeck, M.: Adaptive human-machine interfaces in cognitive production environments. In: Proceedings of IEEE International Conference on Multimedia and Expo, pp. 2246–2249 (2007). doi:10.1109/ICME.2007.4285133
27. Webb, G.: Naïve bayes. In: Sammut, C., Webb, G. (eds.) Encyclopedia of Machine Learning, pp. 713–714. Springer US (2010). doi:10.1007/978-0-387-30164-8_576
28. Wijsman, J., Grundlehner, B., Liu, H., Hermens, H., Penders, J.: Towards mental stress detection using wearable physiological sensors. In: Proceedings of IEEE Engineering in Medicine and Biology Society, pp. 1798–1801 (2011). doi:10.1109/IEMBS.2011.6090512
29. Witten, I.H., Frank, E.: Data Mining: Practical Machine Learning Tools and Techniques, 2 edn. Morgan Kaufmann Publishers Inc., San Francisco (2005)
30. Young, M.S., Stanton, N.A.: Attention and automation: new perspectives on mental underload and performance. Theor. Issues Ergon. Sci. **3**(2), 178–194 (2002). doi:10.1080/14639220210123789

Continuous Real-Time Measurement Method for Heart Rate Monitoring Using Face Images

Daisuke Uchida[✉], Tatsuya Mori, Masato Sakata, Takuro Oya, Yasuyuki Nakata, Kazuho Maeda, Yoshinori Yaginuma, and Akihiro Inomata

Fujitsu Laboratories Ltd., Kawasaki, Kanagawa, Japan
{uchida.daisuke,tatsuya.mori,sakata.masato,oya.takuro,
nakata.yasuyuki,maeda.kazuho,yaginuma,akiino}@jp.fujitsu.com

Abstract. This paper investigates fundamental mechanisms of brightness changes in heart rate (HR) measurement from face images through three kinds of experiments; (i) measurement of light reflection from cheek covered with/without copper film, (ii) spectroscopy measurement of reflection light from face and (iii) simultaneous measurement of face images and laser speckle images. The brightness change of the face skin are found to be caused by both the green light absorption variation by the blood volume changes and the light reflection variation by pulsatory face movements. The Real-time Pulse Extraction Method (RPEM), designed to extract the variation of light absorption by removing motion noise, is corroborated for the robustness by comparing the RPEM with the pulse wave of the ear photoplethysmography. The RPEM is also applied to heart rate measurements of seven participants during office work under non-controlled condition in order to evaluate continuous real-time HR monitoring. RMSE = 6.7 bpm is achieved as an average result of seven participants in five days with the 44% of HR measured rate with respect to the number of reference HRs from the electrocardiogram during face is detected. The result indicates that the RPEM method enables HR monitoring in daily life.

Keywords: Heart rate · Pulse wave · Face images · Real-time remote · Monitoring

1 Introduction

Recently, there has been a growing attention on ICT-enabled personal health services which utilize information on personal health record (PHR) via ubiquitous devices, wireless network and cloud. By continuously monitoring vital signs and activities related to person's health condition, personalized services such as health promotion and disease prevention are expected to be provided. Therefore, the continuous data acquisition in daily life has become an active area of research [1–3]. The research related to human health which focuses on monitoring person finely with low-cost and low-power consumption tools in daily life is now the subject of active investigation. Reginatto, for example, has reported that useful information to health care workers could be provided by acquiring walking motion of older adults with a history of falls by gyro sensors and analyzing the gait [4]. The improved health care using vital signs have been also

© Springer International Publishing AG 2017
A. Fred and H. Gamboa (Eds.): BIOSTEC 2016, CCIS 690, pp. 224–235, 2017.
DOI: 10.1007/978-3-319-54717-6_13

expected. Especially, heart rate (HR) is expected to be utilized widely to keep one's health in good shape. Today, it has been utilized as an index of exercise load. However it is just beginning of utilizations of HR.

The HR has been known as one of the important risk factors for diseases. Dyer reported the HR indicates a prognostic factor for coronary heart disease and mortality [5]. In other reports, the HR has been shown to be a risk factor for diabetes [6]. Therefore, long-term and detailed HR monitoring is expected to be useful for prognostic observation. Monitoring HR is classically done by measuring biopotential with an electrocardiogram (ECG) [7]. A new trend for measuring HR is the photoplethysmogram (PPG), an optical method that retrieves cardiac information [7]. However, these methods require tight contact to the skin, when it needs high accuracy measurement of HR. Such methods make users uncomfortable due to the damage to the skin in the case of long-term monitoring [8]. For that reason, a non-contact measurement method is preferred. In recent years, non-invasive measuring methods of HR using by a depth camera [9], laser [10, 11], or ultrasound [12] are proposed. A method using face images is noted among them. The method doesn't require the special equipments, but a normal RGB camera such as widely used web cameras built into smart phones or laptops, is able to realize non-contact HR measurement [13–18]. Balakrishnan et al. directly detected small head moving amount caused by the blood circulation for measuring HR [17]. Others detected face colour or brightness changes which is also related to blood circulation. In these reports, high accuracy results were obtained under well-controlled conditions. However these methods do not satisfy continuous HR monitoring in daily life. People frequently have various large and small movements, and it makes the extraction of pulse waves from brightness change difficult. Therefore methods which need to accumulate data, such as independent component analysis (ICA) [14, 15] are not suitable because accumulation of data is often interrupted by large motion in daily life, and the method with shorter measurement time is required. In 2013, we demonstrated continuous HR monitoring in daily life by the Real-time Pulse Extraction Method (RPEM) [19]. In this paper, we describe the RPEM and we corroborate the theoretical efficacy of the RPEM by the investigation of brightness change on face images with three fundamental experiments. Continuous real-time HR monitoring with the RPEM in office is also performed as an example of applications.

2   Method for Real-Time Measurement

In this chapter, the framework to measure HR from face images in real time is explained. It has 5 steps. (1) Face images are captured by a RGB camera (webcam) and (2) face detection is performed in each frame. (3) Averaged red, green and blue signals are calculated from region of interest of face images, respectively. (4) Filtering process is performed in order to extract pulse waves due to the blood circulation. (5) Calculation of HR is performed.

In this 4th step of the framework, we focus on the green signal, which is assumed to include pulse components, and remove the noise caused by face movements to obtain the pulse signal. Our method assumes that small head movement affects reflection light from face only in the brightness and not in the colour. Therefore, the intensity ratio between green and red/blue signals stays constant in all frequency range except at the frequency of pulse. We defined the intensity ratio as a, and red and green signals in pulse frequency as g_{signal} and r_{signal}, respectively. We calculate the ratio a in the lower frequency range than the pulse frequency, then we estimate the noise included in green signal g_{signal} by the multiplication of the ratio a and the red signal r_{signal}. We obtain the pulse signal g_{pulse} by subtracting the estimated noise ar_{signal} from the green signal g_{signal} as shown in (1).

$$r_{signal} = g_{signal} - ar_{signal} \qquad (1)$$

With the described method, the noise derived from the larger movement cannot be removed. Thus we also use the confidence indicator with the autocorrelation and remove the HR with small indicator value. The calculation of HR is performed by averaging signals for consecutive 4, 8 or 15 beats with confidence indicators larger than a threshold. With this method, the HR can be measured in several seconds, which is much shorter time than that of conventional HR extraction method such as Discrete Fourier Transform (DFT) method.

3 Fundamental Mechanism

In our method in the Sect. 2, we assumed that green signal has stronger pulsatory component than other colours (hypothesis 1). Also, we assumed that the ratio between green and red signals stays constant in all frequency range except at the pulse frequency (hypothesis 2). In this chapter, we experimentally validate these hypotheses by clarifying the contribution of surface reflection and light absorption caused by blood circulation to brightness change of the face images.

3.1 Light Reflection Measurement

Firstly, the effect of the surface reflection from the face was investigated. The face images (video) were captured and two regions at right and left cheeks were compared. The right cheek was covered with a thin copper film are shown in Fig. 1. The distance between a RGB camera and a participant's face was 80 cm, in addition participants were requested to keep still during the experiment. The face images were recorded by the camera, and the finger PPG signal was also measured as a reference data.

The raw signals of the red, green and blue signals are shown in the upper side of Fig. 2(a) and (b), respectively. The lower side of Fig. 2 shows their frequency characteristics obtained by fast Fourier transform (FFT) in the ROIs at the copper surface and skin surface. In Fig. 2, there are peaks around 75 cycles per minute (cpm) in both of (a) and (b). Note that this peak frequency is the same as the finger pulse rate of 75 bpm simultaneously measured by PPG as shown in Fig. 3.

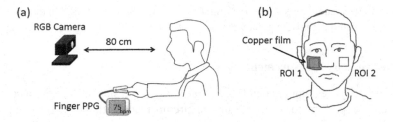

Fig. 1. The experimental setup. (a) A camera is placed at a distance of 80 cm from participant. (b) Region of interests (ROIs). ROI 1 is at a copper film attached on the surface of right cheek.

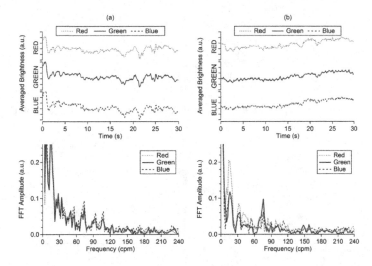

Fig. 2. The raw signals (top) and the frequency characteristics (bottom) of each RGB signal from face images at (a) ROI 1 and at (b) ROI 2. Frequencies are shown as cycles per minute (cpm). (Color figure online)

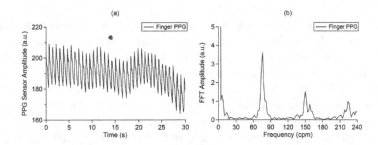

Fig. 3. The finger PPG signal. (a) The raw signal and (b) the frequency characteristics.

Since complete light reflection from the copper surface and no reflection from face skin surface are expected in Fig. 2(a), this peak indicates the contribution of pulsatory movement of the head at 75 cpm caused by the blood circulation. These reflection peaks and profiles in all frequency range are very similar for all RGB signals in (a). On the other hand, the green signal at the peak frequency is stronger than red and blue signals in Fig. 2(b).

3.2 Spectroscopy Measurement

In order to clarify the colour dependency of the signal from skin surface, a spectroscopy experiment was performed. To create similar circumstance with the RGB camera measurement, the distance between a spectroscope and subject's face is about 50 cm and the face was exposed by the intense light using an incandescent lamp (Fig. 4).

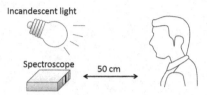

Fig. 4. The experimental setup. A spectroscope is placed at a distance of 50 cm.

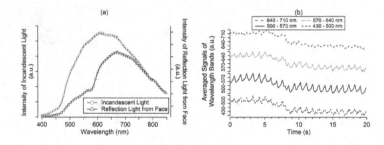

Fig. 5. The results measured with a spectroscope. (a) The spectrum of an incandescent light and a reflection light from face surface, and (b) the time-series variation of averaged signals of each wavelength band.

Figure 5(a) shows the spectra, measured with a spectroscope, of an incandescent lamp and the reflection light from face. The difference of intensity between the incandescent light and the reflection light is due to the loss by scatter and the absorption at the face surface. On the other hand, Fig. 5(b) shows the time-series variation of averaged signals of each wavelength band: 430–500 nm, 500–570 nm, 570–640 nm, and 640–710 nm. It is shown that the waveform in wavelength band of 500–570 nm has higher signal and lower noise than waveforms in other bands. Figure 6 represents the 2D plot of frequency and wavelength dependency of the reflection light measured with spectroscope. Strong amplitude around the frequency of zero cpm is DC component of signals and the relatively large signal around 28 cpm is considered as a component derived from breathing. The characteristic peaks were observed around 540 nm and 570 nm at 68 cpm. These peaks are consistent with the peaks of oxy-haemoglobin absorption at around 540 nm and 570 nm [20]. Since the wavelength of the green light is around 500 nm to 570 nm, the strong peak for green signal in Fig. 2(b) is contributed by the absorption by the oxy-haemoglobin under the face skin. Therefore, it is assumed that the absorption variation by pulsatory blood volume change is causing the strong peak for green signal in Fig. 2(b).

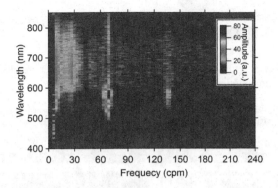

Fig. 6. The frequency and wavelength dependency of the reflection light measured with a spectroscope.

3.3 Laser Speckle Measurement

We also carried out a simultaneous measurement of face images by a RGB camera and blood flow images by a laser speckle imager in order to investigate the relation between the variation of green signal and the blood volume changes [21]. The laser spackle imager is one of the instruments to measure blood flow non-invasively and continuously. The applied principle of the laser speckle imager is that the variation of speckle pattern irradiated by coherent light like laser depends on the velocity of the moving particles, and the velocity can be calculated by the temporal or spatial change on the speckle pattern intensity. In the case of measuring living tissue, the moving particles are erythrocytes (RBCs) [22]. Since light absorption variation depends on blood volume changes, derivative of the light absorption variation means velocity of blood flow. In this experiment, therefore, the changes of blood velocity, measured with laser speckle imager, is expected to be matched with time differential signal of green signal if the light absorption variation, caused by blood volume changes, contributes to the brightness changes of the face images.

The experiment was conducted on the condition that the RGB camera and the laser speckle imager were placed at a distance of 80 cm and 150 cm from participants, respectively. In addition, participants were requested to keep still during the experiment (Fig. 7).

The speckle images were acquired at 10 fps, and the blood flow images of face were calculated by computer. To protect the personal information of the participants, the eyes were covered. Figure 8 shows the ten blood flow images during 1 s. Figure 9 compares the blood flow wave with the time differential green wave, both are obtained simultaneously by averaging signals at the centre area of the face. The phase of differential green wave is in agreement with that of the blood flow wave. The result also indicates blood volume changes contributes the brightness change on face.

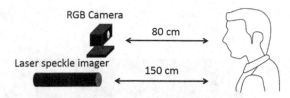

Fig. 7. The experimental setup. A camera and a laser speckle imager are placed at a distance of 80 cm and 150 cm from participant, respectively.

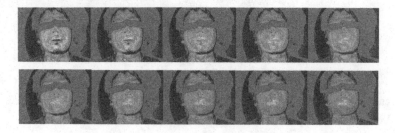

Fig. 8. The ten blood flow images during 1 s. The images are acquired at 10 fps.

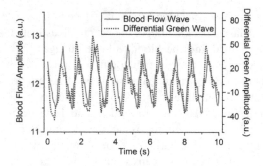

Fig. 9. A comparison of the blood flow wave obtained by a laser speckle imager with the time differential green wave from face images. (Color figure online)

From these experiments and results, the causes of the brightness change on face are combined effects of the oxy-haemoglobin absorption variation by pulsatory blood volume change and the surface reflection variation caused by pulsatory movements. These results validates the hypotheses of our method by the facts that the absorption rate in green is higher than red or blue by the spectroscopy experiment (hypothesis 1), and the influence of movements have no dependence on colour channel in any frequency as shown in Fig. 2(a) (hypothesis 2). Therefore, the RPEM extracts pulse waves due to the blood volume changes from green light by cancelling the effect of head movements.

4   Waves Under Motions

Figure 10 shows a comparison of waveforms when the face is moving. In Fig. 10(a), raw signals of red, green and blue (RGB) channel which are averaged in the region of interest (ROI) are shown. The ROI is determined by choosing a centre part of face detected area. The face movements around at 35.5 s and 43.5 s almost equally affect all RGB signals.

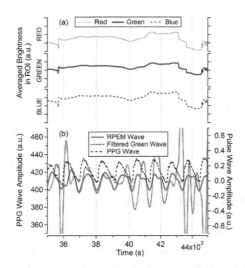

Fig. 10. A comparison of waveforms when the face is moving: (a) the raw RGB signals from face images, and (b) the RPEM, filtered green, PPG wave. (Color figure online)

In Fig. 10(b), an extracted pulse wave by RPEM, filtered green wave and ear PPG wave are shown. The filtered green wave is extracted by a conventional method of infinite impulse response (IIR) filter applied on green signal at frequencies between 50 and 150 bpm. The filtered green wave is largely affected by the motions and the waveform is distorted. On the other hand, the extracted pulse waveform by RPEM is similar to the ear PPG wave without major distortion. From these results, the effectiveness of RPEM for HR measurement is corroborated especially when the filtered green waveform is affected by motions.

To ensure the efficiency of the RPEM statistically, the influence of face movement is estimated with one person's data of about 8 h compared to referential heart rate of electrocardiograph (ECG). Figure 11(a) shows the distribution of averaged velocity of face movement. The result shows that the person has a lot of movement with less than 0.2 cm/s in 8 h. Figure 11(b) shows the relationship of the velocity and the ratio of accurate HR within 3 bpm error. The averaged velocity of the face movement is calculated by dividing the total face movement distance of successive 4 heart beats by the total time of successive 4 beats. The face movement distance is estimated both by the pixels length of face movement in horizontal direction and by the ratio, determined in advance, of pixel width of face detected area to real face width. The ratio of accurate

HR is the ratio of the number of HRs with error of less than 3 bpm to the number of referential HRs. The ratio of accurate HR of filtered green is quite small (>0.2 cm/s). It indicates that the filtered green is not able to extract the pulse wave with subtle movement. On the other hand, accuracy HR rate on our method is about 0.3 at 1 cm/s. It proves that our method has higher resistance to motion.

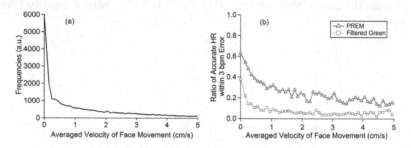

Fig. 11. The influence of face movement on long-term HR measurement for 8 h: (a) The distribution of averaged velocity of face movement and (b) the relationship of the averaged velocity of face movement and the ratio of accurate HR within 3 bpm error.

5   Continuous Heart Rate Monitoring

We applied the RPEM to continuous monitoring of HR during daily office work under non-controlled conditions. In the experiment, seven participants (A, B, C, D, E, F and G) aged from 24 to 55 years old were monitored. Commercially available web cameras were attached on top of the computer display on their desk to capture their face during desk work. Also, an ECG device was on their chest as a reference. All of them were requested to do their work as usual for five days. Face image data for approximately 133 h was obtained in total for seven participants.

Figure 12 shows the HR trend calculated from the results of RPEM for the participant D. The trend of HR in one day is in good agreement with the HR from ECG. The data missing period around noon is because the participant left his desk for lunch, and the large change after the lunch break is due to the effects of running during the break.

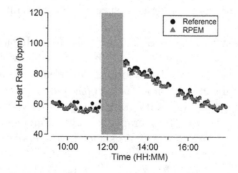

Fig. 12. HR trend during office work compared with HR calculated from ECG (reference).

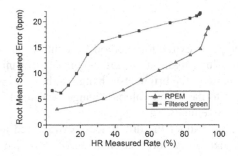

Fig. 13. The trade-off relationship between the HR measured rate and RMSE for RPEM and filtered green. (Color figure online)

During the continuous measurements, the face detection is frequently chopped because people frequently move their face to execute their tasks, such as phone calls, conversation with colleagues, or leaving for lunch or breaks. In one case as an example, only 33% of the sum of the face detection time is for the continuous detections with more than 30 s, and about 90% is for the detections with more than 4 s. Therefore the shorter measurement time is required to increase the chances to measure HR.

Table 1. HR Measured rate and RMSE for seven participants. RPEM is compared with filtered green for different averaging beats.

Participant	HR measured rate (%)				RMSE (bpm)			
	Filtered green	RPEM			Filtered green	RPEM		
	4 beats	4 beats	8 beats	15 beats	4 beats	4 beats	8 beats	15 beats
A	6	40	24	15	11.1	7.2	3.5	2.1
B	19	50	30	16	6.4	4.9	1.9	0.8
C	13	49	30	17	7.9	5.9	2.3	1.7
D	15	45	24	13	18.8	6.7	2.9	1.5
E	9	33	15	7	12.1	9.0	3.3	1.8
F	38	58	40	26	5.1	5.4	3.3	2.8
G	20	37	16	7	7.3	7.6	2.4	0.7

The result of HR measured rate and root mean squared error (RMSE) is shown in Fig. 13. The result is an average for seven participants and the signal averaging is for 4 beats. The HR measured rate is defined as a ratio of the number of beats measured from face images to the number of referential ECG beats while face is detected. The rate can be controlled by changing the threshold of the confidence indicator with autocorrelation.

Smaller RMSE are found at lower HR measured rate, and there is a trade-off relationship. Our method achieves both higher measured rate of HR and higher accuracy than filtered green method. HR measured rate = 44% at the confidence indicator = 0.6 with RMSE = 6.7 bpm are obtained as the mean result of seven participants for five days.

The results of each participant are shown in Table 1. The RPEM result shows 1.5–6.7 times higher HR measured rate with almost equal or higher accuracy than filtered green in 4 averaging beats. By increasing the averaging beats from 4 to 8 or 15, the RMSE improves although HR measured rate decreases.

6   Conclusion

We propose a real-time pulse extraction method for continuous heart rate monitoring from face images. The investigation of fundamental mechanisms experimentally revealed that the main cause of the brightness change of the face image is both the light absorption variation due to the blood volume changes and the face surface reflection generated by pulsatory movements.

Our method enables to extract the differences between red and green absorption derived from oxy-haemoglobin absorption characteristics by cancelling the effect of head movement. The comparison of RPEM with ear PPG under motion ensured the effectiveness of RPEM. We also applied RPEM to HR monitoring in office under non-controlled condition. The HR trend obtained by RPEM is in agreement with the reference ECG result. Our method achieves HR measured rate = 44% with RMSE = 6.7 bpm even in 4 averaging beats measurement. These results indicate that RPEM enables HR monitoring in daily life with high accuracy without losing much data even under non-controlled conditions.

References

1. Pantelopoulos, A., Bourbakis, N.G.: A survey on wearable sensor-based systems for health monitoring and prognosis. IEEE Trans. Syst., Man, Cybern. Part C Appl. Rev. **40**(1), 1–12 (2010)
2. Inomata, A., Yaginuma, Y.: Hassle-free sensing technologies for monitoring daily health changes. Fujitsu Sci. Tech. J. **50**(1), 78–83 (2014)
3. Uchida, D., Nakata, Y., Inomata, A., Shiotsu, S., Yaginuma, Y.: Hassle-free sensing technologies for human health monitoring. In: The IEICE General Conference/The Institute of Electronics, Information and Communication Engineers, S-16 (2015)
4. Reginatto, B., Taylor, K., Patterson, M., Caulfield, B.: Context aware falls risk assessment: a case study comparison. In: The Annual International Conference of the IEEE Engineering in Medicine and Biology Society (EMBC), pp. 5477–5480 (2015)
5. Dyer, A.R., Persky, V., Stamler, J., Paul, O., Shekelle, R.B., Berkson, D.M., Lepper, M., Schoenberger, J.A., Lindberg, H.A.: Heart rate as a prognostic factor for coronary heart disease and mortality: findings in three Chicago epidemiologic studies. Am. J. Epidemio **112**, 736–749 (1980)
6. Jensen, M.T., Suadicani, P., Hein, H.O., Gyntelberg, F.: Elevated resting heart rate, physical fitness and all-cause mortality: a 16-year follow-up in the Copenhagen Male Study. Heart **99**(12), 882–887 (2013)
7. Broeders, J.H., Conchell, J.C.: Wearable electronic devices monitor vital signs, activity level, and more: health monitoring is going wearable. Analog Dialogue **41**(12), 1–6 (2014)
8. Scalise, L.: Non contact heart monitoring. In: TechOpen, p. 84 (2012). www.intechopen.com
9. Yang, C., Cheung, G., Stankovic, V.: Estimating heart rate via depth video motion tracking. In: The Annual International Conference of the IEEE Multimedia and Expo (ICME), pp. 1–6 (2015)
10. Costa, D.: Optical remote sensing of heartbeats. Opt. Commun. **117**(5–6), 395–398 (1995)
11. Parra, J.E., Costa, G.: Optical remote sensing of heartbeats. Proc. SPIE – Int. Soc. Opt. Eng. **4368**, 113–121 (2001)

12. Tanaka, S., Matsumoto, Y., Wakimoto, K.: Unconstrained and non-invasive measurement of heart-beat and respiration periods using a phonocardiographic sensor. Med. Biol. Eng. Comput. **40**, 246–252 (2001)
13. Takano, C., Ohta, Y.: Heart rate measurement based on a time-lapse image. Med. Eng. Phys. **29**, 853–857 (2007)
14. Poh, M.Z., McDuff, D.J., Picard, R.W.: Non-contact, automated cardiac pulse measurements using video imaging and blind source separation. Opt. Express **18**(10), 10762–10774 (2010)
15. Poh, M.Z., McDuff, D.J., Picard, R.W.: Advancements in noncontact, multiparameter physiological measurements using a webcam. IEEE Trans. Biomed. Eng. **58**(1), 7–11 (2011)
16. Kwon, S., Kim, H., Park, S.: Validation of heart rate extraction using video imaging on a built-in camera system of a smartphone. In: The Annual International Conference of the IEEE Engineering in Medicine and Biology Society (EMBC), pp. 2174–2177 (2012)
17. Balakrishnan, G., Durand, F., Guttag, J.: Detecting pulse from head motions in video. In: The IEEE Conference on Computer Vision and Pattern Recognition (CVPR), pp. 3430–3437 (2013)
18. Li, X., Chen, J., Zhao, G., Pietikainen, M.: Remote heart rate measurement from face videos under realistic situations. In: The IEEE Conference on Computer Vision and Pattern Recognition (CVPR), pp. 4321–4328 (2014)
19. Sakata, M., Uchida, D., Inomata, A., Yaginuma, Y.: Continuous non-contact heart rate measurement using face imaging. In: The IEICE General Conference/The Institute of Electronics, Information and Communication Engineers, vol. 1, p. 73 (2013)
20. Steknke, J.M., Shephered, A.P.: Effects of temperature on optical absorbance spectra of oxy-, carboxy-, and deoxyhemoglobin. Clin. Chem. **38**(7), 1360–1364 (1992)
21. Forrester, K.R., Tulip, J., Leonard, C., Stewart, C., Bray, R.C.: A laser speckle imaging technique for measuring tissue perfusion. IEEE Trans. Biomed. Eng. **51**(11), 2074–2084 (2004)
22. Omegawave, Inc. http://www.omegawave.co.jp/en/products/oz/

Algorithm for Temporal Gait Analysis Using Wireless Foot-Mounted Accelerometers

Mohamed Boutaayamou[1,2]([✉]), Vincent Denoël[1], Olivier Brüls[1],
Marie Demonceau[3], Didier Maquet[3], Bénédicte Forthomme[1],
Jean-Louis Croisier[1], Cédric Schwartz[1], Jacques G. Verly[2],
and Gaëtan Garraux[4,5]

[1] Laboratory of Human Motion Analysis, University of Liège (ULg), Liège, Belgium
mboutaayamou@ulg.ac.be
[2] INTELSIG Laboratory, Department of Electrical Engineering
and Computer Science, ULg, Liège, Belgium
[3] Department of Rehabilitation and Movement Sciences, ULg, Liège, Belgium
[4] Movere Group, Cyclotron Research Center, ULg, Liège, Belgium
[5] Department of Neurology, University Hospital Center, Liège, Belgium

Abstract. We present a new signal processing algorithm that extracts
five gait events: heel strike, toe strike, heel-off, toe-off, and heel clearance from only two accelerometers attached on the heels of the subjects
usual shoes. This algorithm first uses a continuous wavelet-based segmentation that parses the signal of consecutive strides into motionless
periods defining relevant local acceleration signals. Then, the algorithm
uses versatile techniques to accurately extract the five gait events from
these local acceleration signals. We validated, on a stride-by-stride basis,
the extraction of these gait events by comparing the results with reference data provided by a kinematic 3D analysis system and a video camera. The accuracy and precision achieved by the extraction algorithm for
healthy subjects, the reduced number of accelerometer units required,
and the validation results obtained, encourage us to further study this
system in pathological conditions.

Keywords: Gait analysis · Wearable accelerometers · Wavelet analysis ·
Validation · Gait segmentation · Gait events · Heel-off · Heel strike · Toe
strike · Toe-off · Heel clearance · Stance time · Swing time · Stride time

1 Introduction

In our aging society, gait disturbance becomes a major concern as it leads to loss
of autonomy and risk of falls. Self-paced walking speed decreases by about 1 to
2% per year from age 60 onwards [1]. A population-based study showed a 35%
prevalence of gait disorders among persons over 70 years [2], and some abnormal
gait parameters can be predictive of the dementia of Alzheimer type, e.g., [3,4].
Gait and balance impairments are associated with falls and loss of mobility,
which both markedly impair the quality of life, e.g., [5]. Thus, it is important to

© Springer International Publishing AG 2017
A. Fred and H. Gamboa (Eds.): BIOSTEC 2016, CCIS 690, pp. 236–254, 2017.
DOI: 10.1007/978-3-319-54717-6_14

develop quantitative methods aimed at monitoring gait disturbances in a natural setting, and to provide the clinical relevance of the extracted features for clinical practitioners and specialized medical care units.

Wearable inertial systems have been proposed to measure gait events and to estimate temporal gait parameters (e.g., [6–10]). Compared to conventional gait analysis techniques, such as optoelectronic motion capture systems and instrumented walkways, these systems are not limited to a controlled laboratory environment; they can handle gait analysis in an ecological environment (e.g., at home, outdoor, or in the medical cabinet) with the possibility to obtain gait parameters over longer walking distances (e.g., [11]). The hardware part of inertial systems, such as accelerometer units, includes low-cost, small, and lightweight sensing units with generally low power consumption, an attractive feature for monitoring over long periods (e.g., [12]). With an appropriate algorithm, these inertial systems are particularly suitable for assessing gait in clinical populations (e.g., [13,14]).

Yet, existing systems often need many sensing units to achieve reasonable accuracy and precision in the extraction of gait events/parameters. furthermore, the challenge for daily clinical use is to find an optimal arrangement of these sensors that minimizes (1) the time needed to attach them on lower limbs and (2) the interference with the movement of the subjects during gait tests.

At the level of the gait event/parameter extraction algorithm in wearable inertial systems, the segmentation of the recorded signal according to gait cycles is of importance since it can provide a stride-by-stride partition of this signal during the gait analysis of normal and pathological subjects. Moreover, once the gait cycles have been identified, a more detailed examination of the gait signals can be performed in order to extract relevant gait parameters that could quantify the gait performance. Some of these extracted gait parameters are indeed known to be of particular interest in, e.g., the development of rehabilitative systems such as functional electrical stimulation during walking (e.g., [6,15]), or remote monitoring of daily activities of elderly people (e.g., [13,16]).

Several techniques have been tested for the segmentation of the recorded signal according to gait cycles using inertial sensor data, such as

- peak detection methods using accelerometer data, e.g., [12], or gyroscope data, e.g., [17],
- sequential stride phase classification algorithms using accelerometer data, e.g., [18], or gyroscope data, e.g., [19],
- hidden Markov models using gyroscope data, e.g., [20], or a combination of accelerometer data and gyroscope data, e.g., [21],
- template-based cross-correlation methods using accelerometer data, e.g., [22], or gyroscope data, e.g., [23],
- dynamic time warping methods using a combination of accelerometer data and gyroscope data, e.g., [24],
- wavelet-based methods using accelerometer data, e.g., [11,25], or a combination of accelerometer data and gyroscope data, e.g., [26].

However, most of these techniques do not allow quantitative measurements of the durations of the stance sub-phases, such as the loading response, mid-stance, and push-off sub-phases. In the context of the gait analysis solely using accelerometer systems, our research deals with the accelerometer-based analysis of normal gait and pathological gait [27,28].

We have previously developed and validated a signal processing algorithm to automatically extract four fundamental gait events in healthy walking–labelled as heel strike (HS), toe strike (TS), heel-off (HO), and toe-off (TO)–from seg-mented accelerometer signals measured by four accelerometer units attached to heels and toes [28]. The algorithm exploited distinctive and remarkable features in these acceleration signals to identify and extract gait events with good accuracy and precision.

In this paper, we present a new signal processing algorithm that extracts the gait events, HS, TS, HO, and TO from only two accelerometers, i.e., one on each shoe, at the level of the heel. Our approach to the aforementioned problem consists therefore in reducing the number of accelerometer units by eliminating the two units on the toes. This algorithm uses a new segmentation method based on the continuous wavelet transform (CWT) that identifies gait patterns by isolating (1) time intervals where the heel acceleration signals are close to zero, from (2) time intervals in which accelerometers (and the heels) are moving. We also extend the previous algorithm to detect a fifth gait event, i.e., the time of heel clearance (HC) which is an important gait event that can refine the swing phase. In addition, we consider the validation on a stride-by-stride basis of the proposed algorithm in a group of healthy people during normal walking. To validate our approach, we compared the results (i.e., measured gait events and calculated temporal gait parameters of interest) to reference data provided by a kinematic 3D system (used as gold standard) and a video camera.

2 Method

2.1 Wearable Accelerometer System

Acceleration signals during walking were recorded by a wireless, wearable, accelerometer-based hardware system which includes small three-axis accelerometer units (2 cm × 1 cm × 0.5 cm), a transmitter module, and a receiver module [12,28] (Fig. 1). This system can measure accelerations up to ±12 g (where $g = 9.81$ m/s^2 is the value of the gravitational acceleration) along its three sensitive axes: x (horizontal), y (transverse), and z (vertical). In this study, two accelerometer units were tightly attached on the right and left feet, i.e., one on each shoe at the level of the heel. The recordings from the right and left accelerometers were synchronized in time. Accelerometers were connected to the transmitter module positioned on the waist. The wires between accelerometers and the transmitter module were tightly strapped around the legs so as to avoid disturbing the subject movements. Acceleration signals were recorded at 200 Hz. All data were analyzed using Matlab 7.6.0 (MathWorks, Natick, MA, USA).

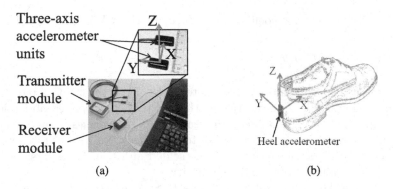

Fig. 1. (a) The wearable accelerometer-based hardware system. (b) Schematic illustration of the placement of a wearable accelerometer (either for right or left foot) and the direction of axes.

2.2 Subjects and Gait Tests Procedure

Gait signals were recorded during walking tests performed by seven young and healthy subjects without any previous lower limbs injury ((mean ± standard deviation) age = 27 ± 2.6 years; height = 181 ± 7 cm; weight = 78 ± 9 kg). All of them provided informed consent. The gait tests procedure was similar to the one reported in [28]. Before we started the measurements, subjects took sufficient time to get used to the instrumentation tools and the experimental procedure. During the tests, they were asked to walk on a 12-m long track, at their preferred, self-selected, usual speed. Each subject performed several gait tests of 60 s duration each. Subjects wore their own regular shoes. All of the walking tests were performed in the Laboratory of Human Motion Analysis, University of Liège, Belgium.

2.3 Wavelet Analysis: Segmentation of Acceleration Signals

We use a segmentation method (i.e., the partition of the considered acceleration signals into separate gait cycles) to identify gait patterns from only heel acceleration signals; thereby reducing the number of wearable accelerometers and allowing for a robust extraction of the gait events/parameters (see Sect. 2.4). This segmentation method is based on the continuous wavelet transform (CWT) to isolate (1) time intervals where the heel acceleration is close to zero (heel flat phases), from (2) time intervals the accelerometers are moving (heel non-flat phases).

The acceleration signals involved in the segmentation method are the x-axis (horizontal) acceleration, y-axis (transverse) acceleration, and z-axis (vertical) acceleration. These acceleration signals are denoted by $\ddot{x}$, $\ddot{y}$, and $\ddot{z}$ respectively. For clarity, we consider only one foot. It is obvious that the segmentation method could be applied in the same way for both feet.

The matrix of wavelet coefficients $C_s(a, b)$ of the CWT of a signal $s(t)$ is defined as

$$C(a, b) = \frac{1}{\sqrt{|a|}} \int_{-\infty}^{+\infty} s(t)\psi^*\left(\frac{t-b}{a}\right) dt, \qquad (1)$$

where $a(a \neq 0, a \in \mathbb{R})$ is the scale parameter, $b(b \in \mathbb{R})$ is the location parameter, ψ^* is the complex conjugate of the mother wavelet function ψ, and t is the time. Compressing a (small values of a) tracks high frequencies changes whereas stretching a (large values of a) tracks low frequencies. $C_s(a, b)$ thus measures the similarity between the signal $s(t)$ and the scaled and translated versions of ψ, with larger values indicating higher similarity.

In order to isolate regions of interest in $s(t)$, we consider an input signal $s_p(t)$ in (1) that preserves the positive values of $s(t) - mean(s(t))$ and sets all others to zero, i.e., $s_p(t) = s(t) - mean(s(t))$ if $s(t) \geq mean(s(t))$ and $s_p(t) = 0$ if $s(t) < mean(s(t))$, where $mean(s(t))$ is the mean of $s(t)$. We denote by MC_p the vector that contains the maximum value in each column of the CWT matrix associated with $s_p(t)$. Each value in MC_p thus corresponds to the best similarity between $s_p(t)$ and ψ at a given scale. We choose here the "Mexican hat" wavelet as the mother wavelet ψ, and we use the integer values 1 to 32 for the scale parameter a.

We apply the CWT separately to each of the following measured signals (as described for the input signal $s_p(t)$ in (1)):

- acceleration signals $\ddot{x}$, $\ddot{y}$, and $\ddot{z}$,
- the magnitude $a_n = \sqrt{\ddot{x}^2 + \ddot{y}^2 + \ddot{z}^2}$,
- the signal defined by $a_g = -\max(\ddot{x}, \ddot{z})$.

From this, we obtain the vector MC_p for each of these signals denoted as $MC_{x,1}$, $MC_{y,1}$, $MC_{z,1}$, $MC_{n,1}$, and $MC_{g,1}$, respectively. An example of the application of the CWT to $\ddot{z}$ is provided in Fig. 2.

We apply the CWT to each of the vectors $MC_{x,1}$, $MC_{y,1}$, $MC_{z,1}$, $MC_{n,1}$, and $MC_{g,1}$ in order to obtain the corresponding vectors $MC_{x,2}$, $MC_{y,2}$, $MC_{z,2}$, $MC_{n,2}$, and $MC_{g,2}$, respectively.

We finally apply the CWT to each of the vectors $MC_{x,2}$, $MC_{y,2}$, $MC_{z,2}$, $MC_{n,2}$, and $MC_{g,2}$ in order to determine the vectors $MC_{x,3}$, $MC_{y,3}$, $MC_{z,3}$, $MC_{(n,3}$, and $MC_{g,3}$, respectively. The summation of these vectors gives the rough envelope (RE) (see an example in Fig. 3(a)) of the 3D acceleration signals $\ddot{x}$, $\ddot{y}$, and $\ddot{z}$, as

$$RE = MC_{x,3} + MC_{y,3} + MC_{z,3} + MC_{n,3} + MC_{g,3}. \qquad (2)$$

The CWT is successively applied in order to determine a best partition of the acceleration signals by, e.g., avoiding intervals with a small size and zero values in the heel non-flat phases.

We then calculate the associated binary function of RE by assigning 0 and 1 to the zero and nonzero elements of the RE, respectively. This binary function is further refined using a function that assigns 1 to intervals with zero values

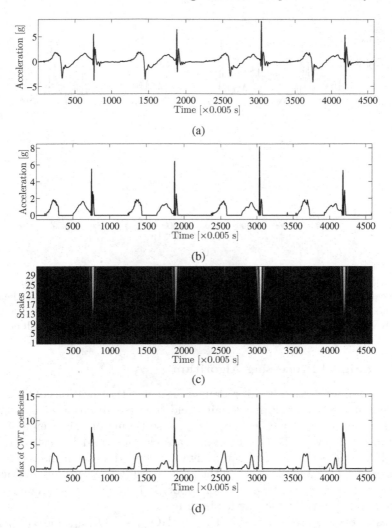

Fig. 2. (a) Example of the z-axis heel acceleration $\ddot{z}$; (b) the corresponding signal $\ddot{z}_p$ that preserves the positive values of $\ddot{z} - mean(\ddot{z})$ and sets all others to zero (i.e., $\ddot{z}_p = \ddot{z} - mean(\ddot{z})$ if $\ddot{z} \geq mean(\ddot{z})$, and $\ddot{z}_p = 0$ if $\ddot{z} < mean(\ddot{z})$); (c) the CWT of $\ddot{z}_p$; and (d) the rough envelope of $\ddot{z}_p$ (i.e., $MC_{z,1}$) defined as the vector that contains the maximum value in each column of the CWT matrix obtained from $\ddot{z}_p$.

in the heel non-flat phases (a typical size of such intervals ranges from 30 ms to 200 ms.) The obtained binary function corresponds to our segmentation that roughly identifies motionless periods in the acceleration signals (Fig. 3(b)). The segmentation method has the advantage that it avoids to look directly for specific gait events. The segmentation only determine rough heel flat/non-flat phases in which gait events of interest can be further extracted with good accuracy.

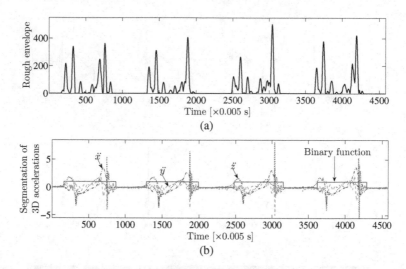

Fig. 3. (a) Example of the total rough envelope obtained by applying successive CWTs to the accelerations $\ddot{x}$, $\ddot{y}$, and $\ddot{z}$ to the magnitude $a_n = \sqrt{\ddot{x}^2 + \ddot{y}^2 + \ddot{z}^2}$, and to the signal $a_g = -max(\ddot{x}, \ddot{z})$. (b) The obtained segmentation of the accelerations; this segmentation is given by calculating the associated binary function of this total rough envelope.

2.4   New Signal Processing Algorithm

In order to estimate precisely gait parameters such as the durations of the stance, swing, and stride phases during a gait cycle (i.e., the duration of a stride), it is necessary to detect, for each foot, the precise timing of gait events of interest within a gait cycle. These gait events are characterized by distinctive and remarkable features in heel acceleration signals. Depending on the nature of these features, a suitable method is used in the present study to accurately extract gait events. For clarity, we consider only one foot. It is obvious that the algorithm could be applied in the same way for both feet.

Times of occurrence of HS_{accel}, TS_{accel}, HO_{accel}, TO_{accel}, and HC_{accel} are identified mainly from the acceleration signals in sagittal plane, i.e., with respect to the x-axis and z-axis accelerations denoted by $\ddot{x}_h$ and $\ddot{z}_h$, respectively. The subscripts $accel$, ref, h, and t refer to our method, to the reference methods, to the heel, and to the toe, respectively.

Gait Events Identification. We now describe the main steps of the detection following the chronological occurrence order of healthy gait events (i.e., not the order that the algorithm follows to extract these events).

(a) The time of the heel strike event: HS_{accel}

In the present study, we adapt the method described in [28] to detect HS_{accel} as follows:

- At HS_{accel}, the heel acceleration signal $\ddot{z}_h$ is subject to abrupt changes (Fig. 4a). To detect HS_{accel}, we only consider the segment defined as the second half of the heel non-flat phase. In this segment, HS_{accel} is identified using the magnitude of $\ddot{z}_h$ filtered with a 4th-order zero-lag Butterworth high-pass filter (cutoff frequency $= 10$ Hz). HS_{accel} is detected as the time of occurrence of the maximum value of the magnitude of this filtered $\ddot{z}_h$ (Fig. 4a). As pointed out in [28], the determination of HS_{accel} is robust with respect to this filtering step, since HS_{accel} occurs rapidly with a frequency larger than 10 Hz.

(b) The time of the toe strike event: TS_{accel}

TS_{accel} can be extracted from the heel acceleration signal as the accelerometer is sensitive enough to measure the acceleration movement of the foot when the toe hits the ground. The main steps to estimate TS_{accel} are as follows:

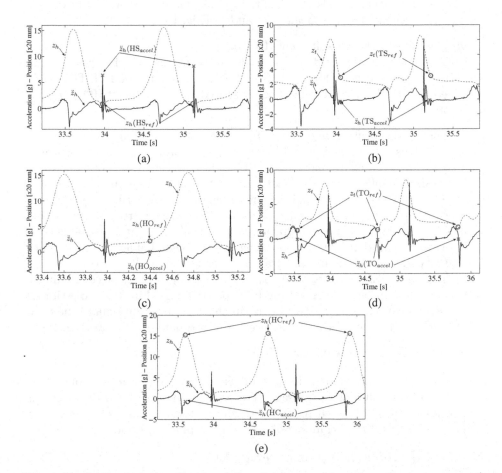

Fig. 4. Vertical heel acceleration signal (i.e., $\ddot{z}_h$ measured by our accelerometer system) and reference kinematic signals (i.e., the vertical heel position z_h and the vertical toe position z_t measured by the Codamotion system). The gait events, i.e., (a) HS, (b) TS, (c) HO, (d) TO, and (e) HC, detected by our method and by reference methods are shown on each signal during typical consecutive strides.

- As TS_{accel} occurs after HS_{accel} and before HO_{accel}, we seek TS_{accel} in the segment $[HS_{accel}, 0.4 \times HS_{accel} + 0.6 \times HO_{accel}]$ (the procedure for extracting HO_{accel} is explained in (c)). TS_{accel} is automatically detected using $\ddot{x}_h$ and $\ddot{z}_h$ restricted to this segment. The resulting local signals are then filtered with a 4th-order zero-lag Butterworth low-pass filter (cutoff frequency $= 20\,Hz$), and integrated twice in order to calculate their associated position signals. The drift related to this double integration is limited since the latter is performed in a small time interval. We then apply a piecewise-linear fitting method to each of these position signals. This method estimates a location of convex curvature in a signal using two linear segments that best fit this signal in the least-square sense (Appendix) [28]. The times of resulting convex curvatures in the two position signals are denoted t_1 and t_2. It is then assumed that TS_{accel} is estimated as the mean of t_1 and t_2 (Fig. 4b).

(c) The time of the heel-off event: HO_{accel}

HO_{accel} is automatically detected in the segment that lies between $125\,ms$ after HS_{accel} and $70\,ms$ before TO_{accel} (the extraction method of TO_{accel} is described in (d)). We adapt the method presented in [28] to detect HO_{accel} from $\ddot{z}_h$ as follows:

- We consider the local signal obtained from the restriction of $\ddot{z}_h$ to the previous segment. This local signal is then filtered with a 4th-order zero-lag Butterworth low-pass filter (cutoff frequency $= 20\,Hz$). This filtering step does not alter the physical significance of the local signal [28]. Since this signal corresponds to a slow movement (some milliseconds before and after HO_{accel}), there is no critical peak to be detected that could be removed erroneously in this filtering step [28]. A double integration of this local acceleration signal is then performed to calculate the corresponding position signal. The drift that could be generated from this double integration is negligible since the latter is carried out in a small time interval. We apply the aforementioned piecewise-linear fitting method twice to the resulting local position signal in order to estimate successive locations of convex curvature in this local position signal. The time of the last location of convex curvature is our estimate of HO_{accel} (Fig. 4c).

(d) The time of the toe-off event: TO_{accel}

- At TO_{accel}, the direction of motion of the ankle joint changes from plantarflexion to dorsiflexion in the sagittal plane [29]. It is assumed that TO_{accel} corresponds to the time when a zero crossing of the vertical heel acceleration signal occurs after the beginning of the non-flat phase (Fig. 4d).

(e) The time of the heel clearance event: HC_{accel}

- HC_{accel} is defined as the moment when the maximum clearance between the heel accelerometer and the ground is achieved during the swing phase. We consider distinctive vertical heel acceleration features that indicate where HC_{accel} can be found in the time and frequency domains. These features are rather sharp negative peaks in $\ddot{z}_h$ (Fig. 4e) involving some mid frequencies. In order to extract HC_{accel}, we apply the CWT (see Sect. 2.3) to the local signal defined as the restriction of $\ddot{z}_h$ to the neighbourhood of these features. The CWT is indeed adapted for identifying HC_{accel} because it allows detection of a specified frequency at a specified time. The previous local signal is then decomposed into wavelet packages. The wavelet "Mexican hat" is used as the mother wavelet to extract HC_{accel} as it is similar to the pattern of the aforementioned features. A typical result is depicted in Fig. 4e.

Extraction of Temporal Gait Parameters. Temporal gait parameters, such as durations of the stance, swing, and stride phases, are calculated on the basis of the previous gait events as follows:

- Right stance duration (time between right HS (HS_{right}) and right TO (TO_{right}) during stride i)

$$Right\ stance = TO_{right}(i) - HS_{right}(i).$$

- Left stance duration (time between left HS (HS_{left}) and left TO (TO_{left}) during stride i)

$$Left\ stance = TO_{left}(i) - HS_{left}(i).$$

- Right swing duration (time between HS_{right} of stride $i+1$ and TO_{right} of stride i)

$$Right\ swing = HS_{right}(i+1) - TO_{right}(i).$$

- Left swing duration (time between HS_{left} of stride $i+1$ and TO_{left} of stride i)

$$Left\ swing = HS_{left}(i+1) - TO_{left}(i).$$

- Right stride duration (time between two consecutive right HSs)

$$Right\ stride = HS_{right}(i+1) - HS_{right}(i).$$

- Left stride duration (time between two consecutive left HSs)

$$Left\ stride = HS_{left}(i+1) - HS_{left}(i).$$

2.5   Stride-by-Stride Validation Method

Reference Data. A kinematic 3D analysis system (Codamotion system; Charnwood Dynamics; Rothley, UK) and a video camera provided reference data to validate, on a stride-by-stride basis, the gait parameters/events determined by our method.

The kinematic system is based on active optical technology; it can accurately measure the 3D positions of active markers placed in the body locations of interest. We collected kinematic data at the level of the heel and the toe of each foot at 400 Hz. The heel marker was placed upon the heel accelerometer.

The video camera (30 fps) was placed close to the track such that the pointing direction is approximately perpendicular to the sagittal plan.

Kinematic data were used to validate, on a stride-by-stride basis, the gait events HS_{accel}, TS_{accel}, TO_{accel}, and HC_{accel}. Reference gait events HS_{ref} and TO_{ref}, were obtained by the kinematic method reported in [30]. HS_{ref} and TO_{ref} were extracted solely from measured heel and toe coordinates during overground walking (Figs. 4a and d). TS_{ref} was extracted from the vertical toe position signal, z_t, in each gait cycle (Fig. 4b) [28]. HC_{ref} was detected as the time when the vertical heel position, z_h, reaches its maximal value after a toe-off (Fig. 4e). The video camera provided HO_{ref}.

Evaluation Method. We evaluated the level of agreement between our method and the reference methods by quantifying the accuracy, precision, absolute error, and intraclass correlation coefficient (ICC). Accuracy and precision were computed for each stride as the mean and standard deviation (std. dev.), respectively, of the differences: $HS_{accel}-HS_{ref}$, $TS_{accel}-TS_{ref}$, $HO_{accel}-HO_{ref}$, $TO_{accel}-TO_{ref}$, and $HC_{accel}-HC_{ref}$. The absolute error was calculated as the mean and std. dev. of absolute values of the previous differences. The ICC evaluates the statistical agreement between our method and the reference methods. A Bland-Altman analysis was also carried out.

3   Results

Table 1 provides a quantitative one-by-one comparison of gait events. Because of the limited number of extracted reference events and the variation in some reference patterns among subjects, the sample size for the compared gait events was not always the same but ranged between 126 and 839 strides. During some gait tests, we observed that some markers — used to record reference kinematic signals — had detached from the shoes. We therefore excluded the associated gait events from the analysis. In addition, we emphasize that HO_{ref} was obtained only by the video camera. The extraction of HO_{ref} is thus limited to one stride during a given gait test. The total number of HO_{ref} (here 126) is therefore much smaller than those of HS_{ref}, TS_{ref}, TO_{ref}, and HC_{ref}. The four last reference data were indeed extracted from consecutive strides.

Timing accuracy and precision of gait events detection ranged from -3.4 ms to 7.2 ms, and 15.7 ms to 27.4 ms, respectively. Given the sampling frequency of 200 Hz of the recorded heel accelerations for both feet, the accuracy and the precision of detection are less than durations of 2 frames (10 ms) and 6 frames (30 ms), respectively.

Figure 5 shows the Bland-Altman plots of gait events differences. We observe small systematic biases in accordance with the accuracy of detection provided in Table 1. The proposed method tends to detect earlier gait events except for HO. In addition, the limits of agreement (i.e., mean ± 1.96 std. dev.) and their associated 95% confidence interval exhibit small variations in the times of gait events (Table 1).

Table 1. The results of our method are compared to the results of reference methods considering several consecutive strides. This evaluation is given as the accuracy (mean of the differences), the precision (std. dev. of the differences), limits of agreement, 95% confidence interval (CI) of the differences, and 95% CI of the lower and upper limits of agreement.

	Accuracy (ms) (precision (ms))	Limits of agreements (ms)	95% CI of the differences (ms)	95% CI of the lower limits (ms)	95% CI of the upper limits (ms)	No. of strides
HS	7.2 (22.1)	[−36.2 50.7]	[5.6 8.8]	[−38.9 − 33.5]	[47.9 53.4]	771
TS	0.7 (19.0)	[−36.6 38.0]	[−0.9 2.3]	[−39.3 − 33.9]	[35.3 40.7]	567
HO	−3.4 (27.4)	[−57.2 50.3]	[−8.2 1.3]	[−66.5 − 49.9]	[43.1 59.7]	126
TO	2.2 (15.7)	[−28.6 33.0]	[1.1 3.3]	[−30.5 − 26.8]	[31.2 34.9]	819
HC	3.2 (17.9)	[−31.9 38.3]	[1.9 4.4]	[−34.7 − 30.5]	[36.9 41.1]	839

Table 2 shows the results of durations of stance, swing, and stride phases calculated by our method and by the reference method (i.e., provided by the Codamotion system) for the right and left feet. These temporal parameters could be estimated with a mean absolute error less than 15 ms. The ICC coefficient was larger than 0.95 for both stance time and stride time, and larger than 0.87 for swing time.

Figure 6 shows the Bland-Altman plots of the temporal parameters for the right and left feet. Most differences of these temporal parameters are within the 1.96 std. dev. lines.

Table 2. Results of right/left stance, swing, and stride phase durations calculated by our method are compared to those obtained by a reference kinematic system, Codamotion, used as gold standard. This comparison is given as the difference of the estimated values (mean error), the mean of the absolute error, and the intraclass correlation coefficient (ICC).

	Foot	Accelero-meters	Codamotion	Mean error	Absolute error	ICC	No. of strides
Stance time (s)	Right	0.670 ± 0.047	0.674 ± 0.055	−0.004 ± 0.016	0.012 ± 0.011	0.95	188
	Left	0.656 ± 0.052	0.662 ± 0.055	−0.006 ± 0.015	0.012 ± 0.010	0.95	220
	Right & left	0.662 ± 0.050	0.668 ± 0.055	−0.006 ± 0.015	0.012 ± 0.010	0.95	408
Swing time (s)	Right	0.404 ± 0.042	0.399 ± 0.035	0.005 ± 0.018	0.014 ± 0.012	0.89	336
	Left	0.418 ± 0.038	0.413 ± 0.035	0.005 ± 0.018	0.014 ± 0.011	0.87	383
	Right & left	0.412 ± 0.041	0.407 ± 0.035	0.005 ± 0.017	0.014 ± 0.011	0.88	719
Stride time (s)	Right	1.080 ± 0.092	1.083 ± 0.092	−0.003 ± 0.016	0.012 ± 0.011	0.98	181
	Left	1.081 ± 0.090	1.089 ± 0.098	−0.008 ± 0.018	0.015 ± 0.013	0.98	227
	Right & left	1.080 ± 0.090	1.087 ± 0.095	−0.006 ± 0.017	0.013 ± 0.012	0.98	408

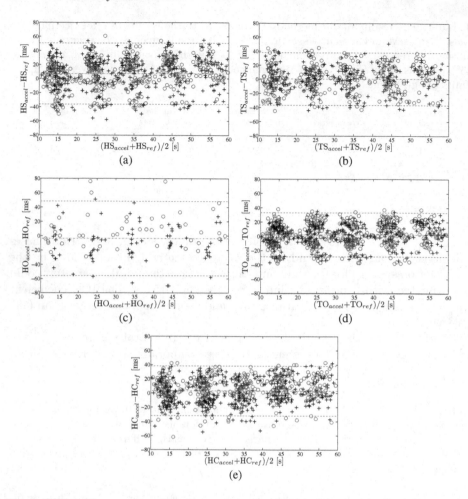

Fig. 5. Bland-Altman plots of the gait events, i.e., (a) HS, (b) TS, (c) HO, (d) TO, and (e) HC, measured using our method and reference methods, with mean (dash-dotted line in the middle) of differences $HS_{accel} - HS_{ref}$, $TS_{accel} - TS_{ref}$, $HO_{accel} - HO_{ref}$, $TO_{accel} - TO_{ref}$, and $HC_{accel} - HC_{ref}$. 95% of these differences are between the lines ± 1.96 std. dev. (dashed lines). (**o**) and (**+**) refer to gait events measured at the right foot and those measured at the left foot, respectively.

4   Discussion

We have presented a new signal processing algorithm that extracts relevant temporal gait parameters/events from only two accelerometers attached to the right and left feet, i.e., one on each shoe at the level of the heel.

The new algorithm is versatile enough to detect gait events. The algorithm is based on the CWT and an original piecewise-linear fitting method. Those methods allow for an automatic and robust extraction of gait events from

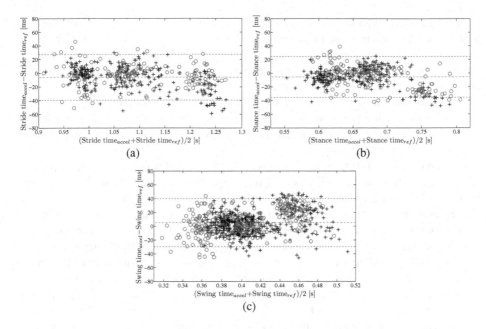

Fig. 6. Bland-Altman plots of the temporal gait parameters, i.e., (a) stance time, (b) swing time, and (c) stride time estimated during consecutive strides by our method and the gold standard method. (o) and (+) refer to right and left gait parameters, respectively.

relevant local acceleration signals. The algorithm was validated by comparing results obtained by our method to those obtained by a kinematic 3D system (used as gold standard) and a video camera. The experimental results show an excellent agreement between our algorithm and the reference, and demonstrate an accurate and precise detection of HS, TS, HO, TO, and HC in a group of healthy people during normal walking. In addition, the algorithm computes the time of stance, swing, and stride phases with a good accuracy and precision.

Previous studies reported results of gait parameters during the normal walk. Compared to stance time calculated in [8] (i.e., 15 ms ± 41 ms), in [31] (i.e., 9 ms ± 69 ms), and in [13] (i.e., 5.9 ms ± 29.6 ms), the accuracy is similar but the precision is improved in our method (i.e., 6 ms ± 15 ms). Similar accuracy and better precision in stride time are also found in our method (i.e., 6 ms ± 17 ms) compared to [31] (i.e., 2 ms ± 68 ms) and to [13] (i.e., 2.2 ms ± 23.2 ms). In addition, the accuracy of swing time in [31] (i.e., 8 ms ± 45 ms) is similar to our results but the precision is improved in our method (i.e., 5 ms ± 17 ms).

Compared to commercial trunk accelerometer systems (e.g., [32]), which only provide global gait features, our system is capable to extract stride-by-stride parameters. The stride-by-stride extraction may be a huge advantage in the gait analysis of some specific population such as Parkinson's disease patients who walk slowly than age-matched controls and experience freezing of gait, a sudden and brief episodic alteration of strides regulation.

Participants did not complain about the hardware system used during the gait tests. They all reported that wires and accelerometers did not interfere with their gait. Since only two accelerometers were attached to heels and wires were behind the legs of the participants during walking, these participants did not notice or complain about the system.

It is noteworthy that all accelerometers of the used hardware system were synchronized. The algorithm can thus extract other important gait parameters such as the times of initial double support, terminal double support, double support, and right/left steps.

Based on TS and HO, the algorithm can extract the durations of the sub-phases of the stance phase, namely: (1) loading response duration (time from HS of one foot to TS of the same foot); (2) foot-flat duration (time from TS of one foot to HO of the same foot); and (3) push-off duration (time from HO of one foot to TO of the same foot). In addition, HC can be used to refine the swing phase duration.

The algorithm is valid in case of a heel strike at initial contact during walking, but might be modified to be more flexible to take into account situations where the heel strike (or other events) is missing (i.e., flat foot) such as in running or in some pathological conditions. Moreover, we assumed a 125 ms and 70 ms criteria in the detection of the heel-off (Sect. 2.4). This criteria was adapted to our data and needs further to be adjusted in case of slower and/or faster walking speeds and in case of pathological gait patterns.

The proposed ambulatory accelerometer system was capable of measuring temporal gait parameters in a very large number of strides without the need of controlled laboratory conditions. We believe that our novel accelerometer-based system offers perspectives for use in a routine clinical practice to deal with abnormal gait (e.g., gait of patients with Parkinson's disease).

5   Conclusion

We presented a new signal processing algorithm that reduces the number of wearable accelerometers for estimating temporal gait parameters. The advantages of this method can be summarized as follows:

- Only two accelerometers are required, i.e., one for each shoe at the level of the heel. This contributes to a simplification of our wearable accelerometer-based system, thus resulting in reducing the costs and time needed to attach the system on body.
- The algorithm is validated for consecutive strides during normal walking. The validation used reference data provided by a kinematic system (used as gold standard) and a video camera.
- Compared to previous studies, the proposed method performs equally well or better in terms of accuracy and precision of detection of temporal gait parameters such as times of swing, stance, and stride phases.

The extension of this method to the study of pathological gait (e.g., gait of patients with Parkinsons disease) is now in progress. The method promises to allow an objective quantification of gait parameters in routine clinical practice.

Acknowledgements. The authors wish to acknowledge the contribution of J. Stamatakis and B. Macq through the design of the accelerometer-based hardware system used in the present study. The authors would like also to thank all the participants to the gait tests of this study.

Appendix

We present the piecewise-linear fitting method used to estimate the locations of the convex curvature in a signal (Sect. 2.4). For this, we consider a given signal $sig = sig(t_1), sig(t_2), \ldots, sig(t_N)$ defined in a time interval $I = t_1, t_2, \ldots, t_N$, where N is the total number of samples of sig. This method first computes the coefficients of piecewise-linear functions with two linear segments that best fit sig in the least-square sense, leading to the computation of least-square errors. The minimum of these least-square errors is then determined and the associated piecewise-linear function provides two linear segments that intersect at the breakpoint $(t_b, sig(t_b))$. The main steps to determine the breakpoint $(t_b, sig(t_b))$ are as follows:

- For each $k = 1, \ldots, N$, one computes the coefficients α_1, α_2, β_1, and β_2 of a piecewise-linear function p_k that best fits sig by minimizing

$$E_k = \sum_{i=1}^{N}(sig(t_i) - p_k(t_i))^2, \tag{3}$$

where

$$p_k(t) = \begin{cases} \alpha_1 * t + \beta_1, & \text{if } t \in [t_1, t_k], \\ \alpha_2 * t + \beta_2, & \text{if } t \in [t_{k+1}, t_N]. \end{cases} \tag{4}$$

This error can be expressed as

$$E_k = ||A\,X_k - B||^2, \tag{5}$$

where

$$X_k = \begin{pmatrix} \alpha_1 \\ \beta_1 \end{pmatrix} \quad A = \begin{pmatrix} t_1 & 1 \\ \vdots & \vdots \\ t_k & 1 \end{pmatrix} \quad B = \begin{pmatrix} sig(t_1) \\ \vdots \\ sig(t_k) \end{pmatrix} \quad if\, t \in [t_1, t_k],$$

and

$$X_k = \begin{pmatrix} \alpha_2 \\ \beta_2 \end{pmatrix} \quad A = \begin{pmatrix} t_{k+1} & 1 \\ \vdots & \vdots \\ t_N & 1 \end{pmatrix} \quad B = \begin{pmatrix} sig(t_{k+1}) \\ \vdots \\ sig(t_N) \end{pmatrix} \quad if\, t \in [t_{k+1}, t_N],$$

The normal equations associated with (5) are

$$A^t A X_k = A^t B. \tag{6}$$

Solving (6) leads to the coefficients α_1, α_2, β_1, and β_2.

- Finally, one obtains the breakpoint $(t_b, sig(t_b))$ by determining the minimum of the least-square errors, i.e.,

$$E_b = \min_{k=1,\ldots,N}(E_k). \tag{7}$$

References

1. Ashton-Miller, J.A.: Age-associated changes in the biomechanics of gait and gait-related falls in older adults. Neurol. Dis. Ther. **73**, 63–100 (2005)
2. Verghese, J., LeValley, A., Hall, C.B., Katz, M.J., Ambrose, A.F., Lipton, R.B.: Epidemiology of gait disorders in community-residing older adults. J. Am. Geriatr. Soc. **54**(2), 255–261 (2006)
3. Verghese, J., Lipton, R.B., Hall, C.B., Kuslansky, G., Katz, M.J., Buschke, H.: Abnormality of gait as a predictor of non-Alzheimer's dementia. New Engl. J. Med. **347**(22), 1761–1768 (2002)
4. Gillain, S., Warzee, E., Lekeu, F., Wojtasik, V., Maquet, D., Croisier, J.-L., Salmon, E., Petermans, J.: The value of instrumental gait analysis in elderly healthy, MCI or Alzheimer's disease subjects and a comparison with other clinical tests used in single and dual-task conditions. Ann. Phys. Rehabil. Med. **52**(6), 453–474 (2009)
5. Stolze, H., Klebe, S., Zechlin, C., Baecker, C., Friege, L., Deuschl, G.: Falls in frequent neurological diseases: prevalence, risk factors and aetiology. J. Neurol. **251**(1), 79–84 (2004)
6. Willemsen, A.T.M., Bloemhof, F., Boom, H.B.: Automatic stance-swing phase detection from accelerometer data for peroneal nerve stimulation. IEEE Trans. Biomed. Eng. **37**(12), 1–8 (1990)
7. Aminian, K., Rezakhanlou, K., Andres, E., Fritsch, C., Leyvraz, P.-F., Robert, P.: Temporal feature estimation during walking using miniature accelerometers: an analysis of gait improvement after hip arthroplasty. Med. Biol. Eng. Comput. **37**(6), 686–691 (1999)
8. Selles, R.W., Formanoy, M.A.G., Bussmann, J.B.J., Janssens, P.J., Stam, H.J.: Automated estimation of initial and terminal contact timing using accelerometers; development and validation in transtibial amputees and controls. IEEE Trans. Neural Syst. Rehabil. Eng. **13**(1), 81–88 (2005)
9. Lee, J.-.A., Cho, S.-H., Lee, J.-W., Lee, K.-H., Yang, H.-K.: Wearable accelerometer system for measuring the temporal parameters of gait. In: Proceedings of the 29th Annual International Conference of the IEEE EMBS, pp. 23–26 (2007)
10. Godfrey, A., Conway, R., Meagher, D., ÓLaighin, G.: Direct measurement of human movement by accelerometry. Med. Eng. Phys. **30**(10), 1364–1386 (2008)
11. Khandelwal, S., Wickström, N.: Identification of gait events using expert knowledge and continuous wavelet transform analysis. In: Proceedings of the International Conference on Bio-inspired Systems and Signal Processing, pp. 197–204 (2014)
12. Stamatakis, J., Crémers, J., Maquet, D., Macq, B., Garraux, G.: Gait feature extraction in Parkinson's disease using low-cost accelerometers. In: Proceedings of the Annual International Conference of the IEEE Engineering in Medicine and Biology Society, pp. 7900–7903 (2011)
13. Salarian, A., Russmann, H., Vingerhoets, F.J.G., Dehollain, C., Blanc, Y., Burkhard, P.R., Aminian, K.: Gait assessment in Parkinson's disease: toward an ambulatory system for long-term monitoring. IEEE Trans. Biomed. Eng. **51**(8), 1434–1443 (2004)

14. Rueterbories, J., Spaich, E.G., Larsen, B., Andersen, O.K.: Methods for gait event detection and analysis in ambulatory systems. Med. Eng. Phys. **32**(6), 545–552 (2010)
15. Pappas, I.P., Popovic, M.R., Keller, T., Dietz, V., Morari, M.: A reliable gait phase detection system. IEEE Trans. Neural Syst. Rehabil. Eng. **9**(2), 113–125 (2001)
16. Moore, S.T., MacDougall, H.G., Gracies, J.-M., Cohen, H.S., Ondo, W.G.: Long-term monitoring of gait in Parkinson's disease. Gait Posture **26**(2), 200–207 (2007)
17. Aminian, K., Najafi, B., Büla, C., Leyvraz, P.F., Robert, P.: Spatio-temporal parameters of gait measured by an ambulatory system using miniature gyroscopes. J. Biomech. **35**, 689–699 (2002)
18. Han, J., Jeon, H.S., Jeon, B.S., Park, K.S.: Gait detection from three dimensional acceleration signals of ankles for the patients with Parkinson's disease. In: Proceedings of the IEEE International Special Topic Conference on Information Technology in Biomedicine, vol. 2628 (2006)
19. Sabatini, A.M., Martelloni, C., Scapellato, S., Cavallo, F.: Assessment of walking features from foot inertial sensing. IEEE Trans. Biomed. Eng. **52**, 486–494 (2005)
20. Mannini, A., Sabatini, A.M.: A hidden Markov model-based technique for gait segmentation using a foot-mounted gyroscope. In: Proceedings of the IEEE Annual International Conference of the Engineering in Medicine and Biology Society, pp. 4369–4373 (2011)
21. Guenterberg, E., Yang, A.Y., Ghasemzadeh, H., Jafari, R., Bajcsy, R., Sastry, S.S.: A method for extracting temporal parameters based on hidden Markov models in body sensor networks with inertial sensors. IEEE Trans. Inf Technol. Biomed. **13**, 1019–1030 (2010)
22. Ying, H., Silex, C., Schnitzer, A., Leonhardt, S., Schiek, M.: Automatic step detection in the accelerometer signal. In: Proceedings of the 4th International Workshop on Wearable and Implantable Body Sensor Networks, pp. 80–85 (2007)
23. Brauner, T., Oriwol, D., Sterzing, T., Milani, T.L.: A single gyrometer inside an instrumented running shoe allows mobile determination of gait cycle and pronation velocity during outdoor running. Footwear Sci. **1**, 25–26 (2009)
24. Barth, J., Oberndorfer, C., Pasluosta, C., Schülein, S., Gassner, H., Reinfelder, S., Kugler, P., Schuldhaus, D., Winkler, J., Klucken, J., Eskofier, B.M.: Stride segmentation during free walk movements using multi-dimensional subsequence dynamic time warping on inertial sensor data. Sensors **15**, 6419–6440 (2015)
25. Sekine, M., Tamura, T., Akay, M., Fujimoto, T., Togawa, T., Fukui, Y.: Discrimination of walking patterns using wavelet-based fractal analysis. IEEE Trans. Neural Syst. Rehabil. Eng. **10**(3), 188–196 (2002)
26. Yuwono, M., Su, S.W., Moulton, B.D., Nguyen, H.T.: Unsupervised segmentation of heel-strike IMU data using rapid cluster estimation of wavelet features. In: Proceedings of the 35th Annual International Conference of the IEEE Engineering in Medicine and Biology Society, pp. 953–956 (2013)
27. Boutaayamou, M., Schwartz, C., Stamatakis, J., Denoël, V., Maquet, D., Forthomme, B., Croisier, J.-L., Macq, B., Verly, J.G., Garraux, G., Brüls, O.: Validated extraction of gait events from 3D accelerometer recordings. In: The International Conference on 3D Imaging, pp. 1–4 (2012)
28. Boutaayamou, M., Schwartz, C., Stamatakis, J., Denoël, V., Maquet, D., Forthomme, B., Croisier, J.-L., Macq, B., Verly, J.G., Garraux, G., Brüls, O.: Development and validation of an accelerometer-based method for quantifying gait events. Med. Eng. Phys. **37**, 226–232 (2015)
29. Whittle, W.: Clinical gait analysis: a review. Hum. Mov. Sci. **15**, 369–387 (1996)

30. Boutaayamou, M., Schwartz, C., Denoël, V., Forthomme, B., Croisier, J.-L., Garraux, G., Verly, J.G., Brüls, O.: Development and validation of a 3D kinematic-based method for determining gait events during overground walking. In: The International Conference on 3D Imaging, pp. 1–6 (2014)
31. Rampp, A., Barth, J., Schülein, S., Gamann, K.-G., Klucken, J., Eskofier, B.M.: Inertial sensor-based stride parameter calculation from gait sequences in geriatric patients. IEEE Trans. Biomed. Eng. **62**(4), 1089–1097 (2015)
32. Auvinet, B., Chaleil, D., Barrey, E.: Analyse de la marche humaine dans la pratique hospitalière par une méthode accélérométrique. Revue du Rhumatisme **66**(7–9), 447–457 (1999)

Online Simulation of Mechatronic Neural Interface Systems: Two Case-Studies

Samuel Bustamante[1]([✉]), Juan C. Yepes[2], Vera Z. Pérez[3], Julio C. Correa[4], and Manuel J. Betancur[5]

[1] Grupo de Automática y Diseño A+D,
Universidad Pontificia Bolivariana, Cir. 1 #73-76, B22, Medellín, Colombia
samuel.bustamante@upb.edu.co
[2] Grupo de Investigaciones en Bioingeniería, Grupo de Automática y Diseío A+D,
Universidad Pontificia Bolivariana, Cir. 1 #73-76, B22, Medellín, Colombia
juancamilo.yepes@upb.edu.co
[3] Facultad de Ingeniería Eléctrica y Electrónica,
Grupo de Investigaciones en Bioingeniería, Universidad Pontificia Bolivariana,
Cir. 1 #73-76, B22, Medellín, Colombia
vera.perez@upb.edu.co
[4] Facultad de Ingeniería Mecánica, Grupo de Automática y Diseño A+D,
Universidad Pontificia Bolivariana, Cir. 1 #73-76, B22, Medellín, Colombia
julio.correa@upb.edu.co
[5] Facultad de Ingeniería Eléctrica y Electrónica,
Grupo de Automática y Diseño A+D, Universidad Pontificia Bolivariana,
Cir. 1 #70-01, Of. 11-259, Medellín, Colombia
manuel.betancur@upb.edu.co

Abstract. Neural interface systems (NIS) are widely used in rehabilitation and prosthetics. These systems usually involve robots, such as robotic exoskeletons or mechatronic arms, as terminal devices. We propose a methodology to assess the feasibility of implementing these kind of neural interfaces by means of an online kinematic simulation of the robot. It allows the researcher or developer to make tests and improve the design of the mechatronic devices when they have not been built yet or are not available. Moreover, it may be used in biofeedback applications for rehabilitation. The simulation makes use of the CAD model of the robot, its Denavit-Hartenberg parameters, and biosignals recorded from a human being. The proposed methodology was tested using surface electromyography (sEMG) signals from the upper limb of a 25-year-old subject to control a kinematic simulation of a KUKA KR6 robot.

It was also used in the design process of an actual lower limb rehabilitation system being developed in our laboratories. The 3D computational simulation of this robot was successfully controlled by means of sEMG signals acquired from the lower limb of a 26-year-old healthy subject. Both real-time and prerecorded signals were used. The tests provided researchers feedback in the design process, looking forward to new iterations in the detailed design and construction phases of the project.

© Springer International Publishing AG 2017
A. Fred and H. Gamboa (Eds.): BIOSTEC 2016, CCIS 690, pp. 255–275, 2017.
DOI: 10.1007/978-3-319-54717-6_15

1   Introduction

In 2011, the World Health Organization reported that there were about one billion people worldwide with some type of disability [29]. In the European Union, almost 45 million people aged between 15 and 64 years reportedly had a disability around that same year, which corresponds to 14.1% of that age group [7]. In the United States of America, approximately 1.7 million people had an amputation in 2008 [21]. Ziegler et al. estimated that each year there are 185,000 new amputees of an upper or lower limb. They also presented an estimation of 3.6 million of people living with the loss of a limb by the year 2050 [31].

Continuous search for engineering solutions with the purpose of helping people experiencing physical disabilities or suffering deficit on the expression of cognitive experiences, has led to the development of artificial neural interfaces [10]. Furthermore, the research to develop systems to help and assist restoration of sensory function, communication and control to impaired humans has brought new branches of experimental neuroscience such as neural prostheses, neural interface systems (NIS) and brain-machine interfaces (BMIs) [11].

NIS are considered bidirectional transduction systems that enable the direct contact between a device and a neurological structure. They are composed of electrodes (or sensors), cables, data acquisition circuitry, and an effector system control unit [10]. The main goal of NIS research is to connect the nervous system to the outside world. This connection can be achieved either by stimulating or by recording electric activity from neural tissue to treat or to assist people with motor, sensory or other neuronal function disabilities. The systems that record electric activity from the neural tissues are called *output NIS*, and are now migrating from research proof of concepts and pilot human clinical trials to useful devices [11]. Some of the most studied devices are neural prostheses, exoskeletons and telemetry robots [10].

Robotic devices, therefore, can be used to improve the quality of life of human beings. Some of these robots can be considered robotic arms, and are similar to serial manipulators. These devices consist of open-loop kinematic chains, *i.e.*, open-loop assemblies of rigid bodies (links) connected by joints [27]. Exoskeletons with three degrees of freedom are examples of these kind of robots.

The scope of this section is to present a four-step methodology that can be used to test a NIS in which the movement of robots is controlled by means of a biosignal. The purpose is to apply this technology in the future, specifically to the design of robotic exoskeletons for rehabilitation and robotic prostheses. The methodology uses a kinematic simulation of a robot, which is useful to test and discuss trajectories, movements, design parameters, and any characteristic of the mechatronic device the researcher may need to check without having an actual robot available. For instance, it can be used to support the design of a new robotic prosthesis. As simulation is a part of the NIS itself, it may be completely controlled by prerecorded or real-time biosignals acquired from a human being.

The first step of the proposed methodology consist of the development of a kinematic simulation of the robot system using its CAD model and Denavit-Hartenberg (D-H) parameters. That kinematic simulation may be implemented on any programming environment that allows communication protocols, such as Octave, Python or Matlab. Given that the access to these environments is relatively easy inside the academic community, the simulation can be considered low-cost. The simulated robot is expected to be a serial manipulator, otherwise the D-H nomenclature can not be used.

The second step consists of acquiring and processing the biosignals of the neural interface being tested, obtaining the real inputs for using later in simulation. The third step is to develop tests with prerecorded signals, which is useful to debug and refine different parameters of the models. By last, the fourth step is to establish real-time communication between the simulation and the acquired biosignal, making the neural interface work with the simulation in the same way it would with a built and working robot.

We present results from two case-studies where the methodology was applied. First, in order to test the four steps, we used a commercial serial manipulator, a KUKA KR6 robotic arm available at our laboratories. It was simulated and controlled by a 25-year-old healthy man through surface electromyography (sEMG). After that, we present results of applying the methodology to the design process of an actual mechatronic device designed for lower limb rehabilitation, which is a powered exoskeleton. Using the tools developed, the current design of this robot was tested using prerecorded and real-time sEMG signals from the lower limb of a 26-year-old healthy woman in order to conduct mechatronic-assisted rehabilitation. The simulation provided useful information to the designers of this NIS, looking forward to later iterations of the design process, and the eventual construction of the robot. An image of both of the robots used can be seen in Figs. 1 and 2, respectively. Both KUKA KR6 and the exoskeleton are serial robots.

This paper is organized as follows. Section 2 explains the proposed methodology with a wide description of each step. Section 3 shows results of the assessment of the four steps with the KUKA robot. Section 4 shows results of the application of the methodology in the detailed design process of the exoskeleton. Finally, Sect. 5 presents conclusions.

2 Proposed Methodology

In this section, the four steps of the previously introduced methodology to assess the feasibility of implementing a NIS are presented.

2.1 Simulation of the Kinematics of a Robot

The purpose of this subsection is to give a brief guide on how to build a simulation of a robot in a programming environment. An example is given in Sect. 3.1.

As mentioned before, the simulation uses the CAD model and the D-H parameters of a serial robot. The D-H parameters of a serial robot refer to the notation introduced in 1955 by Jacques Denavit and Richard Hartenberg to

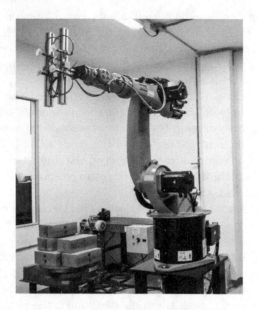

Fig. 1. Commercial robotic KUKA KR6 arm.

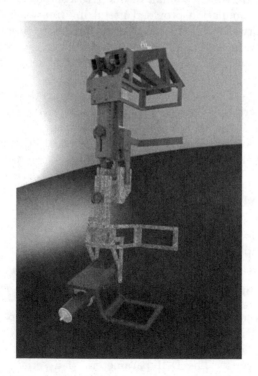

Fig. 2. CAD model of the current design of the lower limb rehabilitation exoskeleton.

describe the geometry of a serial chain of links and joints [2]. Although a complete description of the D-H notation will not be presented in this document, the reader can refer to any robotics text for more information. A simple and systematic methodology to assign the D-H parameters is presented in [3], where more information of the kinematic models described below can be found. We also recommend the detailed description proposed in Sect. 3.1 of [5]. The notation presented here is the same as in that document: The parameters s_i, θ_i, a_{ij} and α_{ij} represent, respectively, the joint offset distances, the joint angles, the link lengths, and the twist angles of the robot.

The CAD model of the robot has to be obtained in STL format. A STL file is a representation of a solid object using small facets, similar to Finite Element Analysis (FEA) meshes. Each triangular facet is explicitly defined in the file by a normal vector and its three vertices [12]. The user must obtain an STL file for each mechanical link of the serial manipulator. Files of additional details, such as motors or wires, can also be used. Considering that an STL file is essentially a text file, it can be generated by virtually any CAD software and easily imported into the computer programming environment used. Then, the vertices can be extracted and plotted using functions for filling polygons.

The plot of all the facets shows a 3D representation of the serial manipulator as if it was inside a CAD environment. Moreover, movement can be given to the model: all the vertices of each part of the robot constitute a cloud of points, susceptible to a homogeneous spatial transformation, such as a rotation or a translation. Each transformation consist of a 4×4 matrix that maps a homogeneous position vector from one coordinate system into another [27]. Assuming that the robot will move through multi-point trajectories, transformation matrices for the cloud of points of each part of the robot must be defined for each individual step of the trajectory. The user will obtain new clouds of points that, when plotted using the same functions as before, will show the robot in new positions. If multiple plots are made, the movement of the robot will be shown in different frames through time, as in an animation. Hence, with the use of these mathematical tools the kinematics of the 3D model of an entire robot can be simulated.

Recall that, in order to apply the procedure described above, the position of each point of each piece of the robot has to be measured in reference to its local origin. However, when the STL file of the robot is imported and plotted, the device appears in the position it was assembled in the CAD software, and all of its points are measured with reference to the absolute origin in the base of the robot instead of their respective local frame. For that reason, a previous step in the process is to apply the inverse kinematic analysis, which is the process of finding the joint angles that satisfy a given position and orientation of the end-effector [9], along with inverse transform matrices to restore each of the pieces of the robot to its local origin. The inverse kinematics of a serial manipulator are not straightforward, and the mathematics vary with each device in particular.

Up to this point, a 3D simulation can be generated for the actual movement of the robot moving smoothly through the demanded positions and following specific trajectories demanded by the user. It is important during the simulation process to test the kinematics of the robot: The user can define a trajectory of their interest and perform a simulation. This desired trajectory can be obtained directly from the motion of the joints of the robot (direct analysis) or by defining the path of its end-effector (inverse analysis).

The last task of this step in the methodology is the creation of a simplified model of the robot. It simulates with straight lines the moving parts of the robot, *i.e.*, the mechanical links. The cloud of points of each part, hence, only has two points, representing the start and the end of the part. Both the direct and the inverse kinematic analysis previously discussed can be applied to the simplified models as well. This is important due to the fact that the simulation of the CAD model may include many details, and the computational resources required to run it in real-time applications are very high. Thus, the simplified model can facilitate the analysis of the performance of the system. However, if the user has a very efficient computational resource it is still possible to use the complete CAD model in real-time applications. In that way the interaction with the system will be very realistic. If the user has a conventional office computer, we strongly recommend the use of the simplified model. In this way, the methodology can be used by any researcher almost independently of the computational resources available.

It is important to note that a kinematic simulation of a serial manipulator in a programming environment using its CAD model was first presented in [4].

2.2 Signal Acquisition, Processing and Control Algorithms

Biosignals obtained from tests such as electroencephalogram, electrocardiogram, electrooculogram or electromyogram can be monitored and measured from human bodies. In order to integrate the kinematic models and the biosignals into a NIS, a signal acquisition device, a signal processing algorithm and a control algorithm must be used.

The idea of the signal acquisition device is to acquire the biosignals for the NIS. For the purpose of the present methodology, this device must be able to record signals in a database and to allow their processing in real-time inside the programming environment being used.

The sensors or electrodes should be connected to the device and located either on a healthy person or on a patient. This process must be developed according to the parameters given by international standards and recommendations for acquiring each specific signal.

The tasks of this steps are the following: First, the user must record in a database all the signals needed to test the NIS. This could be done either in the software provided by the manufacturer of the acquisition device or in a custom made developed software.

Subsequently, the biosignals must be imported in the computational programming environment being used. Then, they should be processed, using one or more signal processing algorithms reported in the literature to develop feature extraction or classification [19], such as filters, amplifiers, envelope detectors [16], peak detectors, artificial neural networks [28], wavelets [17], Hilbert Huang transform [24] Kalman Filters [6,14], and others. These signal processing algorithms should be selected in order to fulfill the requirements of the NIS to be tested, either to enhance the signal, detect the intended movement of human joints, detect human gestures, and so on.

Finally, in order to control the movement of the joints or the end-effector of the kinematic model, a control algorithm must be developed in the programming environment used. It could be a classic or modern algorithm; some examples of them are PID controllers [22], Neuro-Fuzzy Control [13], Computed Torque Control [15], among others. The processed biosignals must be integrated with the control algorithm in order to move the kinematic model as desired in the real NIS.

2.3   Tests with Prerecorded Signals

A test with prerecorded biosignals is very useful before proceeding with actual real-time signals, because it allows the user to check if different elements of the NIS are working properly.

These tests are also useful to check trajectories and discuss design parameters of the moving robot. It is suggested to record several trials with different characteristics, simulating diverse real-life scenarios. If the simulation behaves as expected, the next step of the methodology can be implemented.

2.4   Tests with Real-Time Signals

The signals must be acquired in real-time using the biosignal acquisition device. This device must transmit continuously each sample to the programming environment. The transmission of the data can be either wired or wireless, but the latency of the transmission must be lower than the sample time and the signal processing time, so that it does not affect the algorithms and the whole NIS. Control systems should not create delays perceivable by the user during the operation. There are also reports of real-time constrains on myoelectric control systems when having smooth and continuous controls [1].

Each sample must be processed in real-time with a signal processing algorithm similar to the algorithm evaluated before in the non-real-time tests. The output of the real-time signal processing algorithm, must interact with the control algorithm to compute the movement of the robot.

As explained in Sect. 2.1, to maintain the proposed methodology low-cost, it is recommended to develop this step with the simplified model of the robot rather than with the 3D model of the entire robot, due to the high computational resources needed with the complete model. In these tests, many variables of the NIS can be analyzed such as the working range, specified movements, delays, and so on. If the simulation of the robot behaves in the way the user pretends, the NIS can be developed to further stages.

3   Testing the Methodology with a KUKA KR6 Robot

The methodology proposed in Sect. 2 was applied to a prosthetic-like neural interface consisting of the control of a KUKA KR6 robot by means of the sEMG signals of a 25-year-old healthy male. The reason we selected this particular robot is because an actual prototype is available at our laboratories. KUKA KR6 has six degrees of freedom and a spherical workspace.

3.1   Simulation of the Kinematics

The CAD model of the robot was obtained and successfully imported in a computer programming environment trough the procedure described in Sect. 2.1 using the D-H parameters of the KUKA KR6 [30]. The 3D model obtained is shown in Fig. 3. The quality of the shape and the details can be set when the STL file is exported from a CAD environment such as Solid Edge or Solid Works. Depending on the computational power available, less details mean a faster simulation.

A simplified model of the robot was created as well, in order to optimize real-time tests. This model is shown in Fig. 5. Excepting the base of the robot, each link consists only of two points. The plot, inside the programming environment used, facilitated the creation of straight lines between the points. The red line is the end-effector. Recall that the robot is in the same spatial position in Figs. 3 and 4.

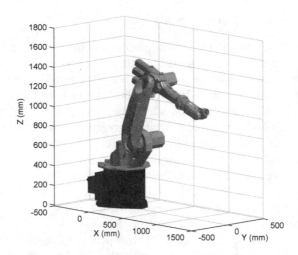

Fig. 3. 3D simulation of the KUKA KR6.

A kinematic test was performed in order to debug and validate the kinematic models. The task was to make the end-effector of the simulated robot to follow smoothly a circular trajectory. As showed in Fig. 5, the simulation performed the task correctly. The computational model of the robot was hence validated.

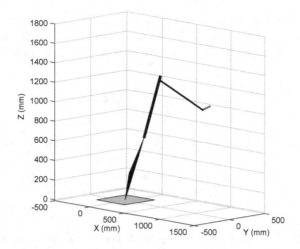

Fig. 4. Simplified simulation of the KUKA KR6. (Color figure online)

3.2 Biosignal Acquisition, Processing and Control Algorithms

The sensors, sensor placement and signal processing methods to acquire and record sEMG signals were based on some of the recommendations of the SENIAM project [26]. According to ISEK Standards for Reporting EMG Data [18] the characteristics of the procedure are shown:

The raw signal was detected using three pairs of 10 mm commercial, disposable and adhesive gel surface electrodes placed in different parts of the forearm of a healthy 25-year-old subject, along with a reference electrode located at the elbow. The electrodes had a disc shape, an area of 3.48 cm^2, and were made in silver-chloride. They were placed with a inter-electrode distance of approximately 2 cm (center point to center point) at the following muscles in order to detect flexion and pronation [8]:

- Biceps brachii, detecting activation when the elbow joint was flexed.
- Brachioradialis, detecting activation when the wrist pronation occured.
- Flexor carpi radialis, detecting activation when the wrist joint was flexed.

The electrodes were placed in parallel direction to the muscle fiber using the dominant middle portion of the muscle belly for best selectivity and avoiding the region of motor points. The signal was acquired using an 8 channel low-cost sEMG signal acquisition device designed and developed in our research labs. The device has a differential configuration, an input impedance of 10 kΩ, a signal-to-noise ratio (SNR) of 112 dB and it was configured with a gain of 12. The biosignal was sampled at 1 kHz. The low-cost sEMG signal acquisition device transmitted the signal through a serial port to a PC, and the programming environment used, recorded or captured in real-time all three differential channels. The signals were obtained one at the time, and they were normalized to the range of $(-1, 1)$.

Subsequently, the signal was rectified with Root Mean Square (RMS) and it was converted into an amplitude envelope as presented in the literature [20]. The RMS was calculated one value at the time using the equation

$$RMS_{emg}(n) = \sqrt{\frac{1}{n} * \sum_{s=1}^{n} sEMG(s)^2},$$ (1)

where $sEMG(s)$ represents the amplitude estimation [25] of the signal in the sth sample and n is the total number of samples in a window from the vector of the signal. The window used was 0.01 s, meaning each window had 10 samples. The signal was not filtered again in the computational software, but the RMS curve was normalized in the range of (0, 1) so it could be useful in the procedure described below.

The normalized RMS curve of the sEMG was the input of the kinematic analysis, in order to move the model of the robot when the subject intended to move his arm. The mathematical relation was defined so that the robot would move whenever the muscle was activated, and in a proportional way to the amplitude of the sEMG signal. If the muscle was not activated, the robot would not move from its home position. When there was a muscle activation, the model would change its position in a proportional way. This movement was done inside the robot position limits, meaning (0) the home position and (1) an arbitrary position for each joint close to its maximum angle with a comfortable safety distance.

The direct kinematic analysis was then effectuated and the movement of the robot was displayed in the screen. It was established that joints 1, 2 and 6 were going to be immobile in the home position of the robot ($0°$, $-90°$ and $0°$ respectively). This indicates that the kinematic model of the robot was used with only three joints, meaning a 3 degrees of freedom (3-DOF) device. Tests were performed with only one signal and its corresponding joint at the time, and the other two joints were set to an arbitrary position.

3.3 Testing Using Pre-recorded Signals

sEMG signals were recorded from the subject during three trials of 20 s using the procedures presented in Sect. 3.2. On each trial the subject was told to perform movements in order to detect activation of each of the three muscles described. On the first trial, he was told to flex his elbow two times. On the second trial, he was told to pronate his wrist four times. And on the third trial, he was told to flex his wrist three times. Muscle activation was measurable when these movements happened.

The sEMG signals obtained from each muscle and their RMS values are shown in Fig. 6. It is important to remark that the results in this Figure are shown before the RMS value was normalized and introduced in the kinematic model. In this figure, the activation and deactivation of the biceps brachii are represented by five letters.

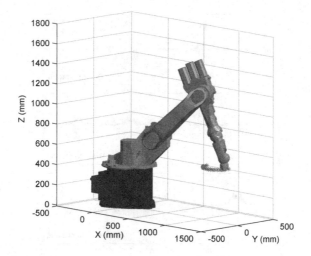

Fig. 5. Circular path executed by the simulated robot.

The output of the NIS is presented in Fig. 7. It shows different positions of the robot during the simulation of the NIS with the data previously recorded from the activation of the biceps brachii. This signal was selected due to the fact that presents the greater contrast between muscular activation-deactivation. Also, it has different levels of contractions producing different levels of signal amplitude, and for that reason is more didactic and visually representative in the simulation.

Each frame of the Fig. 7 corresponds to a position in time. The first frame (a) corresponds to the start of the trial, the second frame (b) to the first activation, which was really weak as can be seen in Fig. 6 section (b), the third frame (c) corresponds to the time between activations, the fourth frame (d) to the second (and strongest) activation as can be seen in Fig. 6 section (d) and the fifth frame (e) to the time after the activation. As these results were satisfactory, the real-time tests were implemented.

3.4 Testing Using Real-Time Signals

In order to assess the feasibility of implementing the designed NIS, the procedures shown in Sect. 2.4 for the signal acquisition, processing, and control were implemented. The signal processing and control algorithms were executed in real-time detecting the activation of the biceps brachii.

The sEMG acquisition device was connected to the subject. It transmitted each sample through serial port protocol at a baud rate of 115.200 through an Universal Serial Bus (USB). The samples of the raw signals were conducted to the signal processing algorithm and then they were mapped in the control algorithm.

The real-time experiments carried out on the healthy subject involved a very simple algorithm to detect the intended movement of the elbow, since the

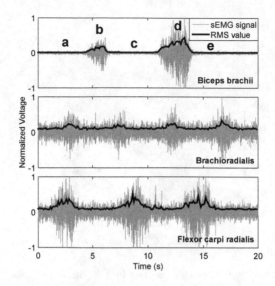

Fig. 6. Biosignals recorded with the low-cost sEMG acquisition device.

purpose of this section is the proposal of the methodology and not presenting a novel algorithm for the detection of the intended movement through sEMG signals. The detection of the elbow flexion was done through a threshold established to the RMS envelope of the biceps brachii signals. If the change in the amplitude envelope exceeded the given threshold, the joint 3 of the simplified kinematic model of robot was put at a maximum value. Otherwise, it stayed at the minimum value. The treshold used was 75% of the range of the RMS envelope signal.

The test consisted on five activations of the muscle separated approximately by two seconds. The simplified model of the robot moved correctly after each activation. Figures 8 and 9 show the status of the robot without and with an activation, respectively.

4   Applying the Methodology to the Design Process of a Powered Exoskeleton

In this section, we present results of applying the methodology in the design process of a powered exoskeleton for lower limb rehabilitation. This mechatronic serial device is currently in the detail design stage. It is conceived as a system that performs rehabilitation tasks in patients with anterior cruciate ligament (ACL) injuries and other pathologies of the lower limb. It is the product of requirements presented by a interdisciplinary group formed by physiotherapists and engineers. It aims to implement an active orthosis controller using EMG signals feedback to conduct assisted therapies for patients. Its mechanical design

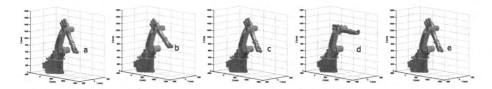

Fig. 7. Movement using prerecorded signals.

consists of a three-link planar mechanism (*i.e.*, a planar workspace) with three degrees of freedom [23].

The methodology was applied using the current design of the exoskeleton, which is being tested before it is manufactured. As the design is an iterative process, we aimed to produce valuable information that may improve the current model of the robot. Some particular design questions were addressed with the tests performed: are the biosignals used robust enough to move a real robot in the way that would be needed in therapy? Are the relative motions between the links safe for patients joints when the simulation of the robot is moving in real-life scenarios? Should any parameter of the robot be changed before the manufacturing process (*e.g.* the length of the links or the eccentricities)? Conclusions extracted from the tests are presented in Sect. 5.

4.1 Simulation of the Kinematics

The simulation of the kinematics of the mechatronic device was performed following the procedure described in Sect. 2.1, and exemplified in Sect. 3.1. The 3D model and the simplified model obtained are shown in Figs. 10 and 11, respectively.

4.2 Biosignal Acquisition, Processing and Control Algorithms

Biosignal acquisition was performed in a similar fashion to the procedure described in Sect. 3.2, signals were acquired from the rectus femoris in order to detect activation when the knee was flexed. The other two joints (hip and ankle) are undergoing further tests, and their results will not be presented in this paper.

Again, sensors, sensor placement and signal processing methods to acquire and record the signals were based on recommendations from SENIAM project [26]. According to ISEK Standards for Reporting EMG Data [18] the characteristics of the procedure are shown:

Raw signal was detected using one pair of 10 mm commercial, disposable and adhesive gel surface electrodes placed in rectus femoris of a healthy 26-year-old female subject, along with a reference electrode located at the knee, additionally a goniometer was localized to acquire the flexo/extension angle.

In this case, electrodes also were placed in parallel direction to the muscle fiber using the dominant middle portion of the muscle belly for best selectivity and avoiding the region of motor points. The signal was acquired using a 16 analog input channels system Powerlab 16/35 and a ML138 Octal Bio Amp. The device used a differential configuration and an input impedance of 200 MΩ. It additionally uses a Fourth-order Bessel Low-pass filter and First-order High-pass filter by hardware, as well as a 10–500 Hz band pass filter by software. The biosignal was sampled at 1 kHz. Powerlab 16/35 uses LabChart software and the captured signals in the different channels were visualized in a friendly interface.

Signals were normalized to the range of (−1, 1). Processing of data was carried out in a similar way as presented in Sect. 3.2, although adjacent windows of 256 ms were used. Equation 1 was applied in order to obtain the RMS value of the signal, and it was normalized and converted into the input of the knee of the robot so that the normalized RMS would be proportional to the movement. The other two joints remained motionless on the home position of the system.

4.3   Testing Using Pre-recorded Signals

Signals were recorded from a subject during one trial of 12 s. The subject was told to extend her knee two times. Force was exerted on the leg by means of a resisted rehabilitation exercise, therefore the amplitude of the signals increased. The biosignal is shown in Fig. 12. As in Sect. 3.3., the activation and deactivation of the muscle are represented by five letters. The corresponding movement of the robot is shown in Fig. 13. Each frame constitutes a position in time, the letters having the exact same meaning as in the movement presented in Fig. 7.

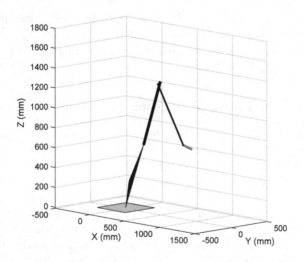

Fig. 8. Simplified KUKA robot 3D model during the real-time experiment with the elbow extended.

In Fig. 14 we show a comparison between the computed angle of movement of the robot and the angle of actual movement of the leg measured with a goniometer during the experiment.

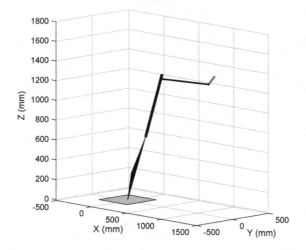

Fig. 9. Simplified KUKA robot 3D model during the real-time experiment with the elbow flexed.

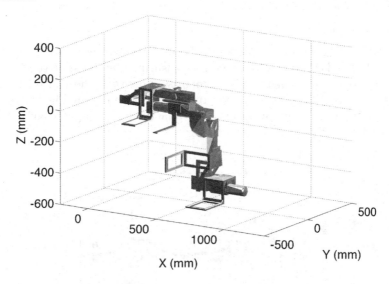

Fig. 10. 3D simulation of the rehabilitation system.

4.4   Testing Using Real-Time Signals

Tests presented in Sect. 3.4 were improved in order to achieve real-time proportional control of the robot instead of on/off control, maintaining the simple processing algorithm used in Sect. 4.3. Due to location issues, signals from the leg were recorded and later simulated in real-time using a Raspberry Pi connected through Ethernet to the computer that was running the simulation of the robot in the programming environment.

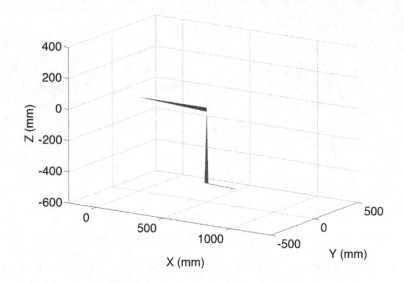

Fig. 11. Simplified model of the rehabilitation system.

One trial of 12 s was performed with the same subject, recording signals from the Rectus Femoris. A goniometer measuring the angle of the knee was used to assess the movement. The comparison between the control signal of the simulation and the angle of the goniometer is shown in Fig. 15. In order to compare the curves the Spearman's correlation was computed and resulted $r_s(54) = .63, p < .001$.

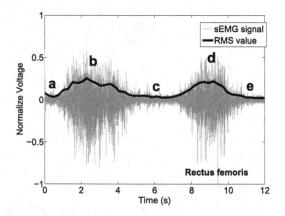

Fig. 12. Biosignals acquired from the lower limb.

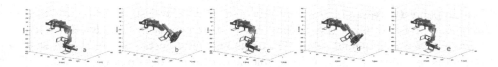

Fig. 13. Movement of the robot using prerecorded signals.

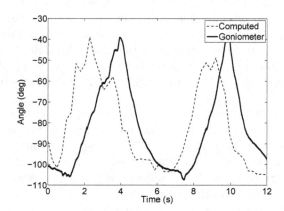

Fig. 14. Computed angle of movement and measured angle of movement of the Neural Interface System.

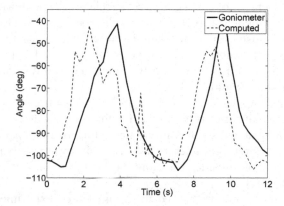

Fig. 15. Computed angle of movement and measured angle of movement after 12 s of real-time testing.

5   Conclusions

In this paper two case-studies were reported. With the first of them we aimed to achieve a validation of the methodology with a simulation of a KUKA KR6, a popular and commercial robot. From the results of Sect. 3 we concluded that the proposed methodology can be implemented in the design stage of a neural interface system (NIS) that involve a robotic device, due to the fact that the simulation of the robot was successfully implemented and controlled by means of biosignals, even with tests in real-time. The movement of the simulation of the robot was similar to the movement of the real robot.

In the second case, however, we had a robot that has not been yet manufactured. We intended to obtain valuable information for the design process of our mechatronic device for rehabilitation. The reliable results obtained in the case of the built robot let to take decisions in the case of an unbuilt device. In both cases the only requirement was to have the CAD files of the systems, which is usual in the design process of many machines nowadays, and convert them to STL format.

The application of the methodology in our rehabilitation robot demonstrated that sEMG signals from the Rectus Femoris are robust enough to move the knee joint of the exoskeleton. Nevertheless, taking into account that the acquisition of signals has been performed in healthy subjects, a new stage of the experiments will include tests with signals from subjects with different lower limbs injuries, and also from muscles from different joints.

In this paper a basic control algorithm was used to map the angles from human joints to robot joints. It is possible to include in the simulations more advanced control algorithms according to the capabilities of the real robot. Some of these algorithms may include dynamic analysis. Therefore, an interesting follow up for the project is the development of force models inside the methodology. Also, although the correlation between the computed angle of movement and the actual angle of movement of the leg may seem relatively low, both curves have a relatively similar shape and the NIS works well (See Fig. 15). The simulation of the robot tends to move every time there is muscular activation, when the limb is starting to move, and therefore the curves seem displaced in time. This is due to the processing algorithm, and from it we concluded that a stronger processing and control algorithm should be developed.

The relative motions between the links of the exoskeleton are safe for patients joints when the simulation of the robot is moving in real-life scenarios, because the range of movement we set has limits according to human body. With the prerecorded tests, different parameters from the robot were visualized in motion, and the authors do not consider that any parameter should be changed before the manufacturing process.

Although sEMG signals were used in the presented experiments, electroencephalogram, electrocardiogram, electrooculogram and even electrocorticogram signals may be used as well as an input for the NIS. It is recommended to follow the proper protocol to acquire the signals.

Finally, we propose that the presented methodology has a potential use in the field of biofeedback for musculoeskeletal and neurologic rehabilitation, since the movement of the robot is an indicator of muscular activation in real-time tests. Nevertheless, in order to assess this proposition clinical tests have to be carried out.

Acknowledgements. The authors would like to thank Cristian D. Martínez for the design and development of the Low-Cost sEMG signal acquisition device. We also would like to thank the physiotherapist Vanessa Montoya for her advisory with the rehabilitation exercises, Álvaro J. Saldarriaga for his support simulating the real-time sEMG signals, and Andrs Orozco-Duque for his support in the acquisition of sEMG signals.

Finally, the authors express gratitude to the Departamento Administrativo de Ciencia, Tecnología e InnovaciÓn Colciencias, from Colombia, for their grant number 121071149736.

References

1. Asghari Oskoei, M., Hu, H.: Myoelectric control systems-a survey. Biomed. Signal Process. Control **2**(4), 275–294 (2007). doi:10.1016/j.bspc.2007.07.009
2. Corke, P.: Robot arm kinematics. In: Corke, P. (ed.) Robotics, Vision and Control: Fundamental Algorithms in MATLAB. Springer Tracts in Advanced Robotics, vol. 73, pp. 137–170. Springer, Heidelberg (2011)
3. Corke, P.I.: A simple and systematic approach to assigning Denavit-Hartenberg parameters. IEEE Trans. Rob. **23**(3), 590–594 (2007). doi:10.1109/TRO.2007.896765
4. Correa, J.C., Ramirez, J.A., Taborda, E.A., Cock, J.A., Gómez, M.A., Escobar, G.A.: Implementation of a laboratory for the study of robot manipulators. In: Proceedings of the ASME 2010 International Mechanical Engineering Congress & Exposition, pp. 23–30 (2010)
5. Crane, C.D., Duffy, J.: Kinematic Analysis of Robot Manipulators. Cambridge University Press, New York (1998)
6. Delis, A.L., Carvalho J.L.A., Rocha, A.F.: Myoelectric Knee Angle Estimation Algorithms for Control of Active Transfemoral Leg Prostheses (2006). Self Organizing Maps - Applications and Novel Algorithm Design, pp. 401–424 (1977). http://cdn.intechopen.com/pdfs/13310/InTech-Myoelectric_knee_angle_estimation_algorithms_for_control_of_active_transfemoral_leg_prostheses.pdf
7. Eurostat: Disability statistics - prevalence and demographics (2014). http://ec.europa.eu/eurostat/statistics-explained/index.php/Disability_statistics_-_prevalence_and_demographics
8. Florimond, V.: Basics of Surface Electromyography Applied to Physical Rehabilitation and Biomechanics, vol. 1, pp. 1–50, March 2010
9. Fu, K.S., Gonzalez, R.C., Lee, C.S.: Robotics: Control, Sensing, Vision and Intelligence. McGraw-Hill, New York (1987)
10. Garcia Quiroz, F., Villa Moreno, A., Castano Jaramillo, P.: Interfaces neuronales y sistemas maquina-cerebro: fundamentos y aplicaciones. Revision. Revista Ingenieria Biomedica (1), 14–22 (2007). http://revistabme.eia.edu.co/numeros/1/art/InterfacesNeuronales.pdf

11. Hatsopoulos, N., Donoghue, J.: The science of neural interface systems. Ann. Rev. Neurosci. **32**, 249–266 (2009). doi:10.1146/annurev.neuro.051508.135241. http://www.ncbi.nlm.nih.gov/pmc/articles/PMC2921719/

12. Hon Wah, W.: Introduction to STL format (1999). http://download.novedge.com/Brands/FPS/Documents/Introduction_To_STL_File_Format.pdf

13. Kiguchi, K., Tanaka, T., Fukuda, T.: Neuro-fuzzy control of a robotic exoskeleton with EMG signals. IEEE Trans. Fuzzy Syst. **12**(4), 481–490 (2004). doi:10.1109/TFUZZ.2004.832525

14. Kyrylova, A., Desplenter, T., Escoto, A., Chinchalkar, S., Trejos, A.L.: Simplified EMG-driven model for active-assisted therapy. In: IROS 2014 Workshop on Rehabilitation and Assistive Robotics: Bridging the Gap Between Clinicians and Roboticists, p. 6 (2014). http://users.eecs.northwestern.edu/~argall/14rar/submissions/kyrylova.pdf

15. Lasso, I.L., Masso, M., Vivas, O.A.: Exoesqueleto para reeducacion muscular en pacientes con IMOC tipodiplejia espastica moderada, pp. 1–88 (2010). http://www.unicauca.edu.co/deic/Documentos/Monograf%EDa%20exoesqueleto.pdf

16. Lenzi, T., Rossi, S.M.M., Vitiello, N., Carrozza, M.C.: Intention-based EMG control for powered exoskeletons. IEEE Trans. Biomed. Eng. **59**(8), 2180–2190 (2012). doi:10.1109/TBME.2012.2198821

17. Lucas, M.F., Gaufriau, A., Pascual, S., Doncarli, C., Farina, D.: Multi-channel surface EMG classification using support vector machines and signal-based wavelet optimization. Biomed. Signal Process. Control **3**(2), 169–174 (2008). doi:10.1016/j.bspc.2007.09.002

18. Merletti, R.: Standards for reporting EMG data (1999). doi:10.1016/S1050-6411(97)90001-8, http://www.isek.org/wp-content/uploads/2015/05/Standards-for-Reporting-EMG-Data.pdf

19. Merletti, R., Parker, P.A.: Electromyography: Physiology, Engineering, and Non-Invasive Applications. Wiley, Hoboken (2004)

20. Mon, Y., Al-Jumaily, A.: Estimation of upper limb joint angle using surface EMG signal. Int. J. Adv. Robot. Syst. 1 (2013). doi:10.5772/56717, http://www.intechopen.com/journals/international_journal_of_advanced_robotic_systems/estimation-of-upper-limb-joint-angle-using-surface-emg-signal

21. National Limb Loss Center Information: Amputation statistics by cause. Limb loss in the United States (2008). http://www.amputee-coalition.org/limb-loss-resource-center/resources-by-topic/limb-loss-statistics/limb-loss-statistics/

22. Pan, D., Gao, F., Miao, Y., Cao, R.: Co-simulation research of a novel exoskeleton-human robot system on humanoid gaits with fuzzy-PID/PID algorithms. Adv. Eng. Softw. **79**, 36–46 (2015). doi:10.1016/j.advengsoft.2014.09.005

23. Patiño, J.G., Bravo, E.E., Perez, J.J., Perez, V.: Lower limb rehabilitation system controlled by robotics, electromyography surface and functional electrical stimulation. In: Pan American Health Care Exchanges, PAHCE 2002 (2013). doi:10.1109/PAHCE.2013.6568341, 6257

24. Revilla, L.M., Delis, A.L., Olaya, A.F.R.: Towards a method to detect movement intention. In: Pan American Health Care Exchanges, PAHCE, pp. 1–6 (2013). doi:10.1109/PAHCE.2013.6568259

25. Ruiz, A.F., Rocon, E., Forner-Cordero, A.: Exoskeleton-based robotic platform applied in biomechanical modelling of the human upper limb. Appl. Bionics Biomech. **6**(2), 205–216 (2009). doi:10.1080/11762320802697380

26. The Seniam Project: SENIAM (2015). http://www.seniam.org

27. Tsai, L.W.: The Mechanics of Serial and Parallel Manipulators. Wiley, New York (1999)
28. Wojtczak, P., Amaral, T.G., Dias, O.P., Wolczowski, A., Kurzynski, M.: Hand movement recognition based on biosignal analysis. Eng. Appl. Artif. Intell. **22**(4–5), 608–615 (2009). doi:10.1016/j.engappai.2008.12.004
29. World Health Organization: World Report on Disability (2011). http://www.who.int/disabilities/world_report/2011/report.pdf
30. Yepes, J.C., Yepes, J.J., Martinez, J.R., Perez, V.Z.: Implementation of an Android based teleoperation application for controlling a KUKA-KR6 robot by using sensor fusion. In: Pan American Health Care Exchanges, PAHCE (2013). doi:10.1109/PAHCE.2013.6568286
31. Ziegler-Graham, K., MacKenzie, E.J., Ephraim, P.L., Travison, T.G., Brookmeyer, R.: Estimating the prevalence of limb loss in the United States: 2005 to 2050. Arch. Phys. Med. Rehabil. **89**(3), 422–429 (2008). doi:10.1016/j.apmr.2007.11.005

Modeling and Clustering of Human Sleep Time Series Using Dynamic Time Warping: Sequential and Distributed Implementations

Chiying Wang[1], Sergio A. Alvarez[2(✉)], Carolina Ruiz[1], and Majaz Moonis[3]

[1] Department Computer Science, Worcester Polytechnic Institute,
Worcester, MA 01609, USA
{wangchiying,ruiz}@wpi.edu
[2] Department Computer Science, Boston College, Chestnut Hill, MA 02467, USA
alvarez@bc.edu
[3] Department Neurology, University Massachusetts Medical School,
Worcester, MA 01655, USA
moonism@ummhc.org

Abstract. We present a new modified dynamic time warping approach for comparing discrete time series that reduces over-warping and maintains the efficiency of global path constraint approaches, without relying on domain-specific heuristics. In a first version, global weighted dynamic time warping, a penalty term for the deviation between the warping path and the path of constant slope is added in a post-processing step, after a standard dynamic time warping computation. A second version, stepwise deviation-based dynamic time warping, incorporates the penalty term into the dynamic programming optimization itself, yielding modified optimal warping paths. Both versions yield modified similarity metrics that we use for time series clustering within the Combined Dynamical Modeling Clustering (CDMC) framework. Additionally, we present a distributed computing implementation of dynamic time warping-based modeling and clustering using CDMC. Experiments over synthetic data, as well as over human sleep data, demonstrate significantly improved accuracy and generative log likelihood as compared with standard dynamic time warping. The distributed computing implementation achieves a reduction in processing time.

Keywords: Dynamic time warping · Deviation · Human sleep · Clustering · Distributed computing

1 Introduction

Many data analysis problems in science and engineering involve data in the form of long time series. Two examples are the analysis of patterns in human sleep data, and the investigation of patterns in web navigation [4]. In the present paper, we focus on the former. Human sleep patterns are closely associated with overall health and quality of life, making the scientific study of sleep an important

© Springer International Publishing AG 2017
A. Fred and H. Gamboa (Eds.): BIOSTEC 2016, CCIS 690, pp. 276–294, 2017.
DOI: 10.1007/978-3-319-54717-6_16

pursuit. Sleep stage transitions [11] and bout durations [6] are essential indicators in characterizing the structure of sleep. Typical patterns of human sleep are known [3], yet sleep microstructure varies across individuals, being affected by age, circadian rhythms [8], and other factors.

Modeling the dynamics of sleep involves many challenges. One substantial challenge is the scarcity of key dynamical events such as stage transitions within individual sleep sequences [3]. This scarcity yields very small samples over which dynamical models are to be trained, leading to high uncertainty in parameter estimates. An approach known as dynamical modeling-clustering (CDMC) was proposed [1] to address this issue of sample size. CDMC selective aggregates instances during a clustering phase that precedes dynamical model estimation. Models are therefore learned over collections consisting of many dynamically similar instances, rather than over individual instances. Pooling dynamically similar instances in this way effectively increases the size of the training sample, reducing the variance of the resulting model. The CDMC framework is described by Algorithm 1. The choice of initial clustering, c_0, measure of clustering similarity, CLUSTERINGSIMILARITY, and specific type of dynamical models to be used in the maximum likelihood learning and clustering procedures, LEARNMLPROTOTYPES and LEARNMLCLUSTERLABELS, are all left open by Algorithm 1, thus providing increased modeling flexibility.

Algorithm 1. Collective Dynamical Modeling-Clustering (CDMC) [1].

Input: An unlabeled time-series dataset $D = \{x = (a_i(x)) \mid i = 1, 2, \ldots, n\}$; a positive integer, k, for the desired number of clusters; an initial guess $c_0 : D \rightarrow \{1, \ldots k\}$ of the cluster label $c_0(x)$ of each instance $x \in D$; parameter values, s, specifying the desired configuration of the models (e.g., number of states); and a real number minSim between 0 and 1 for the minimum clustering similarity required for stopping.

Output: A set $M_1, \ldots M_k$ of generative dynamical models (with configuration parameters s), together with a cluster labeling $c : D \rightarrow \{1 \ldots k\}$ that associates to each data instance, x, the index $c(x)$ of a model $M = M_{c(x)}$ for which the generative likelihood $\prod_{x \in D} P(x \mid M_{c(x)})$ is as high as possible.

CDMC(D, k, c_0, s, $minSim$)

   1. $c(x) = c_0(x)$ for all x in D

   2. $c_{old}(x) = 0$ for all $x \in D$

   3. **while** CLUSTERINGSIMILARITY(c, c_{old}) < minSim

   4. $c_{old} = c$

   5. $(M_1, \ldots M_k) = $ LEARNMLPROTOTYPES(D, k, c, s)

   6. $c = $ LEARNMLCLUSTERLABELS(D, $M_1, \ldots M_k$)

   7. **return** $M_1, \ldots M_k$, c

Dynamic time warping [12] (DTW) is a classical technique that uses dynamic programming to find an optimal alignment of two time series based on comparisons between pairs of data points along a warping path. DTW has been used extensively in speech recognition (e.g., [9, 15]), time series classification [10], and as a measure of similarity for unsupervised time series clustering [13]. Initialization of CDMC using clustering by Dynamic Time Warping (DTW) similarity [13] yields good convergence properties [16].

Despite promising results, standard DTW presents certain problems when used as a similarity measure for unsupervised clustering of time series. One of these is the over-warping problem shown in Fig. 1. Over-warping refers to an unnatural alignment of dissimilar segments in two time series. In Fig. 1, a subsequence of length over 300 in one patient is matched by dynamic time warping to a subsequence of length less than 10 in another. The result is unacceptable due to the extreme distortion introduced in the earlier portion of the data record, yet the standard dynamic time warping distance between the two segments is zero. An explicitly penalized DTW approach has been developed to address the over-warping problem. For example, [7] proposes variable penalty DTW, which reduces nondiagonal moves during alignment. However, this approach is heavily dependent on a user-defined penalty function and thus difficult to apply in practice.

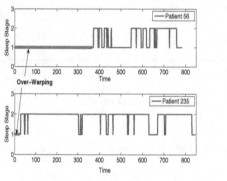

(a) Over-Warping Occurrence in Time Series

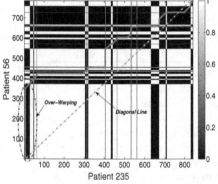

(b) Over-Warping Path in Dynamic Time Warping

Fig. 1. Over-warping of sleep stage sequences using dynamic time warping. (Left) Use of standard dynamic time warping inappropriately matches dissimilar segments (a long one versus a short one) in two sequences. (Right) The same over-warping problem described in terms of warping search area. The area circled by the dashed line indicates a large deviation of the standard warping path from the diagonal path of constant slope. The bar graph on the right indicates local cost measure and the background in the right figure shows the local cost matrix of two discrete time series (patient 56 and patient 235) in dynamic time warping computation (see Sect. 2.1).

An additional concern about DTW is its time complexity. Variants of DTW have been proposed that focus on improving efficiency by globally constraining the warping path to a predefined geometric region such as the Sakoe-Chiba band [15] or the Itakura parallelogram [9]. However, the use of global constraints alone can lead to the over-warping problem described above. Some researchers [14] have argued that the effect of warping band width on the quality of the results is greatly domain dependent and that a narrow band might be valuable. The distribution of optimal warping paths between pairs of sleep time series in Fig. 2 (from the present authors' own work), which is observed to be concentrated around the straight line path between the endpoints of each warping path, likewise suggests that the use of local search constraints would be desirable. This fact motivates the modified dynamic time warping approach that is pursued in the present paper.

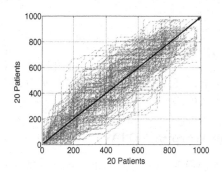

Fig. 2. Optimal time warping paths (dashed lines) between 20 pairs of human sleep recordings. DTW optimal paths are usually close to the diagonal line from bottom-left to top-right in the warping space. The varying boundary of the distribution suggests the desirability of adaptively identifying warping areas locally instead of using a predefined global constraint.

Main Contributions of the Present Paper. The present paper extends our prior work [17]. The main contributions of the extended work appear below. The third item, involving a distributed computing implementation of DTW-based CDMC modeling and clustering, is entirely new.

1. We propose two novel DTW variants, global weighted dynamic time warping (gwDTW) and stepwise deviated dynamic time warping (sdDTW), that penalize deviations of the warping path from the path of constant slope. This overcomes the over-warping issue in Fig. 1, while retaining the efficiency advantages of approaches based on global constraints such as the Sakoe-Chiba band [15] and Itakura parallelogram [9], and without relying on domain dependent specifics as in variable penalty DTW [7].
2. We apply the proposed modified DTW approaches to the task of clustering initialization within the combined dynamical modeling-clustering (CDMC) framework [1] over human sleep time series, and show that this approach better captures the dynamics of human sleep.
3. We present a distributed computing implementation of DTW-based modeling and clustering within the CDMC framework that promises to provide more efficient modeling and clustering performance over large data sets of time series.

Organization of the Paper. Section 2 reviews standard DTW, and describes the proposed deviation-based dynamic time warping approach and its application to time-series clustering. The proposed distributed implementation of modeling-clustering using DTW is also described. Section 3 presents experimental results and analysis on time series clustering using deviation-based dynamic time warping, including time performance of the distributed implementation in the case of standard DTW similarity. Section 4 describes conclusions and future work.

2   Methods

We review standard dynamic time warping in Sect. 2.1, as that technique will serve as a baseline. The proposed deviation-based dynamic time warping approach is described in Sect. 2.2. Section 2.3 describes the proposed distributed computing implementation of dynamic time-warping based modeling and clustering in the CDMC framework of Algorithm 1.

2.1   Dynamic Time Warping (DTW)

Dynamic time warping (DTW) is a classical dynamic programming algorithm for measuring the similarity of two time series (e.g., [12]). It performs an optimal alignment between two time series by nonlinearly warping their time dimensions. DTW has been applied to speech recognition (e.g., [9,15]), time series classification [10], and unsupervised time series clustering [13].

The following are the essentials of standard DTW, as described in [12].

We consider two time sequences $X = (x_1, x_2, \ldots, x_N)$ of length $N \in \mathbb{N}$ and $Y = (y_1, y_2, \ldots, y_M)$ of length $M \in \mathbb{N}$, with individual values x_i, y_j in some feature space $\mathcal{F}$.

A **local cost measure** is a function

$$c : \mathcal{F} \times \mathcal{F} \to \mathbb{R}_{\geq 0} \tag{1}$$

The value of $c(x_i, y_j)$ is small if x_i, y_j are close to each other, and otherwise not. For discrete time series, one can use a cost matrix to define the values $c(x, y)$ for all pairs of values x, y; the simplest possibility is to use the identity matrix, that is, to let $c(x, y) = 0$ if $x = y$, otherwise $c(x, y) = 1$.

A **warping path** between X and Y is a sequence

$$p = (p_1, p_2, \ldots, p_L) \tag{2}$$

where $p_l = (n_l, m_l) \in [1 : N] \times [1 : M]$ for $l \in [1 : L]$ (the diagonal line from bottom-left to top-right in Fig. 1(b)), and $\max\{N, M\} \leq L \leq N + M$. It must satisfy the following three conditions:

- *Boundary condition*: $p_1 = (1, 1)$ and $p_L = (N, M)$ are the start and end points respectively.
- *Monotonicity condition*: horizontal and vertical components increase monotonically: $n_1 \leq n_2 \leq \ldots n_L$ and $m_1 \leq m_2 \leq \ldots m_L$.

– *Step size condition*: for each $l < L$, the difference $p_{l+1} - p_l$ is one of $(1,0)$, $(0,1)$, $(1,1)$.

The **total cost** of a warping path p between X and Y is

$$\Phi_p(X,Y) = \sum_{l=1}^{L} c(x_{n_l}, y_{m_l}), \tag{3}$$

where $c(x_{n_l}, y_{m_l})$ defines the local cost measure in Eq. 1 associated with mapping the data point at n_l in sequence X to the data point at m_l in sequence Y.

An **optimal warping path** p^* is one having minimum total cost $\Phi_{p^*}(X,Y)$ among all warping paths from p_1 to p_L. $\Phi_{p^*}(X,Y)$ is referred to as the *standard DTW distance* between sequences X and Y, as in Eq. 4, where p ranges over all warping paths between X and Y.

$$\Phi_{p^*}(X,Y) = \operatorname*{argmin}_{p=(p_1,p_2,\ldots,p_L)} \sum_{l=1}^{L} c(x_{n_l}, y_{m_l}), \tag{4}$$

2.2   Proposed DTW Approaches

Deviation Measure. We address the standard DTW concerns of over-warping and time complexity described in the Introduction, by penalizing nondiagonal moves in the search for an optimal warping path. This is done by using the measure of deviation discussed below.

Deviation is a measure of the area Δ_p (shaded area in Fig. 3) that is bounded by the warping path p between two time series and the straight-line diagonal path.

Deviation is computed by the procedure described in Algorithm 2. The following are the main steps:

Algorithm 2. Deviation Calculation.

Input: A warping path $p = \{p_1, p_2, \ldots, p_L\}$, each $p_i = (n_i, m_i)$; L: total number of points in path p.
Output: The deviation Δ_p of p from the diagonal line through p_1 and p_L in warping space.
ComputeDeviation(p)
  1. $\Delta_p = 0$
  2. $k = (M-1)/(N-1)$
  3. $b = M - k * N$
  4. $i = 2$
  5. **While**($p_i \neq p_L$)
  6.   if($m_i \neq m_{i-1}$)
  7. $\Delta_p = \Delta_p + |m_i - (k \cdot n_i + b)|$
  8. $i = i + 1$
  9. **return** Δ_p

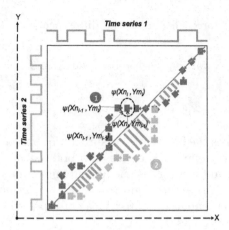

Fig. 3. Deviation (e.g., gray shaded area) of warping path (squares with directional arrows) from path of constant slope (solid red line). Given two warping paths (1 and 2), the path with smaller deviation (the green one) is better. (Color figure online)

- Initialize the deviation Δ_p to be 0. (step 1)
- Calculate the slope k and the intercept b of the diagonal path defined as $y = k * x + b$. (step 2–3)
- Repeat until the end point p_L is reached. (step 5)
    - Add absolute vertical distance between current point p_i and diagonal to the deviation Δ_p, if p_i and p_{i-1} differ vertically. (step 6–7)
    - update the counter, i. (step 8)
- Return the deviation Δ_p (step 9)

Deviation-Based Dynamic Time Warping. This paper proposes the following two distinct approaches to modified dynamic time warping, both based on the measure of deviation computed by Algorithm 2:

- Global weighted dynamic time warping (gwDTW), which adds the deviation as a post-processing penalty term after computation of the standard DTW warping path; and
- Stepwise weighted dynamic time warping (sdDTW), which incorporates the deviation calculation into the dynamic programming computation of the warping path, therefore leading to a modified warping path.

Global Weighted Dynamic Time Warping. The global weighted dynamic time warping $(gwDTW)$ distance between sequences X and Y is defined as

$$gwDTW(X,Y) = \lambda_{gw} \cdot \Phi_{p^*}(X,Y)$$
$$+ (1 - \lambda_{gw}) \cdot \sqrt{\Delta_{p^*(X,Y)}} \qquad (5)$$

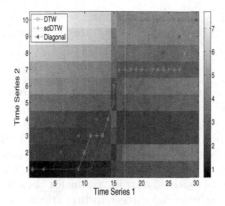

Fig. 4. Optimal warping paths for standard DTW and stepwise deviation-based DTW (sdDTW). Standard DTW allows large deviations in searching for a path of minimum total cost. sdDTW aligns time series closer to the diagonal line. Background shading indicates local cost in sdDTW, which increases with distance to the diagonal.

where $\Phi_{p^*}(X,Y)$ is the standard DTW distance from Eq. 4, and the deviation $\Delta_{p^*(X,Y)}$ (Algorithm 2) is added as a post-processing penalty to standard DTW. $p^*(X,Y)$ is the optimal path between X,Y in standard DTW. λ_{gw} controls the balance between standard DTW cost and the deviation $\Delta_{p^*(X,Y)}$. λ_{gw} ranges from 0 to 1. The best value of λ_{gw} is determined empirically by maximizing the classification accuracy of clustering over labeled synthetic time series data (see Sect. 3.2). The square root of the deviation $\Delta_{p^*(X,Y)}$ is used because it scales linearly with Euclidean distance.

Stepwise Deviation-Based Dynamic Time Warping. The stepwise deviation-based dynamic time warping (*sdDTW*) distance between sequences X, Y is defined as

$$sdDTW(X,Y) = \Psi_{p^*}(X,Y) \tag{6}$$

where $\Psi_{p^*}(X,Y)$ is the minimum total cost of a warping path p^* in Eq. 3 obtained by replacing the local cost measure in Eq. 1 by the modified measure $\varphi(x,y)$ in Eq. 7. Thus, sdDTW optimal warping paths are minimizers of a different cost measure than standard DTW paths.

$$\varphi(x,y) = \lambda_{sd} \cdot c(x,y) + (1 - \lambda_{sd}) \cdot \sqrt{\Delta(x,y)} \tag{7}$$

where $c(x,y)$ denotes the original local cost measure in Eq. 1. $\Delta(x,y)$ is the deviation of a position (x,y) relative to the diagonal path of constant slope for sequences X and Y. See Sect. 2.2. λ_{sd} is a parameter that determines the relative weights of the standard local cost measure and the deviation.

We use dynamic programming as in standard DTW to compute the sdDTW optimal warping path, based on the **modified accumulated cost matrix**

$$\bar{D}(n,m) = \Psi(X(1:n), Y(1:m)) \tag{8}$$

where $X(1:n) = (x_1, \ldots, x_n)$ and $Y(1:m) = (y_1, \ldots, y_m)$. $n \in [1:N]$ and $m \in [1:M]$. That is $X(1:n)$ and $Y(1:m)$ are subsequences of X and Y. The procedure is as follows:

- Initially, $\bar{D}(n,1) = \sum_{k=1}^{n} \varphi(x_k, y_1)$ and $\bar{D}(1,m) = \sum_{k=1}^{m} \varphi(x_1, y_k)$.
- Iteratively, take the minimum accumulated cost from three immediately adjacent directions: $\bar{D}(n-1,m) + \varphi(x_n, y_m)$, $\bar{D}(n, m-1) + \varphi(x_n, y_m)$, $\bar{D}(n-1, m-1) + \varphi(x_n, y_m)$.
- Until the final position (N, M) is reached. $\bar{D}(N, M)$ is the optimal dynamic time warping distance with respect to stepwise deviation-based DTW.

As illustrated in Fig. 4, optimal warping paths for the sdDTW distance metric exhibit less over-warping than the corresponding standard DTW paths.

Deviation-Based Dynamic Time Warping Clustering. Deviation-based dynamic time warping clustering (dDTWC) performs unsupervised agglomerative hierarchical clustering of time series using the deviation-based DTW approaches in Sect. 2.2 to calculate distances. The proposed approach is described in pseudocode in Algorithm 3. The main steps are:

- Initially each time series instance X is in its own cluster (steps 1–2).
- Repeat until only k clusters remain (steps 3–6):
  - Merge the closest clusters, C and C'; the distance between two instances (X and Y) is defined by **gwDTW** or **sdDTW** in Sect. 2.2; the distance between two clusters is the average distance between pairs of instances.
- Return clustering of dataset in k clusters (step 7).

2.3   Distributed CDMC Implementation

In the form in which it is described in Algorithm 1, Collective Dynamical Modeling & Clustering (CDMC) is a sequential procedure. To improve the running time of CDMC on large data sets consisting of long time series, we implement the CDMC framework in Apache Storm [2], a distributed and fault-tolerant real-time computation platform for processing streaming data. In this approach, time series are viewed as data streams. The section that follows describes this distributed computing implementation.

Distributed CDMC System Design. Figure 5 depicts the overall architecture of the distributed collective dynamical modeling clustering framework. The general idea is to avoid sequential processing of the collection of input time series and decompose CDMC into a parallel architecture. One processing node (a "bolt" in Storm terminology) per cluster label is used to store clustering results, and one processing node per cluster label is used for data modeling. For simplicity, we use the term "processor" here synonymously with "processing node" (bolt), with the caveat that a node does not necessarily correspond to a physical processor.

Algorithm 3. Deviation-Based DTW Clustering (dDTWC).

Input: An unlabeled time series dataset $D = \{X \mid X$ is a time series$\}$; a positive integer, k, the desired number of clusters; a predefined local cost measure $c : \mathcal{F} \times \mathcal{F} \to \mathbb{R}_{\geq 0}$ where $\mathcal{F}$ is the feature space in which the time series in D take their values. **dDTW** denotes the total cost measure associated with c, which is defined by Eq. 5 in the case of gwDTW and by Eq. 6 in the case of sdDTW.
Output: A partition of D into k clusters
dDTWC(D, k, d)

1. for each i, let $C_i = $ a cluster that contains only the i-th time series in D
2. $s = $ the number of time series in D (initial number of clusters)
3. **while** $s > k$
4. $(i^*, j^*) = \arg \min\limits_{i,j \in \{1,\ldots,s\}} \bar{c}(C_i, C_j)$ (where $\bar{c}$ is mean cost for instance pairs in the two clusters)
$$= \arg \min\limits_{i,j \in \{1,\ldots,s\}} \left\{ \frac{\sum_{X \in C_i, Y \in C_j} \mathbf{dDTW}(X,Y,c)}{|C_i| \cdot |C_j|} \right\}$$
5.   Merge C_{i^*} and C_{j^*} to reduce the number of clusters to $s - 1$
6. $s = s - 1$
7. **return** $\{C_1, \ldots, C_k\}$

The distributed collective dynamical modeling & clustering system implementation has six main components, as listed in Table 1.

The following are a few important features of the core components:

- For the data source, a desired number (e.g., n) of time series are cached before starting the data modeling and clustering procedures. When n time series are ready, the data source "spout" (streaming data source) assigns each time series to a processor based on its cluster label. Initially, cluster labels for an input time series are randomly specified. After that, the clustering results produced by clustering step in CDMC, Algorithm 1, are applied to the existing time series.
- In the data partition step, time series with the same cluster label are stored in the same processor. Additional steps such as discretization and compression of time series could be performed there before passing the data to the next processor.

Table 1. Components of the distributed system.

Distributed CDMC components	
Core components	Initial data assignment
	Data partition
	Data modeling
	Evaluation
External components	Data source
	Parameter configuration

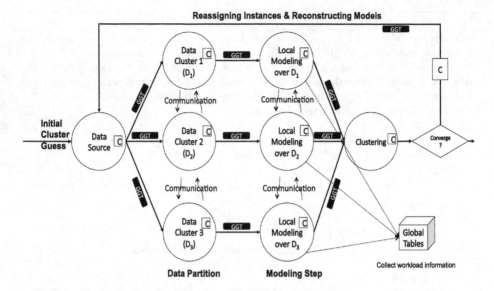

Fig. 5. Distributed Collective Dynamical Modeling & Clustering Framework Implementation. Take three models and three clusters as an example. Incoming time series emerging from the data source node are assigned to clustering processing nodes according to initial cluster label guesses. Each cluster is then fed to a dedicated processing node that extracts a single model of the cluster. After finishing this modeling process, all instances are reassigned to clusters according to the new models. If two consecutive clusterings are close enough to each other, the clustering-modeling procedure stops; otherwise the procedure continues until a stable clustering result is produced or a maximum number of iterations is reached.

- In the modeling step, each processor receives tuples (time series) from the corresponding node in the clustering step, then builds a model on the collection of time series available to it. After that, the node produces model parameters or takes models as input to the next clustering processing node.
- In the clustering step, time series are grouped into clusters in terms of a predefined criterion such as maximum likelihood of the input data, using the current dynamical models.

Note that when a new time series comes in, it is temporarily put in the buffer of the data source node in Fig. 5. Unless updated clustering results arrive in the processing node, the new time series is added to the dataset with a randomly selected cluster label. To extend the parallelism of the CDMC framework, it is possible to parallelize the local modeling procedure, provided that the modeling process can be divided into several independent parts.

Factors Contributing to the Performance of the Distributed System.
Compared with a sequential system, the main factors that can lead to efficiency gains due to parallelization are as follows:

1. Caching of intermediate results, including state transition and state duration statistics for individual instances. The cached information reduces the processing time in subsequent rounds, as it avoids the need for recalculation.
2. Processing multiple tasks in parallel and allocating resources dynamically (memory, hard disk, registers in processors) in a way that optimizes parallel performance.

3 Experimental Evaluation

Deviation-based DTW clustering as described in Sect. 2.2 was compared with clustering using the standard DTW distance metric of Eq. 4. For all dynamic time warping computations, the local cost measure in Eq. 1 was defined as $c(x, y) = 1$ if the elements x and y are different, otherwise $c(x, y) = 0$. The weight values λ_{gw} and λ_{sd} in Eqs. 5 and 6, respectively, were determined empirically in order to maximize mean accuracy over a sample of labeled synthetic data generated as in Sect. 3.2 (but separate from the synthetic data sample used for performance evaluation in Sect. 3.2): λ_{sd} was set to 0.67. λ_{gw} was set to 0.83. All experiments were performed in MATLAB® (*The MathWorks*, 2015).

Two sets of experiments were carried out, corresponding to synthetic data and human sleep data, respectively. Details specific to each of these are described in Sects. 3.2 and 3.3 below. Experiments were also carried out to evaluate the time performance of the sequential and distributed versions of DTW-based CDMC over human sleep data, and are described separately, in Sect. 3.4.

3.1 Statistical Significance

Pairwise comparisons of median classification accuracy values (see Sect. 3.2) of gwDTW and sdDTW clustering against the accuracy of standard DTW clustering (for synthetic data) and of negative log likelihood values of gwDTW and sdDTW clustering against that of standard DTW clustering (for human sleep data) were carried out by a non-parametric two-sided Wilcoxon rank sum test, since a Lilliefors normality test rejected normality at the $p < 0.05$ significance level in each case. A Bonferroni correction was performed jointly on the accuracy and log likelihood Wilcoxon p-values to ensure a familywise error rate less than 0.05.

3.2 Synthetic Markov Mixture Data

Dataset Generation. A synthetic dataset of discrete sequences was generated as in [1], from two distinct Markov models, each with two states. The two models differ in their transition probability matrices. Self-transition probabilities of 0.6 in one model and 0.8 in the other were selected. The probabilities of transitioning between states were 0.4 and 0.2, respectively. One of the two models is selected randomly and used to generate a sequence of the desired length, L. This process is iterated until a predetermined number of sequences, N, is obtained. The present paper uses the values $N = 100$ and $L = 300$ in all trials.

Experimental Procedure

Clustering Classification Accuracy. Supervised classification via clustering was performed with the generating model label as the classification target. Each cluster was associated with the class c that occurs most frequently among its members. The evaluation metric was classification accuracy, equal to the fraction of labeled instances (X, c) that are assigned to a cluster in which c is the majority class. Statistical hypothesis testing was performed using a Wilcoxon rank sum test to compare median accuracies as described in Sect. 3.1. Experimental procedure was as in the following pseudocode:

Experimental Procedure, Synthetic Data Classification:
begin
    for $i := 1$ to $TrialNum$
        SD = generateSyntheticDataset(N, L);
        Accuracy(1, i) = evaluateByDTW(SD);
        Accuracy(2, i) = evaluateBygwDTW(λ_{gw}, SD);
        Accuracy(3, i) = evaluateBysdDTW(λ_{sd}, SD);
    end
    Perform Wilcoxon rank sum test over Accuracy.
end

Notes:

– *generateSyntheticDataset* followed the description in Sect. 3.2.
– *evaluateBy***DTW** refers to the clustering procedure in Sect. 2.2 and clustering evaluation by classification accuracy (see above); likewise for *evaluateBy***gwDTW** and *evaluateBy***sdDTW**.
– The total number of sequences, N, was set to 100.
– The length of a sequence, L, was set to be 300.
– The number of trials, $TrialNum$, was set to 100.

Synthetic Data Results. This section evaluates performance of clustering over synthetic (labeled) data using globally weighted DTW (gwDTW) (Eq. 5) or stepwise deviation-based DTW (sdDTW) as the similarity measure, as compared with standard DTW similarity.

As in Sect. 3.2, synthetic data were generated from two distinct hidden Markov models, each with two states, but with different transition probability matrices and state duration statistics. To determine the best weight value (e.g., λ_{gw}), an exhaustive search over values $0 < \lambda < 1$ was conducted. The value of λ leading to the highest average accuracy over the labeled data was selected.

Clustering accuracies over the synthetic dataset appear in Fig. 6. Both gwDTW and sdDTW perform significantly better than standard DTW, proving the benefit of incorporating deviation into the DTW computation for clustering of synthetic time series data. Median accuracies appear in Table 2.

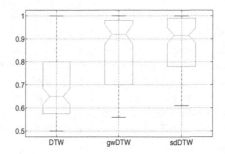

Fig. 6. Clustering accuracies using standard DTW, gwDTW, and sdDTW as similarity measures over hidden Markov mixture data. Non-overlapping notches indicate significant difference in medians ($p < 0.05$). gwDTW and sdDTW are significantly more accurate than standard DTW.

Table 2. Median accuracies of clusterings based on DTW, gwDTW, and sdDTW. Asterisks denote Bonferroni-corrected statistical significance of differences with standard DTW in Wilcoxon rank sum test ($p < 0.05$).

DTW	gwDTW	sdDTW
0.65	0.92*	0.91*

3.3 Human Sleep Data

Datasets. A collection of 244 fully anonymized human polysomnographic recordings was extracted from polysomnographic overnight sleep studies performed in the Sleep Clinic at Day Kimball Hospital in Putnam, Connecticut, USA. Each polysomnographic recording is split into 30-second epochs. Lab technicians staged each 30-second epoch into one of the sleep stages Wake, stage 1, stage 2, stage 3, and REM (Rapid Eye Movement). Three versions of the human sleep dataset are considered, depending on whether these stage labels are grouped in some way:

- (W5) uses the five standard stage labels Wake, 1, 2, 3, REM.
- (WNR) uses the three stage labels Wake, NREM (stages 1, 2, and 3), REM.
- (WDL) uses the three stage labels Wake, Deep (stage 3), Light (stages 1, 2, REM).

Experimental Procedure. Unsupervised clustering was performed over the human sleep datasets described in Sect. 3.3. The collective dynamic modeling clustering algorithm [1] was used for clustering, with two-state hidden semi-Markov chain models as the dynamical models. Initial cluster labels were computed by deviation-based DTW clustering as described in Algorithm 3, with either gwDTW (Eq. 5) or sdDTW (Eq. 6) as the distance metric. Clustering driven by the standard DTW distance metric was used as a basis for comparison.

Generative negative log likelihood was used to measure the quality of model fit for unsupervised clustering. Given a hidden semi-Markov model, M, built over a group of sequences such as human sleep sequences, the generative negative log-likelihood $-\log(P(s|M))$ of a sequence, s, is a measure of the probability that the sequence, s, would be produced by the model, M. Lower negative log-likelihood values (higher generative probabilities) imply a better model fit. The goal of clustering was to minimize the generative negative log likelihood. Comparison of median negative log likelihoods for different models was measured by a Wilcoxon rank sum test as described in Sect. 3.1.

Experimental Procedure, Sleep Data Clustering:
```
begin
    D1, D2, D3 = W5, WNR, WDL datasets (3.3)
    m1, m2, m3 = DTW, gwDTW, sdDTW
    for j = 1 to 3
        for k = 1 to 3
            (M, nlogll(Dj, mk, s1...s244)) = CDMC(Dj, mk)
        end
        Perform pairwise Wilcoxon rank sum tests on
            nlogll(Dj, m1...m3, s1...s244)
    end
end
```

Notes:

- The W5, WNR, WDL datasets are as in Sect. 3.3.
- DTW refers to clustering using standard DTW as the similarity metric; gwDTW and sdDTW refer to the deviation-based clustering techniques described in Sect. 2.2.
- $CDMC(Dj, mk)$ refers to CDMC clustering [1] with semi-Markov cluster models, using the given method, mk, for clustering initialization, and is assumed to return a set of dynamical models together with negative generative log likelihoods nlogll(Dj,mk,sl) for all input sequences, sl, $l = 1, \cdots 244$.

Human Sleep Data Results. Figure 7 shows the two CDMC clusters (circles and triangles) with coordinates equal to the Weibull scale and shape parameters for the wake stage in the WNR dataset. gwDTW clusters better capture the boundary between the natural Weibull dynamical clusters in the human sleep dataset, as compared with standard DTW clusters.

Model fit was significantly better for both global weighted DTW (gwDTW) clustering and stepwise deviation-based DTW (sdDTW) clustering as compared with standard DTW-driven clustering, as shown in Table 3. This shows that deviation-based DTW is superior to standard DTW as a similarity metric for initialization of CDMC clustering over human sleep data, as well as for stand-alone clustering over synthetic data as shown in Sect. 3.2.

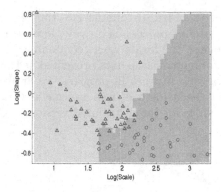

Fig. 7. Visualization of clusters over human sleep dataset using gwDTW as similarity measure. Coordinates are Weibull shape and scale parameters for Wake stage. Red circles and blue triangles denote gwDTW clusters; background colors represent DTW clusters. (Color figure online)

Table 3. Median negative log likelihoods of gwDTW, sdDTW, and standard DTW clusterings over WNR, WDL, and W5 human sleep datasets in Sect. 3.3. Asterisks indicate Bonferroni-corrected significance of differences with standard DTW in Wilcoxon rank sum test ($p < 0.05$).

	DTW	gwDTW	sdDTW
WNR	150.7	148.2*	147.9*
WDL	159.4	157.9*	158.1*
W5	194.1	192.3*	191.9*

3.4 Time Performance of Sequential and Distributed DTW-Based CDMC

Experimental Procedure. A standard sequential implementation of the CDMC algorithm (Algorithm 1) was compared with a distributed implementation using Storm, as in Sect. 2.3. The standard DTW-based similarity metric was employed in both cases (with a Sakoe-Chiba band constraint on the warping paths). In other words, clustering was carried out by using the standard DTW metric from Eq. 4 as the total cost measure **dDTW** in Algorithm 3. The W5 human sleep data set described in Sect. 3.3 was used. Semi-Markov chains were selected as the dynamical models, with five states (one per sleep stage), each using a Weibull duration distribution to model bout durations for the corresponding stage.

The sequential system uses a single machine with one processor and a 1-Gigabyte memory and 100-Gigabyte disk space to store data instances, data models, and clustering results. The distributed system uses three machines with one processor per machine and the same size of 1-Gigabyte memory and 100-Gigabyte disk space to hold data instances as well as models and clusters. Total running time, as a function of the number of input instances, was used as the performance measure.

Running Time Results. Figure 8 summarizes time performance for human sleep data. As seen, distributed processing provides lower run times. However, the advantage is within a factor of 2 for data sets containing up to 1000 instances or so. Limited availability of human sleep data prevented us from performance comparisons for larger data sets. In work in progress, we apply distributed CDMC to a larger data set in the domain of web use navigation behavior, and find considerably greater savings for the distributed version of CDMC.

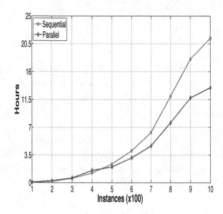

Fig. 8. Running time of distributed DTW-based CDMC over human sleep data.

4   Conclusions

This paper proposes two versions of a modified dynamic time warping (DTW) approach for comparing discrete time series such as human sleep stage sequences: global weighted dynamic time warping (gwDTW) and stepwise deviation-based dynamic time warping (sdDTW). Both versions penalize deviations from the path of constant slope in the warping space, yielding the efficiency advantages of DTW approaches based on global constraints such as the Itakura parallelogram or the Sakoe-Chiba band, while better accounting for local deviations. gwDTW adds a deviation-based term to the standard DTW distance metric. sdDTW adds a deviation term into the local cost function that drives the DTW dynamic programming optimization itself, yielding an improved warping path together with a similarity metric. When used within the collective dynamical modeling and clustering (CDMC) framework [1], both gwDTW and sdDTW lead to significantly better clustering results than DTW in a classification task over labeled synthetic semi-Markov data, as well as in unsupervised clustering of human sleep data. The authors learned of an interesting "salient feature" approach to constrained DTW [5] after completing the work reported in the present paper. The salient feature approach extracts features of the input sequences that are then used to define locally adaptive constraints on the warping path. In future work, it would be desirable to pursue a performance comparison of the salient feature approach of [5] with that of the present paper. The present paper also introduced a distributed computing implementation of the DTW-based modeling and clustering in the CDMC framework, using the standard DTW metric for similarity. The distributed implementation reduces the running time for large data sets of long time series. Future work will explore the performance gains achieved by the distributed version as compared with the sequential version, for data sets that are larger than the sleep data set considered in the present paper, and for modified DTW similarity metrics such as those proposed in the present paper.

Acknowledgements. The authors thank the anonymous referees for comments that helped improve the legibility of the paper, and for making us aware of [5].

References

1. Alvarez, S.A., Ruiz, C.: Collective probabilistic dynamical modeling of sleep stage transitions. In: Proceedings Sixth International Conference on Bio-Inspired Systems and Signal Processing (BIOSIGNALS 2013), Barcelona, Spain, pp. 209–214 (2013)
2. Apache Software Foundation. Apache Storm (2013)
3. Bianchi, M.T., Cash, S.S., Mietus, J., Peng, C.-K., Thomas, R.: Obstructive sleep apnea alters sleep stage transition dynamics. PLoS ONE **5**(6), e11356 (2010)
4. Cadez, I.V., Heckerman, D., Meek, C., Smyth, P., White, S.: Model-based clustering and visualization of navigation patterns on a web site. Data Min. Knowl. Disc. **7**(4), 399–424 (2003)
5. Candan, K.S., Rossini, R., Wang, X., Sapino, M.L.: sDTW: computing DTW distances using locally relevant constraints based on salient feature alignments. Proc. VLDB Endowment **5**(11), 1519–1530 (2012)
6. Chu-Shore, J., Westover, M.B., Bianchi, M.T.: Power law versus exponential state transition dynamics: application to sleep-wake architecture. PLoS ONE **5**(12), e14204 (2010)
7. Clifford, D., Stone, G., Montoliu, I., Rezzi, S., Martin, F.-P., Guy, P., Bruce, S., Kochhar, S.: Alignment using variable penalty dynamic time warping. Anal. Chem. **81**(3), 1000–1007 (2009)
8. Dijk, D.J., Lockley, S.W.: Invited review: integration of human sleep-wake regulation and circadian rhythmicity. J. Appl. Physiol. **92**(2), 852–862 (2002)
9. Itakura, F.: Minimum prediction residual principle applied to speech recognition. IEEE Trans. Acoust. Speech Sig. Process. **23**(1), 67–72 (1975)
10. Jeong, Y.-S., Jeong, M.K., Omitaomu, O.A.: Weighted dynamic time warping for time series classification. Pattern Recogn. **44**(9), 2231–2240 (2011)
11. Kishi, A., Struzik, Z.R., Natelson, B.H., Togo, F., Yamamoto, Y.: Dynamics of sleep stage transitions in healthy humans and patients with chronic fatigue syndrome. Am. J. Physiol. Regul. Integr. Comp. Physiol. **294**(6), R1980–R1987 (2008)
12. Müller, M.: Dynamic time warping. In: Information Retrieval for Music and Motion, pp. 69–84. Springer, Heidelberg (2007)
13. Oates, T., Firoiu, L., Cohen, P.R.: Using dynamic time warping to bootstrap HMM-based clustering of time series. In: Sun, R., Giles, C.L. (eds.) Sequence Learning. LNCS, vol. 1828, pp. 35–52. Springer, Heidelberg (2000). doi:10.1007/3-540-44565-X_3

14. Ratanamahatana, C.A., Keogh, E.: Making time-series classification more accurate using learned constraints. In: Proceedings of SIAM International Conference on Data Mining (SDM 2004), pp. 11–22 (2004)
15. Sakoe, H., Chiba, S.: Dynamic programming algorithm optimization for spoken word recognition. IEEE Trans. Acoust. Speech Sig. Process. **26**(1), 43–49 (1978)
16. Wang, C., Alvarez, S.A., Ruiz, C., Moonis, M.: Semi-Markov modeling-clustering of human sleep with efficient initialization and stopping. In: Proceedings Seventh International Conference on Bio-Inspired Systems and Signal Processing (BIOSIG-NALS 2014), Barcelona, Spain (2014)
17. Wang, C., Alvarez, S.A., Ruiz, C., Moonis, M.: Deviation-based dynamic time warping for clustering human sleep. In: Proceedings of Ninth International Conference on Bio-Inspired Systems and Signal Processing (BIOSIGNALS 2016), Rome, Italy, pp. 21–23 (2016)

Voice Restoration After Laryngectomy Based on Magnetic Sensing of Articulator Movement and Statistical Articulation-to-Speech Conversion

Jose A. Gonzalez[1]([⊠]), Lam A. Cheah[2], James M. Gilbert[2], Jie Bai[2], Stephen R. Ell[3], Phil D. Green[1], and Roger K. Moore[1]

[1] Department of Computer Science, University of Sheffield, Sheffield, UK
{j.gonzalez,p.green,r.k.moore}@sheffield.ac.uk
[2] School of Engineering, University of Hull, Kingston upon Hull, UK
{l.cheah,j.m.gilbert,j.bai}@hull.ac.uk
[3] Hull and East Yorkshire Hospitals Trust, Castle Hill Hospital, Cottingham, UK
srell@doctors.org.uk

Abstract. In this work, we present a silent speech system that is able to generate audible speech from captured movement of speech articulators. Our goal is to help laryngectomy patients, i.e. patients who have lost the ability to speak following surgical removal of the larynx most frequently due to cancer, to recover their voice. In our system, we use a magnetic sensing technique known as Permanent Magnet Articulography (PMA) to capture the movement of the lips and tongue by attaching small magnets to the articulators and monitoring the magnetic field changes with sensors close to the mouth. The captured sensor data is then transformed into a sequence of speech parameter vectors from which a time-domain speech signal is finally synthesised. The key component of our system is a parametric transformation which represents the PMA-to-speech mapping. Here, this transformation takes the form of a statistical model (a mixture of factor analysers, more specifically) whose parameters are learned from simultaneous recordings of PMA and speech signals acquired before laryngectomy. To evaluate the performance of our system on voice reconstruction, we recorded two PMA-and-speech databases with different phonetic complexity for several non-impaired subjects. Results show that our system is able to synthesise speech that sounds as the original voice of the subject and also is intelligible. However, more work still need to be done to achieve a consistent synthesis for phonetically-rich vocabularies.

Keywords: Silent speech interfaces · Speech rehabilitation · Speech synthesis · Permanent magnet articulography

1 Introduction

People whose larynx have been surgically removed following throat cancer, trauma or destructive throat infection normally find themselves struggling with

© Springer International Publishing AG 2017
A. Fred and H. Gamboa (Eds.): BIOSTEC 2016, CCIS 690, pp. 295–316, 2017.
DOI: 10.1007/978-3-319-54717-6_17

oral communication after losing their voice. This often has a severe impact on people's lives and can lead to social isolation, feelings of loss of identity and, sometimes, clinical depression [2,3,7]. To make things worse, existing methods for voice restoration are far from ideal [13,18]. The 'gold-standard', the tracheo-oesoephageal valve, requires frequent replacement every 3–4 months due to biofilm growth and, for this reason, is an expensive treatment [11,12,22]. The electro-larynx, on the other hand, despite being relatively easy to use and safe, produces a robotic voice. Finally, oesophageal speech, a technique in which air is injected into the mouth and then it is released in a controlled manner to make the oesophagus vibrate to create speech, sounds gruff and masculine and, also, is difficult to learn. Other methods such as Augmentative and Alternative Communication (AAC) devices, where the user types words and the device synthesises them, are only suitable for short conversations due to their slow manual text input [15].

In an attempt to surpass the limitations of current existing methods, we propose in this work a completely different approach for post-laryngectomy voice rehabilitation. Our proposed method can be seen as a Silent Speech Interface (SSI), which is a system that enables oral communication in the absence of audible speech [9]. Somewhat akin to lip reading, the most common way a SSI works is by first decoding the message encoded in other biosignals associated with speech production using Automatic Speech Recognition (ASR) software and then synthesising the recognised text using a Text-To-Speech (TTS) synthesiser. Many different SSIs have been proposed so far, mainly differing in the type of biosignal they rely on. Thus, we can find SSIs that exploit the electrical signals generated by the neurons in the brain [23] or in the articulator muscles [31,42,49] or the movement of the speech articulators themselves [9,14,18,21,26, 29,40,44]. In our work we use a magnetic sensing technique known as Permanent Magnet Articulography (PMA) [13,18] for capturing the movement of the speech articulators. In brief, the principle of PMA is that articulator movement can be captured by attaching a set of permanent magnets to the articulators (typically the lips and tongue) and then sensing the variations of the resultant magnetic field generated while the person articulates words with sensors located close to the mouth. Compared to other techniques for articulator motion capture, such as Electromagnetic Articulography (EMA) or surface Electromyography (sEMG), PMA has the potential advantage of being unobtrusive, since there are no wires coming out of the mouth or electrodes attached to the skin.

The recognise-then-synthesise approach outlined above is the most common way of synthesising speech from articulator movement. This approach, however, is not exempt from problems [21]. Firstly, due to the variable delay introduced by the ASR and TTS systems, speech articulation and the corresponding acoustic feedback produced by the SSI are disconnected (i.e. speech is not generated in real time). An analogy for this would be like having an interpreter: the person 'mouths' words, waits and, after a while, the SSI generates the corresponding acoustic signal. Another limitation is that speech can only be synthesised for the language and vocabulary of the ASR and TTS systems. Finally, as another limitation, the non-linguistic information embedded in the articulatory signal, such as emotion or speaker identity, is normally lost after speech recognition. To address these shortcomings, we investigate in this work an alternative approach for SSI-based voice restoration known as *direct speech synthesis*.

In the direct synthesis approach, a parametric transformation is applied to the captured articulatory data to obtain a sequence of speech parameter vectors from which a time-domain speech signal is synthesised. Thus, no intermediate speech recognition step is performed. To enable this method, we present a statistical approach in which simultaneous recordings of articulator movement and speech data captured before laryngectomy are used to learn the parameters of the transformation. In particular, in our statistical framework the transformation adopts the form of a joint probability distribution represented as a Mixture of Factor Analysers (MFA) [17]. Once this transformation is learned, it can be used in conversion time to compute the speech-parameter posterior distribution given the articulatory data which, in turn, allows us to estimate the speech parameters associated with the measured articulator gesture. Because the transformation is estimated from recordings of the patient's voice, our method has the potential to synthesise speech that sounds as the original voice. Moreover, if the conversion can be done in real-time, the voice will sound spontaneous and natural and it will enable the patient to receive real-time acoustic feedback of her/his articulation.

This chapter is organised as follows. First, the details of the PMA technique for capturing articulator movement are provided in Sect. 2. In Sect. 3, we describe the proposed technique for synthesising audible speech from PMA data. Then, in Sect. 3.3, we discuss some practical implementation issues. The details about the experimental evaluation of the proposed technique and the results obtained can be found in Sect. 5. Finally, we summarise this work and outline some future research in Sect. 6.

2   Articulator Motion Capture Based on Magnetic Sensing

As commented above, in our work we use PMA, a magnetic sensing technique, for capturing the movements of the speech articulators, more typically the lips and tongue [5,13,18]. The principle of PMA is simple as illustrated in Fig. 1: small magnets are attached to the speech articulators whose movement we wish to capture and magnetic sensors located close to the mouth are employed for measuring the magnetic field generated by the magnets. In the current set-up magnets are temporarily attached to the flesh using Histoacryl surgical tissue adhesive (Braun, Melsungen, Germany), but eventually the magnets will be surgically implanted for long term usage. A total of six magnets are used: four are attached to the lips (1 mm diameter × 5 mm height), one to the tongue tip (2 mm × 4 mm), and one to the tongue blade (5 mm × 1 mm). As shown in Fig. 1, an external headset is used to hold the four magnetic sensors employed in the current prototype, though other arrangements such as an intra-oral device similar to a dental retainer has been and are currently being investigated [4,5].

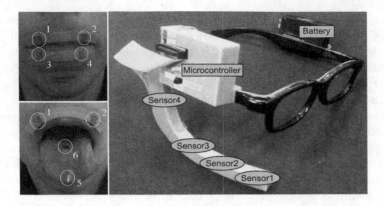

Fig. 1. *Upper-left and lower-left*: placement of magnets used for measuring lips and tongue movements. *Right*: components of the PMA headset: micro-controller, battery and magnetic sensors used to detect the variations of the magnetic field generated by the magnets.

From the four sensors in Fig. 1, only the first three sensors (Sensor1-Sensor3), the ones which are closer to the mouth, are actually used for data acquisition, while the last one (Sensor4) is used as a reference sensor for Earth's magnetic field cancellation in the data captured by Sensor1-Sensor3 [5]. Each sensor provides three channels of data for the 3D spatial components of the magnetic field at the sensor location, thus making 9 channels of data in total.

The data recorded by the sensors may then be used to determine the speech associated with the articulatory gestures, either by performing ASR on the PMA data [18,26] or by transforming the data to an acoustic signal, as we do in this work. It should be noted that contrary to other mechanisms for capturing articulator movement data such as EMA, in PMA the exact coordinates of the magnets in the mouth are unknown since the magnetic field sensed by the sensors is a composite of the fields generated by all the magnets. Nevertheless, as each articulatory gesture generates a recognisable pattern in the captured articulatory data, we can resort to statistical approaches for modelling the relationship between the PMA data and the corresponding acoustics. This is developed in the next section.

3   Speech Synthesis from Articulator Movement

In this section we describe the technique for synthesising audible speech from articulator movement data. The goal of the proposed technique is to model the mapping $\boldsymbol{y}_t = \boldsymbol{f}(\boldsymbol{x}_t)$ between source parameter vectors $\boldsymbol{x}_t$ derived from the PMA signal and target parameter vectors $\boldsymbol{y}_t$, which correspond to a low-dimensional, parametric representation of the audio signal. Because the positions of the magnets cannot be easily inferred from PMA data (inference of the Cartesian coordinates of the magnets from the magnetic field captured by the sensors is an inverse problem), we cannot resort here to approaches such as that proposed in [45,47] where 2-dimensional vocal tract shapes are directly computed from the measured positions of the EMA sensors and, from the vocal tract shapes, speech is synthesised by using an articulatory synthesis method [35]. In contrast, we propose a data-driven approach for modelling the mapping function $y_t = f(x_t)$. The proposed approach assumes the existence of a parallel dataset containing simultaneous recordings of PMA and acoustic data acquired before laryngectomy. From this parallel dataset, the parameters of the mapping function, which is represented here as a Mixture of Factor Analysers (MFA) [17], are estimated during an initial training stage. Later, in conversion time, the learned transformation is used to convert PMA parameter vectors into speech parameter ones. More details about the training and conversion phases are given in the next sections.

3.1 Training Phase

Instead of trying to directly model the mapping function $y_t = f(x_t)$, we assume that x_t and y_t are the outputs of a stochastic process whose state v_t is not directly observable. We also assume that the dimensionality of v_t is much less than that of x_t and y_t, such that the latent space offers a more parsimonious representation of the observable data. Under these assumptions, we have the following model:

$$x_t = f_x(v_t) + \epsilon_x, \tag{1}$$
$$y_t = f_y(v_t) + \epsilon_y, \tag{2}$$

where ϵ_x and ϵ_y are Gaussian-distributed noise processes with zero mean and diagonal covariances Ψ_x and Ψ_x, respectively.

In general, f_x and f_y will be non-linear and, hence, difficult to model. To represent them, a piece-wise linear regression approach is adopted in which the functions are approximated by a mixture of K local factor analysis models, each of which has the following form,

$$x_t^{(k)} = W_x^{(k)} v_t + \mu_x^{(k)} + \epsilon_x^{(k)}, \tag{3}$$
$$y_t^{(k)} = W_y^{(k)} v_t + \mu_y^{(k)} + \epsilon_y^{(k)}, \tag{4}$$

where $k = 1, \ldots, K$ is the model index; $W_x^{(k)}$, $W_y^{(k)}$ are the factor loadings matrices; $\mu_x^{(k)}$, $\mu_y^{(k)}$ are bias vectors that allow the data to have a non-zero mean, and $x_t^{(k)}$, $y_t^{(k)}$ are respectively local approximations of x_t and y_t around the means $\mu_x^{(k)}$, $\mu_y^{(k)}$. This model can be written more compactly as,

$$z_t^{(k)} = W_z^{(k)} v_t + \mu_z^{(k)} + \epsilon_z^{(k)}, \tag{5}$$

where $z_t = [x_t^{(k)^\top}, y_t^{(k)^\top}]^\top$, $W_z^{(k)} = [W_x^{(k)^\top} W_y^{(k)^\top}]^\top$, $\mu_z^{(k)} = [\mu_x^{(k)^\top}, \mu_y^{(k)^\top}]^\top$, and $\epsilon_z^{(k)} \sim \mathcal{N}(0, \Psi_z^{(k)})$, with $\Psi_z^{(k)}$ being the following diagonal covariance matrix,

$$\Psi_z^{(k)} = \begin{bmatrix} \Psi_x^{(k)} & 0 \\ 0 & \Psi_y^{(k)} \end{bmatrix}. \tag{6}$$

From (5) we see that the conditional distribution of the observed variables given the latent ones is $p(z|v, k) = \mathcal{N}(z; W_z^{(k)} v + \mu_z^{(k)}, \Psi_z^{(k)})$. By assuming that the latent variables are independent and Gaussian with zero mean and unit variance (i.e. $p(v|k) = \mathcal{N}(0, I)$), the k-th component marginal distribution of the observed variables, i.e.

$$p(z|k) = \int p(z|v, k)p(v|k)dv, \tag{7}$$

also becomes normally distributed as $p(z|k) = \mathcal{N}(z; \mu_z^{(k)}, \Sigma_z^{(k)})$, where

$$\Sigma_z^{(k)} = \Psi_z^{(k)} + W_z^{(k)} W_z^{(k)^\top} \tag{8}$$

is the reduced-rank covariance matrix.

The generative model is completed by adding mixture weights $\pi^{(k)}$ for each mixture component ($\sum_k \pi^{(k)} = 1$). Then, the joint distribution $p(z) \equiv p(x, y)$ finally becomes the following mixture model,

$$p(z) = \sum_{k=1}^{K} \pi^{(k)} p(z|k). \tag{9}$$

To learn the parameters $\{\langle \pi^{(k)}, \mu_z^{(k)}, W_z^{(k)}, \Psi_z^{(k)} \rangle, k = 1, \ldots, K\}$ of the MFA model in (9) we use a slightly modified version the expectation-maximization (EM) algorithm proposed in [17] in which the noise covariances $\Psi_z^{(k)}$ are cluster dependent. Let $\{z_i = [x_i^\top, y_i^\top]^\top, i = 1, \ldots, N\}$ be the parallel dataset used for training. Then, the E-step and M-step of the EM algorithm are as follows:

(1) *E-step:* Using the MFA parameters estimated in the previous iteration, compute the component posterior probabilities $\gamma_i^{(k)} = P(k|z_i)$ and the expectations $\langle v_{ik} \rangle$ and $\langle v_{ik} v_{ik}^\top \rangle$ for the hidden variables:

$$\gamma_i^{(k)} = \frac{p(z_i|k)\pi^{(k)}}{\sum_{k'=1}^{K} p(z_i|k')\pi^{(k')}}, \tag{10}$$

$$\langle v_{ik} \rangle = S_k^{(k)} \left(z_i - \mu_z^{(k)} \right), \tag{11}$$

$$\langle v_{ik} v_{ik}^\top \rangle = I - S^{(k)} W_z^{(k)} + \langle v_{ik} \rangle \langle v_{ik} \rangle^\top, \tag{12}$$

with $S^{(k)} = W_z^{(k)^\top} \Sigma_z^{(k)^{-1}}$ and $\Sigma_z^{(k)}$ as given by (8).

(2) M-step: To simplify the derivation of the updating equations, we define the following augmented variables,

$$\langle \tilde{\boldsymbol{v}}_{ik} \rangle = \begin{bmatrix} \langle \boldsymbol{v}_{ik} \rangle \\ 1 \end{bmatrix}, \tag{13}$$

$$\langle \tilde{\boldsymbol{v}}_{ik} \tilde{\boldsymbol{v}}_{ik}^{\top} \rangle = \begin{bmatrix} \langle \boldsymbol{v}_{ik} \boldsymbol{v}_{ik}^{\top} \rangle & \langle \boldsymbol{v}_{ik} \rangle \\ \langle \boldsymbol{v}_{ik} \rangle^{\top} & 1 \end{bmatrix}, \tag{14}$$

$$\tilde{\boldsymbol{\Sigma}}_{v|z}^{(k)} = \langle \tilde{\boldsymbol{v}}_{ik} \tilde{\boldsymbol{v}}_{ik}^{\top} \rangle - \langle \tilde{\boldsymbol{v}}_{ik} \rangle \langle \tilde{\boldsymbol{v}}_{ik} \rangle^{\top}. \tag{15}$$

Then, the updated MFA parameters are obtained as follows,

$$\hat{\pi}_z^{(k)} = \frac{1}{N} \sum_{i=1}^{N} \gamma_i^{(k)}, \tag{16}$$

$$\hat{\boldsymbol{\Lambda}}_z^{(k)} = \left[\sum_{i=1}^{N} \gamma_i^{(k)} \boldsymbol{z}_i \langle \tilde{\boldsymbol{v}}_i \rangle^{\top} \right] \left[\sum_{i=1}^{N} \gamma_i^{(k)} \langle \tilde{\boldsymbol{v}}_{ik} \tilde{\boldsymbol{v}}_{ik}^{\top} \rangle \right]^{-1}, \tag{17}$$

$$\hat{\boldsymbol{\Psi}}_z^{(k)} = \text{diag} \left(\hat{\boldsymbol{\Lambda}}_z^{(k)} \tilde{\boldsymbol{\Sigma}}_{v|z}^{(k)} \hat{\boldsymbol{\Lambda}}_z^{(k)^{T}} + \frac{\sum_{i=1}^{N} \gamma_i^{(k)} \tilde{\boldsymbol{\varepsilon}}_i^{(k)} \tilde{\boldsymbol{\varepsilon}}_i^{(k)^{\top}}}{\sum_{i=1}^{N} \gamma_i^{(k)}} \right). \tag{18}$$

with $\tilde{\boldsymbol{\varepsilon}}_i^{(k)} = \boldsymbol{z}_i - \hat{\boldsymbol{\Lambda}}_z^{(k)} \langle \tilde{\boldsymbol{v}}_i \rangle$. The updated values for the factor loadings $\hat{\boldsymbol{W}}_z^{(k)}$ and mean vectors $\hat{\boldsymbol{\mu}}_z^{(k)}$ are obtained from the augmented factor loadings matrix $\hat{\boldsymbol{\Lambda}}_z^{(k)} = [\hat{\boldsymbol{W}}_z^{(k)} \hat{\boldsymbol{\mu}}_z^{(k)}]$.

3.2 Conversion Phase

The conversion phase involves two steps. First, the sequence of speech parameter vectors $\boldsymbol{Y} = (\boldsymbol{y}_1, \ldots, \boldsymbol{y}_T)$ associated with the articulatory gesture captured by the PMA device is estimated under the probabilistic framework presented above. Then, a parametric synthesis algorithm is used to generate the final time-domain signal from the estimated speech parameters. To estimate the speech parameter vectors, we employ a frame-by-frame procedure based on the well-known Minimum Mean Square Error (MMSE) estimator:

$$\hat{\boldsymbol{y}}_t = \mathbb{E}[\boldsymbol{y}|\boldsymbol{x}_t] = \int \boldsymbol{y} p(\boldsymbol{y}|\boldsymbol{x}_t) d\boldsymbol{y}. \tag{19}$$

where $\hat{\boldsymbol{y}}_t$ is the estimate for the speech parameters at time t, $\mathbb{E}[\cdot]$ represents the expected value, and $p(\boldsymbol{y}|\boldsymbol{x}_t)$ is the speech parameter posterior distribution. This distribution is derived from the joint distribution $p(\boldsymbol{x}, \boldsymbol{y})$ in (9) as

$$p(\boldsymbol{y}|\boldsymbol{x}_t) = \sum_{k=1}^{K} P(k|\boldsymbol{x}_t) p(\boldsymbol{y}|\boldsymbol{x}_t, k), \tag{20}$$

where

$$P(k|\boldsymbol{x}_t) = \frac{\pi^{(k)}\mathcal{N}\left(\boldsymbol{x}_t; \boldsymbol{\mu}_x^{(k)}, \boldsymbol{\Sigma}_{xx}^{(k)}\right)}{\sum_{k'=1}^{K} \pi^{(k')}\mathcal{N}\left(\boldsymbol{x}_t; \boldsymbol{\mu}_x^{(k')}, \boldsymbol{\Sigma}_{xx}^{(k')}\right)}, \tag{21}$$

$$p(\boldsymbol{y}|\boldsymbol{x}_t, k) = \mathcal{N}\left(\boldsymbol{y}; \boldsymbol{\mu}_{y|x_t}^{(k)}, \boldsymbol{\Sigma}_{y|x}^{(k)}\right). \tag{22}$$

The parameters of the k-th component conditional distribution $p(\boldsymbol{y}|\boldsymbol{x}_t, k)$ are derived from those of the joint pdf $p(\boldsymbol{x}, \boldsymbol{y}|k)$ in (7). As already mentioned, the latter distribution is Gaussian with mean $\boldsymbol{\mu}_z^{(k)}$ and covariance matrix $\boldsymbol{\Sigma}_z^{(k)}$. Then, using the standard properties of the joint Gaussian distribution, we can derive the parameters of the conditional distribution as follows,

$$\boldsymbol{\mu}_{y|x_t}^{(k)} = \boldsymbol{\mu}_y^{(k)} + \boldsymbol{\Sigma}_{yx}^{(k)} \boldsymbol{\Sigma}_{xx}^{(k)^{-1}} \left(\boldsymbol{x}_t - \boldsymbol{\mu}_x^{(k)}\right), \tag{23}$$

$$\boldsymbol{\Sigma}_{y|x}^{(k)} = \boldsymbol{\Sigma}_{yy}^{(k)} + \boldsymbol{\Sigma}_{yx}^{(k)} \boldsymbol{\Sigma}_{xx}^{(k)^{-1}} \boldsymbol{\Sigma}_{xy}^{(k)}, \tag{24}$$

where the marginal means $\boldsymbol{\mu}_x^{(k)}$, $\boldsymbol{\mu}_y^{(k)}$ and covariance matrices $\boldsymbol{\Sigma}_{xx}^{(k)}$, $\boldsymbol{\Sigma}_{yy}^{(k)}$, $\boldsymbol{\Sigma}_{xy}^{(k)}$ are obtained by partitioning $\boldsymbol{\mu}_z^{(k)}$ and $\boldsymbol{\Sigma}_z^{(k)}$ into their x and y components.

Finally, by substituting the expression of the conditional distribution $p(\boldsymbol{y}|\boldsymbol{x}_t)$ in (20) into (19), we reach the following expression for the MMSE estimation of the speech parameter vectors,

$$\hat{\boldsymbol{y}}_t = \sum_{k=1}^{K} P(k|\boldsymbol{x}_t) \int \boldsymbol{y} p(\boldsymbol{y}|\boldsymbol{x}_t, k) d\boldsymbol{y}$$

$$= \sum_{k=1}^{K} P(k|\boldsymbol{x}_t) \left(\boldsymbol{A}^{(k)}\boldsymbol{x}_t + \boldsymbol{b}^{(k)}\right), \tag{25}$$

with $\boldsymbol{A}^{(k)} = \boldsymbol{\Sigma}_{yx}^{(k)} \boldsymbol{\Sigma}_{xx}^{(k)^{-1}}$ and $\boldsymbol{b}^{(k)} = \boldsymbol{\mu}_y^{(k)} - \boldsymbol{A}^{(k)}\boldsymbol{\mu}_x^{(k)}$ as can be deduced from (23).

For comparison purposes, we also evaluate a fast, approximate version of the above estimator, which we will refer to as fast MMSE (fMMSE), in which only the most likely Gaussian component k^* is involved:

$$k^* = \underset{1 \le k \le K}{\operatorname{argmax}} P(k|\boldsymbol{y}_t).$$

Then, the fMMSE estimate is defined as:

$$\hat{\boldsymbol{y}}_t \approx \boldsymbol{A}^{(k^*)}\boldsymbol{x}_t + \boldsymbol{b}^{(k^*)}. \tag{26}$$

3.3   Recalibration Procedure

The principle of the direct synthesis technique is that the mapping between articulator movement and the corresponding acoustics can be estimated from a parallel dataset containing simultaneous recordings of PMA and speech signals.

Ideally, this dataset should be recorded soon after it has been agreed that a laryngectomy will be performed to the patient. During the recording session, the patient's voice and corresponding PMA data are acquired using adhesively attached magnets. In addition, the information on the location of the magnets is documented, so that they can be later surgically implanted accordingly. From the collected data the PMA-to-acoustic transformation is learned as described in Sect. 3.1, so it can be readily available to be used by the patient soon after the laryngectomy, as described in Sect. 3.2.

In certain conditions, however, the above training procedure might fail. For example, it is highly unlikely that the magnet positions can be exactly replicated during surgical implantation. Any magnet misplacement will inevitably lead to discrepancies between the PMA data used for training and the data captured during the use of the system, hence leading to the degradation of the speech quality. Furthermore, variations of the relative positions of the magnets with respect the head-frame used to hold the magnetic sensors (see Fig. 1) will also lead to mismatches. Therefore, in most of the practical cases it would be necessary to recalibrate the system to compensate for any magnet misplacement with respect to their original positions used for acquiring the training data. In the following we present a data-driven recalibration procedure to this end.

We will assume that the positions of the magnets before (magnets glued) and after magnet implantation only vary slightly. In this case, the mismatch between the articulatory data captured for the same articulatory gesture pre- and post-magnet implantation can be approximately modelled as,

$$\boldsymbol{x}_t = \boldsymbol{h}(\tilde{\boldsymbol{x}}_t), \tag{27}$$

where $\boldsymbol{x}_t$ and $\tilde{\boldsymbol{x}}_t$ denote the data for the pre- and post- magnet implantation arrangements, respectively, and $\boldsymbol{h}$ is the mismatch function.

We propose the following procedure to estimate the mismatch function $\boldsymbol{h}$. First, after magnet implantation, the patient has to attend another recording session in which he/she is asked to mouth along to some of the utterances recorded during the first recording session. In this case, however, only PMA data is acquired since the patient has already lost their voice. Furthermore, only a small fraction of the data recorded during the first session needs to be recorded during the second session, as the aim of it is not to estimate the full PMA-to-acoustic mapping (as in the first recording session), but to learn the mismatch produced by the magnet misplacement. Next, as the durations of the PMA signals obtained for the same sentence in both sessions may be different, the PMA data for both recording sessions are time-aligned using the Dynamic Time Warping (DTW) technique [41]. From the time-aligned signals the mismatch function $\boldsymbol{h}$ is estimated. Here, we investigate two alternative methods for modelling this function. First, the function is represented as a simple linear mapping:

$$\boldsymbol{x}_t = \boldsymbol{C}\tilde{\boldsymbol{x}}_t + \boldsymbol{d}, \tag{28}$$

with $\boldsymbol{C}$ and $\boldsymbol{d}$ being estimated by least squares regression from the aligned data.

Alternatively, a Multilayer Perceptron (MLP) is used to model $x_t = h(\tilde{x}_t)$. The input to the MLP are the PMA parameter vectors $\tilde{x}_t$ and it tries to predict the corresponding vectors x_t used for training the MFA model. More details about the MLP architecture and its training are given in Sect. 5.6.

After h is estimated (either as a linear operator or a neural network), it is used in a second round of the recalibration procedure to improve the alignment of the PMA data. Thus, the PMA data recorded after magnet implantation is first compensated using the estimated transformation and then DTW-aligned with the original data (recorded with magnets glued). Next, the alignments obtained for the compensated data are used to estimate a more accurate transformation between the PMA data captured in both sessions. This procedure is repeated several times until convergence.

4   Related Work

The direct synthesis technique presented in the previous section shares some similarities with other recently proposed methods. In this section we discuss the relationships between our proposal and those methods, pointing out the similarities and differences.

First, regarding our own previous work, we reported in [13,18,26] that not only speech recognition from PMA data is possible, but also that the performance obtained is on par with that obtained using audio, at least on isolated words and connected digits recognition tasks. This study was later successfully extended to multiple subjects in [24]. With respect to the direct synthesis approach, in [25] we carried out a feasibility study on prediction of the first two speech formants (F1 and F2) from the sensor data using a simple linear transformation. Though promising, the results showed that a more powerful transformation is required for modelling the non-linear mapping between sensor data and acoustic parameters. In [19], a more powerful conversion technique was investigated: a statistical mapping based on shared Gaussian process dynamical models [6,50], which are non-parametric models providing a shared low-dimensional embedding of the articulatory and acoustic data as well as a dynamic model in the latent space. Results reported for isolated-digit synthesis showed the superiority of this approach for modelling the PMA-to-acoustic mapping. Finally, in [21], we proposed a conversion system based on mixture of factor analysers, similar to the one described in this work, and showed the viability of voice reconstruction from PMA data for continuous speech.

In addition to ourselves, other authors have also made important contributions to the field of silent speech interfaces. A good introduction to this subject can be found in [9]. In general, approaches for direct speech synthesis from sensor data can be classified into two categories: model-based and data-driven approaches. Model-based approaches are those in which the sensor data provides interpretable information about the position of the speech articulators. From this information, the shape of the vocal tract can be recovered and speech can be synthesised by using an articulatory synthesiser. For example, in [36,37] the articulatory synthesiser is driven by Magnetic Resonance Imaging (MRI) and X-ray images, respectively, while in [45, 47] EMA data is used instead. In data-driven approaches, on the other hand, the relationship between the sensor data and the acoustics is learned from parallel datasets, as with our method. Various techniques have been investigated in the past to model this relationship: Gaussian Mixture Models (GMMs) [38, 44], Hidden Markov Models (HMMs) [27, 28], neural networks [10, 48], support vector regression [46], and a concatenative, unit-selection approach [51]. In the next section, we will compare the performance of our MFA-based conversion technique with the well-known GMM-based technique proposed in [43, 44].

5   Experimental Evaluation

In this section, we evaluate the performance of our silent speech system on a voice reconstruction task for non-impaired speakers. Although our system is thought to help laryngectomy patients to recover the voice, at this initial stage of the development our priority is to assess performance for normal speakers and then, once the system is robust, it can be tested with real patients. More details about the evaluation framework are given in the following.

5.1   Vocabulary and Data Recording

We recorded two parallel databases with different phonetic coverage to evaluate our system. The first one is based on the TIDigits speech database [34] and consists of sequences of up to seven connected English digits. The vocabulary is made up of eleven words: the digits from 'one' to 'nine' plus 'zero' and 'oh'. The second database consists of utterances selected at random from the CMU Arctic corpus of phonetically balanced sentences [32]. Parallel data was then recorded for the two databases by adult speakers with normal speaking ability. For the TIDigits database, four male speakers (M1 to M4) and a female speaker (F1) recorded 308 sentences (385 sentences for M2) comprising 7.2, 10.5, 8.0, 9.7 and 8.5 min of data, respectively. Speaker M1 also recorded a second dataset with 308 sentences (7.4 min of data) in a different recording session with the aim of evaluating the recalibration procedure described in Sect. 3.3. The magnet arrangement in the first recording session was documented and replicated in the second session. Despite this, as will be discussed below, small variations in the magnet positions and/or orientations unintentionally occurred. For the Arctic database, parallel data was recorded for two male speakers: M1 (same as in the TIDigit database) and M5. M1 recorded 420 utterances comprising 22 min of data and M5 recorded 509 sentences comprising 26 min.

The audio and 9-channel PMA signals were recorded simultaneously at sampling frequencies of 16 kHz and 100 Hz, respectively, using an AKG C1000S condenser microphone and the PMA device shown in Fig. 1. Background cancellation was later applied to the PMA signals to mitigate the effect of the Earth's magnetic field on the articulatory data [5]. Finally, all data were endpointed in the audio domain using an energy-based algorithm to prevent modelling silence parts, as the speech articulators may adopt any position during the silence parts.

5.2   Feature Extraction

The source x_t and target y_t parameter vectors are computed as follows in the proposed system. The PMA signals are first segmented into overlapping frames using a 20 ms analysis window with 10 ms overlap. Next, to better model contextual phonetic information, sequences of ω consecutive frames, with a single-frame displacement, are concatenated and the Partial Least Squares (PLS) technique [8] is applied to reduce the dimensionality of the resultant frames. In PLS, the number of principal components retained are those explainng the 95% of the variance in the target speech features. The audio signals are represented in this work as 25 Mel-Frequency Cepstral Coefficients (MFCCs) [16] obtained at the same frame rate as that for PMA. Neither F_0 nor voicing information are extracted from the audio signals because of the limited ability of PMA to model this aspect of speech articulation [20]. Rather, the audio signals are re-synthesised without voicing as whispered speech. Finally, the PMA and speech parameter vectors are converted to z-scores with zero mean and unit variance to improve statistical training.

5.3   Objective Evaluation of Voice Reconstruction Accuracy

In this work we use objective quality measures to evaluate the performance of the techniques under different conditions. As we have access to the original speech signals recorded by the subjects, we can compare the speech signals predicted from articulator movement with the original ones to evaluate reconstruction accuracy. In particular, we use the well-known Mel-Cepstral Distortion (MCD) measure [33] between the MFCCs extracted from the original audio signals, c, and the ones predicted from PMA data, $\hat{c}$, with smaller values indicating better results:

$$\text{MCD[dB]} = \frac{10}{\ln 10} \sqrt{2 \sum_{d=1}^{D} (c_d - \hat{c}_d)^2}. \tag{29}$$

Results reported in this work are obtained using a 10-fold cross-validation scheme. Hence, the available data for each subject is randomly divided into ten sets of same length and, in each round, 9 sets are used for training and the remaining one for testing. The MCD results reported in the following sections correspond to the average MCD result for the 10 rounds.

5.4   Results on the TIDigits Database

The left and middle panels in Fig. 2 show contour plots with the average MCD results across the five subjects for the two conversion algorithms introduced in Sect. 3.2: MMSE and fast MMSE (fMMSE). The results are presented as a

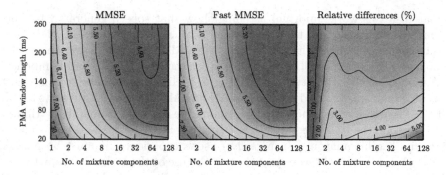

Fig. 2. Mel-cepstral distortion results on the TIDigits database. *Left and middle*: Average MCD results across all the speakers for the MMSE and fMMSE conversion systems as a function of the number of mixture components used in the MFA model and the length of the PMA-frame window. *Right*: Relative degradation of the fMMSE system with respect to the MMSE-based one.

function of the number of mixture components in the MFA model (i.e. K in (20)) and the length of the sliding window used to extract the PMA parameter vectors. As can be seen, results greatly improve when more mixture components and longer windows are used for modelling the PMA-to-acoustic mapping. Using more mixtures means that this mapping, which is known to be highly non-linear [1,39], is more finely represented. For example, the mapping is approximated by a linear transformation when using 1-mixture models, while a piece-wise linear approximation is employed for $K > 1$. For a PMA frame window of $\omega = 200$ ms, the relative MCD reduction when using $K = 64$ components, which is the optimum number of mixtures as can be seen in the figure, with respect to just using a single mixture is 30.33% for the MMSE system and 29.08% for the fMMSE-based one. Increasing the length of the PMA frame window is also beneficial for the mapping because it reduces its uncertainty by taking into account more contextual information about the temporal evolution of the PMA signal. In particular, for $K = 64$ mixtures, the relative reduction in MCD achieved when using a window of $\omega = 200$ ms instead of $\omega = 20$ ms is 13.52% and 17.28% for the MMSE and fMMSE systems, respectively.

The right panel in Fig. 2 shows the relative differences (expressed in percent) between the MMSE method and its fast version fMMSE. As it can be seen, both methods perform almost equally except when $K > 1$ or when short windows are used. In those cases, the performance of the fMMSE method degrades because the mapping uncertainty is higher when using short windows and, hence, it is more difficult for the fMMSE method to choose the 'correct' mixture component for performing the mapping. For example, for $K = 128$ and $\omega = 20$ ms, the fMMSE algorithm is 7.21% worse than MMSE. Conversely, the differences are almost insignificant ($\leq 2\%$ of degradation) for $K = 1$ or when long windows are used (e.g. $\omega = 200$ ms). In terms of speech intelligibility, though not formally evaluated in this work, informal listening show that speech generated by both methods is intelligible and that the speaker's voice is clearly identifiable[1].

The detailed MCD results obtained by the MMSE conversion system for each of the five subjects in the TIDigits database are shown in Fig. 3. A 64-component MFA model, which is the best model in Fig. 2, is chosen. As can be seen in Fig. 3, the best results are obtained for subjects M1 and M4 and the worst results for M2 and F1. The differences in performance among speakers can be mainly attributed to two factors: the user's experience in using the PMA device and how well the device fits her/his anatomy. In regard of user's experience, it must be pointed out that M1, M3 and M4 were proficient in the use of the PMA device, while for M2 and F1 the data recording session was also the first time they used the PMA device. With respect to the second reason, the current PMA device prototype was specifically designed for M1, so it is reasonable to think that articulatory data is more accurately captured for him than for the other subjects.

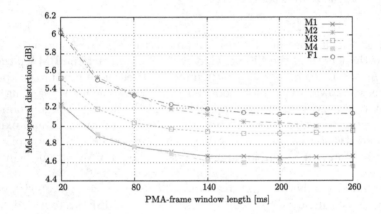

Fig. 3. Mel-ceptral distortion results obtained by the MMSE conversion system for each speaker in the TIDigits database.

Next, we compare our proposal with the well-known GMM-based conversion technique proposed by Toda et al. in [43, 44]. For a fairer comparison, both methods are evaluated using the MMSE-based mapping algorithm. Also, we evaluate our proposal using different dimensions for the latent space variable v_t in (5). The dimensions are 5, 10, 15, 20, and 25, the latter being the dimensionality of the speech feature vectors. Results are shown in Fig. 4 for both systems. It can be seen that both methods perform almost equally except when the dimensionality of the latent space in the our system is very small (i.e. 5 or 10). In this case, the quality of synthetic speech is slightly degraded due to the inability of properly capturing the correlations between the acoustic and PMA spaces in such low-dimensional latent spaces. For dimensions greater than 15, both approaches

---

[1] Several speech samples are available in the Demos section of http://www.hull.ac.uk/ speech/disarm.

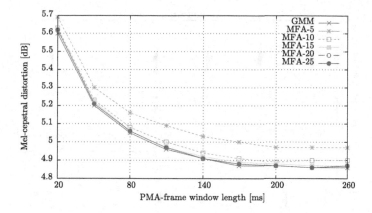

Fig. 4. Comparison between the proposed approach for articulatory-to-acoustic conversion using MFAs and Toda's et al. approach using GMMs [44]. For our proposal, the conversion accuracy using different latent space dimensions (i.e. 5, 10, 15, 20, and 25) for v_t in (5) is evaluated.

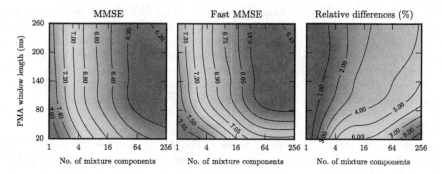

Fig. 5. Mel-cepstral distortion results on the Arctic database. *Left and middle*: Average results across all subjects for the MMSE and fMMSE systems as a function of the number of mixture components used in the MFA model and the length of the PMA-frame window. *Right*: Relative degradation of the fMMSE system with respect to the MMSE-based one.

report more or less the same results, with the benefit that our proposed approach is more computationally efficient because of the savings of carrying out the computations in the reduced-dimension space.

5.5   Results on the Arctic Database

Figure 5 shows the MCD results obtained by the MMSE and fMMSE systems on the Arctic database (left and middle contour plots), as well as the relative degradation of the fast MMSE algorithm w.r.t. the MMSE algorithm (right contour plot). It can be seen that the results for the Arctic database are not as good as the results for the TIDigit database in Fig. 2. This is due to greater phonetic

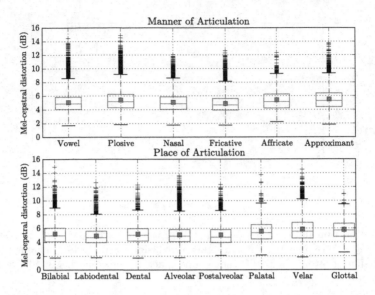

Fig. 6. Box plot of the distributions of the MCD results for different phone categories on the Arctic database. Bottom and top edges of the boxes are the first (Q1) and third (Q3) quartiles of the data. Red bands inside the boxes represents the median, while the means are represented with small, red boxes. Whiskers extend up to 1.5 times the interquartile range (i.e. Q3-Q1) and the outliers are plotted with black crosses. Phone categories are those in the IPA chart [30]. *Top*: Results when considering the manner of articulation of the phones. *Bottom*: Results for the place of articulation.

variability of the Arctic sentences. In fact, the Arctic corpus was designed for phonetic balance. This greater complexity results in a more complex PMA-to-acoustic mapping in the case of the Arctic sentences. Apart from that, another reason the MCD results are worse on the Arctic database is the limitations of the current PMA device for modelling some areas of the vocal tract (e.g. sounds articulated at the back of the mouth) [20,21], as discussed below. Since the Arctic sentences contain more phones articulated in those areas than the digits vocabulary, this harms the overall reconstruction performance achieved on the Arctic database. Regarding the two conversion algorithms, the MMSE-based system outperforms its fast version (fMMSE) again. Nevertheless, the differences between both systems are small, particularly when both high number of mixtures and long windows are used for the mapping. The best overall results are obtained using 256 mixtures in the MFA model and an analysis window spaning 80 ms. This is configuration used in the rest of this section.

To investigate in detail the performance of our system, we conduct a second analysis at the phone level in which the distortions are computed for each phone and the resultant distributions are represented as a box plot. For the sake of clarity, the MCD results of the phones sharing similar articulation properties are grouped together rather than presenting the results for each individual phone. Here, phones are grouped according to their manner and place of articulation.

For segmenting the speech signals into phones, we force-aligned their word-level transcriptions using a cross-word, triphone-based speech recogniser adapted to each subject. The phone-level transcriptions with timing information provided by the forced-alignment procedure are then used to segment the original and estimated speech signals. The results of this second analysis are shown in Fig. 6. When considering the manner of articulation, we can see that the plosive, affricate and approximant consonants tend to be synthesised less accurately than other sound classes due to their more complex articulation and dynamics. When considering the place of articulation, it can seen that sounds articulated at the middle and back of the mouth (palatal, velar, and glottal consonants) are, on average, less well reconstructed than other phones. This is due to the limitations of the current PMA prototype for modelling those areas of the vocal tract [20,21].

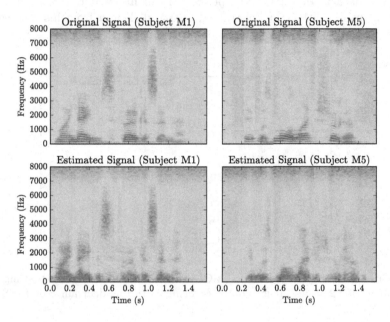

Fig. 7. *Top-left and bottom-left*: Spectrograms of natural speech (top), and speech estimated from PMA data (bottom) corresponding to the sentence "My name is Ferguson" spoken by the subject M1. *Top-right and right-left*: Spectrograms of the sentence "It was more like sugar" spoken by M5.

Finally, in Fig. 7, a comparison between the spectrograms computed from natural speech and from the speech synthesised by our system is presented for the two subjects in the Arctic database. As can be seen, our system is able to predict with high level of accuracy the speech formants and, in general, the spectral envelope of the signals. Spectral detail, however, is lost due to over-smoothing when training the MFA models [52] and the limited information provided by PMA about the articulation process. It also can be seen that the estimated signals are synthesised with no voicing. This is also due to the limited ability of the current PMA prototype for capturing the movement of the vocal folds.

5.6   Cross-Session Synthesis Results

So far it has been assumed that there is no mismatch between the data used for training and that used for testing. However, as already discussed in Sect. 3.3, this is not always true. Variations in the positions of the magnets pre- and post-implantation as well as variations in the relative position of magnets with respect to the head-frame used to hold the magnetic sensors (see Fig. 1), will inevitably lead to mismatches that will degrade the quality of speech synthesised from sensor data. In this section, we evaluate the performance of the direct synthesis technique in one scenario which introduces such mismatch: speech is synthesised from PMA data recorded by the speaker M1 in his second recording session (Session 2) using a MFA model trained on parallel data from his first session (Session 1).

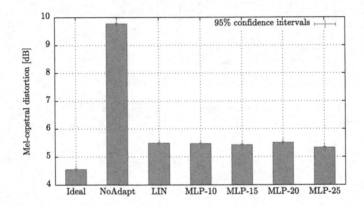

Fig. 8. Cross-session synthesis results: MCD results obtained when synthesising PMA data from Session 2 in the TIDigits database using a MFA model trained on the Session 1 dataset.

Figure 8 shows the MCD results obtained for the above experiment when a 64-component MFA model and a PMA-frame window of 200 ms are used. In the figure, Ideal refers to the ideal case in which there is no mismatch between training and testing (i.e. parallel data from Session 2 is used for training and testing within the cross-validation scheme), the NoAdapt system directly convert the sensor data from Session 2 using the model trained on data from Session 1 with no compensation, and the remaining results are for the compensation technique proposed in Sect. 3.3: LIN models the mismatch function as a linear transformation, while MLP uses a multilayer perceptron with 10, 15, 20 and 25 sigmoid units in the hidden layer.

In the figure, the best results are obtained in the Ideal case where there is no mismatch between training and testing. Even though magnet placement was documented to avoid misplacement between sessions, we see from the NoAdapt results that even small changes between sessions are catastrophic in terms of the synthesised speech quality. This is greatly alleviated, however, by the proposed compensation technique. In this case, the results are only slightly worse than the result obtained in the ideal case. Regarding the different approaches for mismatch compensation, it can be seen that the best results are obtained using a MLP with 25 hidden units due to the greater modelling flexibility allowed by this model. Nevertheless, a simple linear transformation (LIN) also achieves very similar results to MLP-25 with the benefit of LIN being more computationally efficient.

6   Conclusions

In this chapter we have described a system for synthesising speech from motion data captured from the lips and tongue using magnetic sensing. Preliminary evaluation of the system via objective performance metrics show that the proposed system is able to generate speech of sufficient quality for some vocabularies. However, problems still remain to scale up the system to work consistently for phonetically rich vocabularies. It has also been reported that one of the current limitations of our sensing technique, that is, the differences between the articulatory data captured in different sessions, can be greatly reduced by applying a pre-processing technique to the sensor data before the conversion. This brings us closer to being able to apply our voice reconstruction system in a realistic treatment scenario. These results encourage us in pursuing our goal of developing a system that will ultimately allow laryngectomy patients to recover their voice. In order to reach this point, a number of questions will need to addressed in future research such as improving the performance for phonetically rich vocabularies, ways of predicting the prosodic information from PMA data, and extending the technique to impaired subjects.

Acknowledgements. This is a summary of independent research funded by the National Institute for Health Research (NIHR)'s Invention for Innovation Programme. The views expressed are those of the authors and not necessarily those of the NHS, the NIHR or the Department of Health.

References

1. Atal, B.S., Chang, J.J., Mathews, M.V., Tukey, J.W.: Inversion of articulatory-to-acoustic transformation in the vocal tract by a computer-sorting technique. J. Acoust. Soc. Am. **63**(5), 1535–1555 (1978)
2. Braz, D.S.A., Ribas, M.M., Dedivitis, R.A., Nishimoto, I.N., Barros, A.P.B.: Quality of life and depression in patients undergoing total and partial laryngectomy. Clinics **60**(2), 135–142 (2005)

3. Byrne, A., Walsh, M., Farrelly, M., O'Driscoll, K.: Depression following laryngectomy. A pilot study. Brit. J. Psychiat. **163**(2), 173–176 (1993)
4. Cheah, L.A., Bai, J., Gonzalez, J.A., Gilbert, J.M., Ell, S.R., Green, P.D., Moore, R.K.: Preliminary evaluation of a silent speech interface based on intra-oral magnetic sensing. In: Proceedings BioDevices, pp. 108–116 (2016)
5. Cheah, L.A., Bai, J., Gonzalez, J.A., Ell, S.R., Gilbert, J.M., Moore, R.K., Green, P.D.: A user-centric design of permanent magnetic articulography based assistive speech technology. In: Proceedings BioSignals, pp. 109–116 (2015)
6. Chen, J., Kim, M., Wang, Y., Ji, Q.: Switching Gaussian process dynamic models for simultaneous composite motion tracking and recognition. In: Proceedings IEEE Conference Computer Vision and Pattern Recognition, pp. 2655–2662 (2009)
7. Danker, H., Wollbrück, D., Singer, S., Fuchs, M., Brähler, E., Meyer, A.: Social withdrawal after laryngectomy. Eur. Arch. Oto-Rhino-L **267**(4), 593–600 (2010)
8. De Jong, S.: SIMPLS: an alternative approach to partial least squares regression. Chemom. Intell. Lab. Syst. **18**(3), 251–263 (1993)
9. Denby, B., Schultz, T., Honda, K., Hueber, T., Gilbert, J., Brumberg, J.: Silent speech interfaces. Speech Commun. **52**(4), 270–287 (2010)
10. Desai, S., Raghavendra, E.V., Yegnanarayana, B., Black, A.W., Prahallad, K.: Voice conversion using artificial neural networks. In: Proceedings ICASSP, pp. 3893–3896 (2009)
11. Ell, S.R.: Candida: the cancer of silastic. J. Laryngol. Otol. **110**(03), 240–242 (1996)
12. Ell, S.R., Mitchell, A.J., Parker, A.J.: Microbial colonization of the groningen speaking valve and its relationship to valve failure. Clin. Otolaryngol. Allied Sci. **20**(6), 555–556 (1995)
13. Fagan, M.J., Ell, S.R., Gilbert, J.M., Sarrazin, E., Chapman, P.M.: Development of a (silent) speech recognition system for patients following laryngectomy. Med. Eng. Phys. **30**(4), 419–425 (2008)
14. Freitas, J., Teixeira, A., Bastos, C., Dias, M.: Towards a multimodal silent speech interface for European Portuguese. In: Speech Technologies, vol. 10, pp. 125–150. InTech (2011)
15. Fried-Oken, M., Fox, L., Rau, M.T., Tullman, J., Baker, G., Hindal, M., Wile, N., Lou, J.S.: Purposes of AAC device use for persons with ALS as reported by caregivers. Augment Altern. Commun. **22**(3), 209–221 (2006)
16. Fukada, T., Tokuda, K., Kobayashi, T., Imai, S.: An adaptive algorithm for Mel-cepstral analysis of speech. In: Proceedings ICASSP, pp. 137–140 (1992)
17. Ghahramani, Z., Hinton, G.E.: The EM algorithm for mixtures of factor analyzers. Technical report CRG-TR-96-1, University of Toronto (1996)
18. Gilbert, J.M., Rybchenko, S.I., Hofe, R., Ell, S.R., Fagan, M.J., Moore, R.K., Green, P.: Isolated word recognition of silent speech using magnetic implants and sensors. Med. Eng. Phys. **32**(10), 1189–1197 (2010)
19. Gonzalez, J.A., Green, P.D., Moore, R.K., Cheah, L.A., Gilbert, J.M.: A nonparametric articulatory-to-acoustic conversion system for silent speech using shared Gaussian process dynamical models. In: UK Speech, p. 11 (2015)
20. Gonzalez, J.A., Cheah, L.A., Bai, J., Ell, S.R., Gilbert, J.M., Moore, R.K., Green, P.D.: Analysis of phonetic similarity in a silent speech interface based on permanent magnetic articulography. In: Proceedings Interspeech, pp. 1018–1022 (2014)
21. Gonzalez, J.A., Cheah, L.A., Gilbert, J.M., Bai, J., Ell, S.R., Green, P.D., Moore, R.K.: A silent speech system based on permanent magnet articulography and direct synthesis. Comput. Speech Lang. **39**, 67–87 (2016)

22. Heaton, J.M., Parker, A.J.: Indwelling tracheo-oesophageal voice prostheses post-laryngectomy in Sheffield, UK: a 6-year review. Acta Otolaryngol. **114**(6), 675–678 (1994)
23. Herff, C., Heger, D., de Pesters, A., Telaar, D., Brunner, P., Schalk, G., Schultz, T.: Brain-to-text: decoding spoken phrases from phone representations in the brain. Front. Neurosci. **9**, 217 (2015)
24. Hofe, R., Bai, J., Cheah, L.A., Ell, S.R., Gilbert, J.M., Moore, R.K., Green, P.D.: Performance of the MVOCA silent speech interface across multiple speakers. In: Proceedings Interspeech, pp. 1140–1143 (2013)
25. Hofe, R., Ell, S.R., Fagan, M.J., Gilbert, J.M., Green, P.D., Moore, R.K., Rybchenko, S.I.: Speech synthesis parameter generation for the assistive silent speech interface MVOCA. In: Proceedings Interspeech, pp. 3009–3012 (2011)
26. Hofe, R., Ell, S.R., Fagan, M.J., Gilbert, J.M., Green, P.D., Moore, R.K., Rybchenko, S.I.: Small-vocabulary speech recognition using a silent speech interface based on magnetic sensing. Speech Commun. **55**(1), 22–32 (2013)
27. Hueber, T., Bailly, G.: Statistical conversion of silent articulation into audible speech using full-covariance HMM. Med. Eng. Phys. **36**, 274–293 (2016)
28. Hueber, T., Bailly, G., Denby, B.: Continuous articulatory-to-acoustic mapping using phone-based trajectory HMM for a silent speech interface. In: Proceedings Interspeech, pp. 723–726 (2012)
29. Hueber, T., Benaroya, E.L., Chollet, G., Denby, B., Dreyfus, G., Stone, M.: Development of a silent speech interface driven by ultrasound and optical images of the tongue and lips. Speech Commun. **52**(4), 288–300 (2010)
30. International Phonetic Association: The international phonetic alphabet (2005)
31. Jou, S.C., Schultz, T., Walliczek, M., Kraft, F., Waibel, A.: Towards continuous speech recognition using surface electromyography. In: Proceedings Interspeech, pp. 573–576 (2006)
32. Kominek, J., Black, A.W.: The CMU Arctic speech databases. In: Fifth ISCA Workshop on Speech Synthesis, pp. 223–224 (2004)
33. Kubichek, R.: Mel-cepstral distance measure for objective speech quality assessment. In: Proceedings of IEEE Pacific Rim Conference on Communications, Computers and Signal Processing, pp. 125–128 (1993)
34. Leonard, R.: A database for speaker-independent digit recognition. In: Proceedings of ICASSP, pp. 328–331 (1984)
35. Maeda, S.: A digital simulation method of the vocal-tract system. Speech Commun. **1**(3), 199–229 (1982)
36. Mullen, J., Howard, D.M., Murphy, D.T.: Waveguide physical modeling of vocal tract acoustics: flexible formant bandwidth control from increased model dimensionality. IEEE Trans. Audio Speech Lang. Process. **14**(3), 964–971 (2006)
37. Murphy, D.T., Jani, M., Ternström, S.: Articulatory vocal tract syntheis in supercollider. In: Proceedings of International Conference on Digital Audio Effects, pp. 1–7 (2015)
38. Nakamura, K., Toda, T., Saruwatari, H., Shikano, K.: Speaking-aid systems using GMM-based voice conversion for electrolaryngeal speech. Speech Commun. **54**(1), 134–146 (2012)
39. Neiberg, D., Ananthakrishnan, G., Engwall, O.: The acoustic to articulation mapping: non-linear or non-unique? In: Proceedings Interspeech, pp. 1485–1488 (2008)
40. Petajan, E.D.: Automatic lipreading to enhance speech recognition (speech reading). Ph.D. thesis, University of Illinois at Urbana-Champaign (1984)
41. Sakoe, H., Chiba, S.: Dynamic programming algorithm optimization for spoken word recognition. IEEE Trans. Acoust. Speech Sig. Process. **26**(1), 43–49 (1978)

42. Schultz, T., Wand, M.: Modeling coarticulation in EMG-based continuous speech recognition. Speech Commun. **52**(4), 341–353 (2010)
43. Toda, T., Black, A.W., Tokuda, K.: Voice conversion based on maximum-likelihood estimation of spectral parameter trajectory. IEEE Trans. Audio Speech Lang. Process. **15**(8), 2222–2235 (2007)
44. Toda, T., Black, A.W., Tokuda, K.: Statistical mapping between articulatory movements and acoustic spectrum using a Gaussian mixture model. Speech Commun. **50**(3), 215–227 (2008)
45. Toutios, A., Maeda, S.: Articulatory VCV synthesis from EMA data. In: Proceedings Interspeech (2012)
46. Toutios, A., Margaritis, K.G.: A support vector approach to the acoustic-to-articulatory mapping. In: Proceedings Interspeech, pp. 3221–3224 (2005)
47. Toutios, A., Narayanan, S.: Articulatory synthesis of French connected speech from EMA data. In: Proceedings Interspeech, pp. 2738–2742 (2013)
48. Uria, B., Renals, S., Richmond, K.: A deep neural network for acoustic-articulatory speech inversion. In: Proceedings of NIPS 2011 Workshop on Deep Learning and Unsupervised Feature Learning (2011)
49. Wand, M., Janke, M., Schultz, T.: Tackling speaking mode varieties in EMG-based speech recognition. IEEE Trans. Bio-Med. Eng. **61**(10), 2515–2526 (2014)
50. Wang, J.M., Fleet, D.J., Hertzmann, A.: Gaussian process dynamical models for human motion. IEEE Trans. Pattern Anal. Mach. Intell. **30**(2), 283–298 (2008)
51. Zahner, M., Janke, M., Wand, M., Schultz, T.: Conversion from facial myoelectric signals to speech: a unit selection approach. In: Proceedings Interspeech, pp. 1184–1188 (2014)
52. Zen, H., Tokuda, K., Black, A.W.: Statistical parametric speech synthesis. Speech Commun. **51**(11), 1039–1064 (2009)

Health Informatics

Connecting Multistakeholder Analysis Across Connected Health Solutions

Noel Carroll[1(✉)], Marie Travers[2], and Ita Richardson[2]

[1] ARCH – Applied Research in Connected Health Technology Centre, University of Limerick,
Limerick, Ireland
noel.carroll@lero.ie
[2] Lero – the Irish Software Research Centre, University of Limerick, Limerick, Ireland
{marie.travers,ita.richardson}@lero.ie

Abstract. Connected Health can be described as a patient- or consumer-centred socio-technical healthcare management model which exploits the use of information and communication technology (ICT) during clinical or wellness decision-making tasks. It facilitates the connectivity of information sources and extends healthcare services and processes beyond traditional healthcare institutions. However, while much of the emphasis has been on developing the technology to facilitate Connected Health, there are few efforts that have established an evaluation model to encapsulate and assess the value and potential impact of Connected Health solutions, particularly from multiple stakeholders' perspectives. Many information systems (IS) and health information systems (HIS) models are narrow in their evaluation focus. We present the Connected Health Evaluation Framework (CHEF), which offers a generic approach that encapsulates a holistic view of a Connected Health evaluation process. It focuses on four key domains: end-user perception, business growth, quality management and healthcare practice. We present a case study on the Irish Blood Transfusion Service (IBTS) and describe the multi-stakeholder analysis challenge which this presents from an evaluation perspective. We also explore how CHEF could have guided more successful outcomes though our evaluation process.

1 Introduction

Societal and demographic changes, coupled with economic challenges, have driven the need for us to reconsider how we deliver health and social care in our community [40]. Healthcare places considerable financial burdens on both public purse and personal finance. In addition, due to demographic shifts, there is a growing demand for care to be delivered in a more personalised context, delivering 'smart' solutions via technological devices. Connected Health is an emerging and rapidly developing field which has the potential to transform healthcare service systems by increasing its safety, quality and overall efficiency.

While considered a disruptive technological approach in healthcare, Connected Health is used by different industries in various sector contexts (for example, healthcare, social care and the wellness sector). Thus, various definitions exist with different

© Springer International Publishing AG 2017
A. Fred and H. Gamboa (Eds.): BIOSTEC 2016, CCIS 690, pp. 319–339, 2017.
DOI: 10.1007/978-3-319-54717-6_18

emphasis placed on healthcare, business, technology and support service providers, or any combination of these.

Within the research community, Connected Health is not well defined and remains an ambiguous concept. The ECHAlliance [13] group promotes the concept of Connected Health to act as "the umbrella description covering digital health, eHealth, mHealth, telecare, telehealth and telemedicine". In addition, Caulfield and Donnelly [4] defines of Connected Health as "a conceptual model for health management where devices, services or interventions are designed around the patient's needs, and health related data is shared, in such a way that the patient can receive care in the most proactive and efficient manner possible". The key here is the connectedness and the manner in which technological solutions enable healthcare solutions. In addition, the FDA (2014) describes Connected Health as "electronic methods of health care delivery that allow users to deliver and receive care outside of traditional health care settings. Examples include mobile medical apps, medical device data systems, software, and wireless technology". Thus, as technological solutions seek to enable new healthcare relationships and partnerships, there is a growing interest in examining information and communications technology (ICT) to support the development of Connected Health. Connected Health has been defined by Richardson [39] as "patient-centred care resulting from process-driven health care delivery undertaken by healthcare professionals, patients and/or carers who are supported by the use of technology (software and/or hardware)". Therefore Connected Health can be considered to be a socio-technical healthcare model that extends healthcare services beyond healthcare institutions. We capture this in the term 'ecosystem'. A Connected Health Ecosystem implies that we need to strike a balance between the various requirements and dynamics associated with different stakeholder groups in a modern healthcare sector. This can include primary care, secondary care, payers, policy makers, pharmacies, clinicians, patients, family members, innovators, public officials, patient groups, academics and entrepreneurs, all/any of whom can collaborate to experiment, develop protocols and tests, and evaluate new Connected Health service solutions.

As technological solutions seek to enable such connectivity between healthcare stakeholders [21], there is a growing interest in examining how ICT enables Connected Health solutions. If healthcare technology is not designed, developed, implemented, maintained, or used properly, it can pose risks to patients [2]. Therefore, a continuous evaluation lifecycle is critical for various stages of the service lifecycle. However, healthcare technology, such as the case with Connected Health, lags behind in presenting evidence-based evaluation on the contribution of ICT in supporting healthcare services (for example, [14, 20, 31, 45]).

This paper offers an overview of some of the key evaluation frameworks in e-health and information systems (IS) and investigates how these can contribute towards the evaluation of Connected Health. Bridging these efforts, we propose a *Connected Health Evaluation Framework* (CHEF). CHEF also plays on the fact that we need to evaluate all of the 'ingredients' before we can learn of the potential impact of Connected Health technology.

2   Objective and Approach

Connected Health is emerging as a solution which offers significant promise in how healthcare can deliver accessible care with improved safety and patient outcomes. Connected Health encompasses terms such as wireless, digital, electronic, mobile, and tele-health and refers to a conceptual model for health management where devices, services or interventions are designed around the patient's needs.

Considering the emerging nature of Connected Health, there are few attempts to develop evaluation frameworks to guide how to investigate the impact of Connected Health technologies. To address this gap, we formulate the following research question: *What are the main evaluation categories to improve the ability to holistically assess the impact of Connected Health innovation?*

To explore this question, we undertook a literature review with a particular emphasis on information systems (IS) and healthcare IS (HIS) evaluation literature.

3   IS and HIS Evaluation Models

The process of evaluation serves a number of fundamental objectives. Within a healthcare context, evaluating the impact of IS is important to understand the dynamic nature of technology and its ability to improve clinical performance, patient care, and service operations [30]. Therefore, evaluation offers us the ability to learn from past and present performance [15] with a view to improving processes, care [28], economics [7, 47] and healthcare satisfaction for the future [27, 46].

Identifying various methods of evaluation throughout the IS literature enables us to build on the current knowledge and identify techniques to improve healthcare systems [49] which support the emergence and evidence-base of Connected Health innovation. We build on the work of O'Leary et al. [34] in adopting a generic approach to untangle the complexity of evaluating Connected Health innovation. We also extend on our research on CHEF [3] with a view to demonstrate the need for multi-stakeholder analysis though our evaluation process.

There have been several well-cited evaluation models across the IS and healthcare field which we examined and which have influenced the development of CHEF. For example, various evaluation approaches on IS were developed with different outlooks including technical, sociological, economic, human and organisational. A number of frameworks also explicitly focus on HIS evaluation.

Our selection criteria were based on the search for information system evaluation models which adopts multiple perspectives of assessment. The majority of the IS models focus on a business context but offer interesting characteristics which could be employed in a Connected Health context. We summarise these perspectives as follows:

- **Clinical:** medical practice, based on observation, interaction and treatment of patients;
- **Technical:** the application of hardware and software devices to connect healthcare service operations in a more efficient manner;

- **Economic:** understanding the processes that govern the production, distribution and consumption of goods and services which impact on healthcare;
- **Human:** training, personnel attitudes, ergonomics and regulations affecting employment and patient experience in healthcare. This can also examine the evolution of social behaviour and development through the influence of both internal (e.g. attitudes, emotion, or health status) and external factors (e.g. service availability or economics of care);
- **Organisational:** the nature of the healthcare organisation, its structure, culture and politics affect an evaluation;
- **Regulation:** a mechanism to sustain and focus control which is often exercised by a public agency over activities that are valued by the healthcare community and its stakeholders.

We examine these key factors in a number of HIS and IS evaluation models and summarise their primary focus in Table 1. We discovered that many of the models were too narrow in focus and only address a specific element of IS which would not be suitable for the generic nature of Connected Health.

Table 1. Summary of IS Evaluation Frameworks.

Framework	Clinical	Technical	Economic	Human	Organisational	Regulation
4Cs Model	✓	✗	✗	✓	✓	✗
CHEATS Model	✓	✓	✗	✓	✓	✗
TEAM	✗	✓	✗	✓	✓	✗
ITAM	✗	✓	✗	✓	✓	✗
IS Success Model	✗	✓	✓	✓	✓	✗
TAM	✗	✓	✗	✓	✗	✗
HOT-fit Model	✗	✓	✗	✓	✓	✗
Integrated Model	✓	✓	✗	✓	✗	✗
RATER Model	✗	✓	✗	✓	✓	✗
Search Engine Success Model	✗	✓	✓	✓	✓	✗

This indicates that there is a lack of a holistic evaluation approach on healthcare which must be addressed in Connected Health to deliver innovative and perhaps 'disruptive' solutions [5, 43]. Regardless, there have been some efforts to evaluate HIS including clinical decision support systems. We present the most suitable IS models which were considered to influence the development of CHEF.

3.1   4Cs Model

The 4Cs Evaluation Framework steers away from the technical issues of evaluation and using a social interactionist perspective, it examines how human, organisational and social issues are important for service design, development and deployment. The 4Cs framework examines issues associated with communication, care, control, and context based on medical informatics [25, 26].

3.2   CHEATS Model

Another model that evaluates the use of ICT in healthcare includes the CHEATS framework [44]. It evaluates healthcare through six core areas:

- **Clinical:** focusing on issues such as quality of care, diagnosis reliability, impact and continuity of care, technology acceptance, practice changes and cultural changes;
- **Human and Organisational:** focusing on issues such as the effects of change on the individual and on the organisation;
- **Educational:** focusing on issues such as recruitment and retention of staff and training;
- **Administrative:** focusing on issues such as convenience, change and cost associated with health systems;
- **Technical and Social:** focusing on issues such as efficacy and effectiveness of new systems and the appropriateness of technology, usability, training and reliability of healthcare technology.

3.3   TEAM

The Total Evaluation and Acceptance Methodology (TEAM) is also used evaluate HIS. It offers an approach based on systemic and model theories [18] and identifies three key IS evaluation dimensions in biomedicine:

- **Role:** evaluates IS from the designer, specialist user, end user and stakeholder perspective;
- **Time:** identifies four main phases which provide relative stability of the IS;
- **Structure:** distinguishes between strategic, tactical or organisational and operational levels.

3.4   ITAM

Adopting a similar outlook on technology evaluation, Dixon [12] presents a socio-technical evaluation model which examines the behavioural aspects of technology using the IT Adoption Model (ITAM). ITAM provides a framework for using implementation strategies and evaluation techniques from an end-user's perspective (i.e. fit for purpose, user perceptions of innovation usefulness and ease of use, and adoption and utilisation). Related research also focuses on consumer health behaviours and their adoption of medical technologies. For example, Wilson and Lankton [48] examines consumer acceptance of HIS to support patients in managing chronic disease.

3.5 IS Success Model

From an IS perspective, there are also several well cited evaluation frameworks which we examined. For example, the IS Success Model [10, 11] examines the success of IS from a number of different perspectives and classifies them into six categories of success [10]. The model adopts a multidimensional framework which measures interdependencies between the various categories (Fig. 1):

1. Information
2. System and service quality
3. Use (intention to)
4. User satisfaction
5. Net benefits

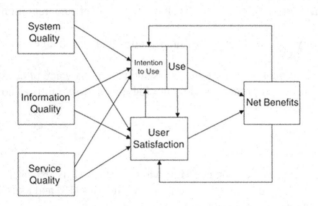

Fig. 1. IS Success Model [10].

These dimensions suggest that there is a clear relationship between the six categories and influences the success of the IS (i.e. net benefits). The net benefits influence user satisfaction and use of the information system.

3.6 TAM

The Technology Acceptance Model (TAM) examines how users accept the use of technology though a number of important influential factors [8]. Among these factors are (see Fig. 2):

- The perceived usefulness (U) of the technology;
- The perceived ease-of use (E) of the technology.

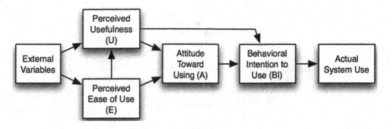

Fig. 2. Technology Acceptance Model [8].

TAM suggests that these factors determine people's intention to use a technology. While TAM provides an excellent approach to examining people's acceptance of technology, it is limited in explanatory terms [19] of technological 'value'.

3.7 Hot-Fit Model

Yosof et al. [49] proposed the Human, Organization and Technology-fit (HOT-fit framework) which was developed from a literature review on HIS evaluation studies. Our literature review revealed that specific instances of the evaluation of healthcare technology does exist [29, 35]. However, there is no evidence of a generic evaluation model which can be applied to Connected Health to provide a holistic view of its potential impact.

3.8 Integrated Model

Wilson and Lankton [48] integrated the use of TAM to extend the model into the Integrated Model (Fig. 3). Their Model merges the perception of technology's usefulness (PU) with extrinsic motivation (EM) in a PU-EM scale and perception of a technology's ease of use (PEOU) scales. The key factors of this model evaluate healthcare technology by examining the:

- Perception of a technology's usefulness (PU);
- Perception of a technology's ease of use (PEOU);
- Behavioural intention (BI) to use the technology;
- Intrinsic motivation (IM);
- Extrinsic motivation (EM) to determine BI.

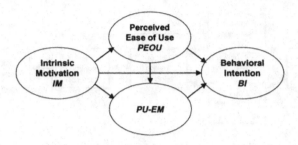

Fig. 3. Integrated Model [48].

The five dimensions identified using the Integrated Model can also provide a useful lens to understand the impact of technology in Connected Health, particularly the influential factors on IT-enabled innovation and the adoption of solutions. Identifying gaps in health service sectors is important to enhance the overall quality of the service delivery and identify how Connected Health solutions can address these gaps.

3.9   RATER Model

There are a number of methods which evaluate the quality of services with a view of identifying areas to prioritise service improvements. For example, the RATER Model [50] offers a simplified version of the SERVQUAL model [36] using five key customer service issues, shown in Table 2.

Table 2.   Key Dimensions within the RATER Model.

Dimension	Description
Reliability	Ability to provide dependable service, consistently, accurately, and on-time
Assurance	The competence of staff to apply their expertise to inspire trust and confidence
Tangibles	Physical appearance or public image of a service, including offices, equipment, employees, and the communication material
Empathy	Relationship between employees and customers and the ability to provide a caring and personalised service
Responsiveness	Willingness to provide a timely, high quality service to meet customers' needs

They focus on five dimensions to analyse and improve service offerings. The five key dimensions can also support the development of a service plan to improve service delivery and are particularly apt in Connected Health solutions. There have been several well-cited evaluation models across the IS and healthcare field which we examined and which have influenced the development of CHEF. For example, various evaluation approaches on IS were developed with different outlooks including technical, sociological, economic, human and organisational. A number of frameworks also explicitly focus on HIS evaluation.

3.10   Search Engine Success Model

In a similar vein, Carroll [1] extends the IS Success Model to develop the Search Engine Success Model and examines the complex task of evaluating the impact of search engine technology on users. The interdependencies between the components builds upon DeLone and McLean IS Success Model but include a more comprehensive view of the value co-creation relationship between the organisation and end-user. From a Connected Health perspective, this model illustrates the cyclical nature of establishing trust to generate and sustain net benefits. The model adopts a multidimensional framework which measures interdependencies between the various categories (Fig. 4):

- Information
- System and service quality
- Use (intention to)
- Technological capabilities
- Quality of experience
- User expectation
- User satisfaction
- Cognitive reasoning
- Knowledge generation
- Net benefits through a co-creation relationship

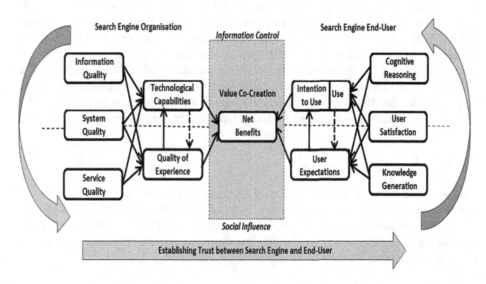

Fig. 4.   Search Engine Success Model (Carroll 2014).

3.11   Research Gap

From our literature review, we can conclude that evaluating the value of HIS is a complex task. This is also confirmed by a recent report on 'The Value of Health Information Technology: Filling the Knowledge Gap' [42] which draws similar conclusions in that the majority of evaluation articles are limited. They state that evaluation articles use "incomplete measures of value and fail to report the important contextual and implementation characteristics that would allow for an adequate understanding of how the study results were achieved", and provide a conceptual framework using three key principles for measuring the value of healthcare IT as follows:

- Value includes both costs and benefits;
- Value accrues over time;
- Value depends on which stakeholder's perspective is used.

These principles suggest that a core focus of an evaluation strategy ought to focus on 'value' and how this can be represented from various stakeholders' perspectives. Other models discussed above referred to this as 'net benefits' or 'value co-creation'. In summary, while the frameworks explored in this report evaluate various aspects of HIS and IS, they do not provide a holistic view of healthcare technology and cannot be successfully applied to support the board nature of Connected Health.

To address this gap, we posed the following research question: what are the main evaluation categories to improve the ability to holistically assess the impact of Connected Health innovation?

Through our literature search and experience with Connected Health companies and with the aim of developing a more universally adoptable framework for multiple perspectives of Connected Health, we propose the Connected Health Evaluation Framework (CHEF). The need for such an approach was also highlighted by [42] who raise concerns regarding evaluation in healthcare: "unfortunately, we have found that few studies include both costs and benefits in their definitions of value. Most studies look at only short-term time horizons, which ignore many of the downstream benefits of the HIT, and many studies don't even explicitly state to whom the value is accruing." We explain how CHEF sets out to address this gap and our research question.

4   CHEF

This section presents the Connected Health Evaluation Framework (CHEF). The development of CHEF (Fig. 5) is influenced by both the strengths of current HIS/IS models and the limitations of these models which emerged from the literature review. In addition, although economics and regulation often shape innovation, both have been largely overlooked in many of the evaluation models we identified.

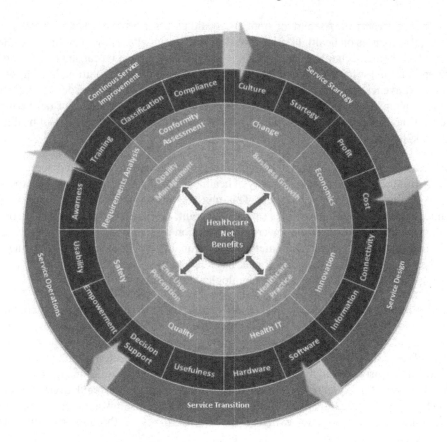

Fig. 5. Connected Health Evaluation Framework (CHEF).

4.1 CHEF Structure

'Healthcare net benefits' are presented at the core of CHEF. It comprises four main layers for Connected Health, broadly addressing clinical, business, users and systems with a view to determine how these co-create value. Each of the categories supports specific Connected Health operations across all service lifecycle stages, ultimately generating healthcare net benefits.

Business Growth. As part of the overall healthcare service strategy phase, Business Growth focuses on driving change and economics in healthcare and organisational market share. Particular emphasis on evaluation focuses on the cultural and strategy change for the introduction of Connected Health innovations. While introducing Connected Health innovation, an economic evaluation should be undertaken to examine the business case for Connected Health innovation, potential profits and costs associated its implementation.

Healthcare Practice. As part of both the healthcare service design and transition phases, this focuses on health IT and innovation and how it alters practice/clinical pathways [33]. From a technological perspective, an evaluation must be carried out on both the hardware and software capability to deliver a Connected Health solution. In addition, the innovativeness of altering healthcare practice is evaluated from a socio-technical and ethnography viewpoint. This allows the examination of the impact of delivering information in a new format and whether it enhances the overall connectivity of healthcare stakeholders.

End-User Perception. As part of both the healthcare service transition and operations phases, this focuses on safety and quality of healthcare innovation from a user's perspective (e.g. a doctor, a patient or carer). This phase evaluates the safety and quality of Connected Health solutions. From a safety viewpoint, an evaluation may be carried out on the usability and level of empowerment a solution may provide in order to provide a balance in empowerment and safety. From a quality viewpoint, we can evaluate whether Connected Health technologies have led to improved healthcare decision-making and enhanced usefulness of technological innovations.

Quality Management. As part of both the healthcare service operations and continuous service improvement phases, quality management focuses on technical and regulation requirements and conformity assessment. This phase can evaluate the requirements of healthcare stakeholders to generate awareness of Connected Health innovation and to support users through improved training programmes. In addition, an evaluation may also assess the organisation's conformity with medical device regulations in terms of technology classification and compliance. This also informs how an organisation can realign their service strategy evolving the service lifecycle through a continuous improvement philosophy.

4.2   Service Lifecycle

Within each of these subcategories, key metrics should be identified [41] which are associated with the evaluation of Connected Health solutions. As part of our future work, we will identify operational key metrics for each category and its components to support Connected Health innovation. The outer layer of CHEF comprises of various service lifecycle stages and highlights the need to identify value points in each of the service lifecycle phases.

The service lifecycle phases play a critical role in aligning the service development process and the market opportunities (Fig. 6). The Connected Health environment addresses healthcare technology requirements to enhance the level of healthcare service offerings. Connected Health can potentially address unfulfilled needs in healthcare as a result of external forces and various demographic drivers. Many of these drivers are also opening new market opportunities that enable Connected Health solutions to improve healthcare service maturity through enhanced service performance. The value of Connected Health solutions includes an improved quality of experience and usefulness in technological solutions to deliver healthcare.

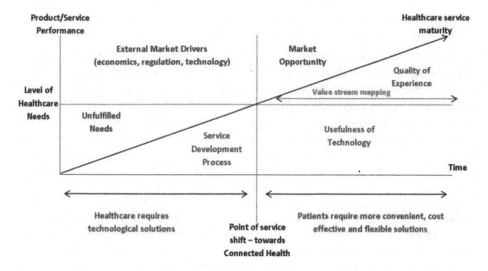

Fig. 6. Connected Health Environment.

4.3 Evaluation Using CHEF

While acknowledging that technology can provide healthcare solutions, it is equally important to question at each phase of the service lifecycle, for example "what problem does information solve?" [37] and "what is the problem to which this technology is a solution?" [38]. Postman's question applies equally well to the Connected Health field as a basic evaluation question. Building on this, it is critical that, as a starting point, and before we can successfully identify value in Connected Health, the current healthcare system is modelled, for example, actor interaction, value stream mapping, resource exchange, service bottlenecks, workflows, organisational structures and the healthcare solutions market landscape.

CHEF offers an approach to guide the evaluation process. Thus, the two key aspects as we move forward in Connected Health evaluation can be derived in:

- Ensuring the systems, devices and services meet the health and social needs of users through evidence-based research;
- Developing innovative patient-centred technological solutions to effectively manage their health and wellness in the home and community [9].

In addition, from a Connected Health perspective, evaluation must be conducted to assess its impact across the broad spectrum of care services. The scope of CHEF explicitly acknowledges the existence of different stakeholders. CHEF will facilitate evaluations through an assessment process designed to provide:

1. A holistic view of a healthcare system;
2. Tailored analysis of healthcare service lifecycle;
3. Performance metrics on service operations and patient-focused analytics;

4. Scorecard and benchmark tools to assess healthcare technological integrations, healthcare interventions and healthcare providers.

CHEF can also promote innovation by guiding evaluation at all stages of the health ICT product lifecycle and encouraging organisations to consider the complex socio-technical ecosystem in which healthcare products are developed, implemented, and used. Particular interests include the quality systems in place to govern Connected Health data management, access to clinical information, stakeholder communication, knowledge management and patient privacy. Regulations and conformity assessment supports the technology evaluation processes from a health and safety perspective. CHEF will also support organisations in examining potential risks posed by Connected Health functionality and in comparing them to the potential net benefits, for example, developing a benefit-risk profile. In addition, by meeting the regulatory evaluation of a medical device, conformity assessment will evaluate whether they present challenges to Connected Health innovation. CHEF promotes the need to incorporate Connected Health evaluation at various stages using quality management principles, adopt continuously revised standards and harness a learning and continual improvement environment to improve patient safety.

CHEF will enable organisations to identify poorly designed healthcare solutions, assess performance requirements, monitor human interaction (end-user) and identify potential gaps within a business strategy. In addition, CHEF offers a first step towards employing evaluation to extend the evidence-based foundation for Connected Health through the assessment of best practice and by identifying interventions and opportunities for improvement based on the CHEF evaluation and evidence gathered.

5   Case Study: Irish Blood Transfusion Service

We applied CHEF to understand a connected health issue which recently arose within the Irish Blood Transfusion Service (IBTS). IBTS is a national organization responsible for collecting, processing, testing and distributing blood and blood products. Voluntary donors provide donations to ensure a consistent supply of blood and blood components to patients [22].

Before a potential donor can donate blood with the IBTS a number of pre-checks are completed. One of these pre-checks includes testing the potential donor's haemoglobin level (Hb) [24]. This test is carried out to make sure the potential donor is not anaemic and will not suffer adversely from donating [32]. If the Hb level is high enough to donate, the potential donor will be asked to read the information provided about donating blood and to complete a questionnaire [24]. Only when satisfied with the donor's ability to give blood, will the IBTS take the donation.

Prior to July 2014 a Hb test was carried out by taking a finger stick sample of blood and analysing it. At this time, the IBTS changed the way potential donors Hb levels were checked. The new testing method used a medical device to check the Hb level without needing the finger stick blood sample. Using white xenon light to measure redness of blood in the small blood vessels of a potential donor's finger, this new test provided a non-invasive testing method. The IBTS identified the following advantages of this new test [23]:

- It is non-invasive;
- There is less discomfort for donors;
- Reduced delays at the clinics;
- The test is more robust;
- The test is less sensitive to fluctuations from warm weather (this had been a problem in previous summers with some donors being turned away during summer months).

The testing device was purchased from a German company and is also in use in other countries [6, 32]. Before the rollout of this new testing device, the IBTS carried out 945 tests with the new device which was then compared to the gold standard test i.e. the finger stick sample of blood test. These were in addition to the previous testing by the company who developed the device, allowing them to achieve European Union regulatory compliance. Following these internal IBTS tests, between July and November, 2015, the new device was used to test 180,000 potential donors in Ireland. However, in November 2015, iron deficiency anaemia in female blood donors was identified when a general practitioner (GP) identified that a female patient who was allowed to donate blood was she was anaemic [6, 32].

Urgently and immediately, the IBTS has to implement emergency measures. The device was removed from use. They suspended taking blood donations from women who had given blood in the previous 18 months until a blood sample from each donor was checked for sufficient Hb levels [16]. Every person who had donated blood within the previous five months were offered the opportunity to have a blood test taken to check iron levels – this test was paid for the IBTS. They also stated that several hundred donors, mainly women, could have become iron deficient and anaemic from blood donation during that time, as the new testing device was subsequently found to have missed a subset of women who were anaemic [6, 16]. Furthermore, the IBTS plan to introduce new software to reanalyse all the electronic results from all donors tested with discrepant results reported to the relevant donors.

A concern is that the IBTS was not made aware of previous issues with the testing device in other countries [6]. Cullen further pointed out that in order for device to pass the regulatory standards in Europe it needed to perform only 2,000 tests.

To return to the previous finger-prick test meant that the devices had to be taken out of storage, sterilised and re-prepared for use. Staff had to quickly revert to the previous way of testing which involved a finger stick blood sample test. Staff who had recently joined the service needed to be trained in the use of the older device.

What potentially could have been different if IBTS had used an evaluation model such as CHEF? Before the Hb testing device was deployed a more extensive evaluation of the potential and the risks of the device would have been implemented. CHEF is a model that can facilitate an evaluation of the potential impact of a Connected Health technology. CHEF focuses on four key domains, all of which were impacted on through the IBTS case, for example:

- End-user perception
  - IBTS credibility was tarnished in the eyes of the public who donate;

- Business growth and cost of the device
  - Extra costs were incurred by the IBTS e.g. provision of blood test to donors; retrieving equipment from storage;
- Quality management
  - Temporarily excluding women from donation put a greater burden on the already limited blood stocks and IBTS had to appeal to more men than ever to donate as blood stock levels were low;
- Healthcare practice
  - Returning to the old test meant reverting to an invasive testing practice.

To further illustrate the usefulness of the CHEF model we will discuss one domain/quadrant from the CHEF model namely Healthcare practice (Fig. 7) in further detail. In Healthcare practice the focus is on:

- Health IT: Hardware, Software;
- Innovation: Information, Connectivity.

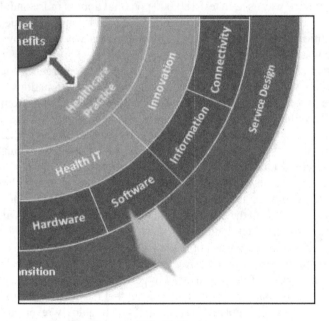

Fig. 7. Connected Health Environment excerpt.

Within Health IT the question to be asked is; how are practice/clinical pathways altered through the introduction of new healthcare devices? In this case study the new Hb testing device reduced delays at the clinics and as it was non-invasive there was less risk to IBTS staff carrying out the test. This was a clear advantage over the old test that involved taking a blood sample to test the Hb levels. From a technical point of view, an evaluation must be carried out on the hardware and software capability to deliver a Connected Health solution. In this case study the small number of only 2000 tests that were required for regulatory compliance of the Hb testing device was not identified as a potential risk which should have clearly been considered as part of the service design phase.

As part of the innovation evaluation, assessing how technologies alter healthcare practice from a socio-technical and ethnography viewpoint should have been a critical concern. In this case study, as the new Hb testing device was non-invasive, it was less distressing to donors. The new Hb testing device impacted the donors and IBTS staff carrying out the test directly by improving the way the test was carried out. However, it impacted on healthcare practice since they failed to account the connectivity between patient history and patient outcomes.

CHEF examines potential risks posed by Connected Health functionality, comparing them to the potential net benefits. The IBTS case study demonstrates that, had a benefit-risk profile been completed using CHEF or a similar model, existing problems with the new Hb testing device in other counties would have been identified. The IBTS did carry out some testing on the device – but obviously this number was not large enough.

CHEF encourages the use of quality management principles at the various stages of a Connected Health evaluation. Key metrics associated with the evaluation of Connected Health solutions must be identified. The outer layer of CHEF has various service life-cycle stages value points required in each of the service lifecycle phases. The service lifecycle aligns the service development process with market opportunities. Key metrics are useful to set and measure against targets following which appropriate action can be taken. Such metrics would have been useful in this case study to facilitate a more proactive approach rather than the reactive approach which ensued when the inaccuracies of the Hb testing device were identified. An additional point to note is the inaccuracies were not found by the IBTS but by a patient presenting to a GP with severe anaemia.

We conclude that it would have been very useful to complete a CHEF evaluation before implementing this Connected Health solution. The unanticipated situation to which the IBTS had to react put pressure on the donor system and cost money to resolve. If the CHEF model were used, there would have been proactive measures available to evaluate the potential of the new Hb testing device.

6   Discussion

Throughout our evaluation research, we also discovered that the concept of connectedness through Information and Communications Technology (ICT)-enabled healthcare is a complex socio-technical environment which is also impacted on various geography, socio-economic status, and technological competence – often influencing attitudes to Connected Health innovation. Technology therefore plays a key role in fostering healthcare relationships, giving healthcare stakeholders a sense of being interconnected. Through evaluation processes, if we can develop a better understanding of the Connected Health network structure, we can begin to further evaluate the impact of ICT innovation on a healthcare ecosystem.

As an extension to this research, we have also been studying Design Thinking techniques [2]. Design Thinking provides a formal process to capture people's various needs or pain-points. Therefore, this is particularly apt to guide our research in identifying healthcare innovation requirements. Such guidance is vital since innovation enabled through healthcare software development has much at stake, most notably patient safety. Our research suggests that Design Thinking moves beyond requirements gathering and being constrained by preconceptions of software solutions in isolation [17]. It supports us to guide innovation through: (1) empathising to fully understand the experience of the users, (2) defining a wide variety of possible Connected Health solutions, (3) ideating creative healthcare solutions, (4) prototyping ideas into tangible form, and (5) testing to refine and examine the value/impact of technological solutions. This fosters a learning lifecycle through various actions about both the solution and informing how we can bridge our understanding of healthcare needs and the software design process. More importantly, we have identified the need to consider the evaluation process within the design phase rather than considering the evaluation process as a separate task. However, this can be challenging, but the Design Thinking process offers a formal approach for practical, creative resolution of barriers which can guide improved healthcare software solutions. This approach also supports Connected Health solutions development and can reduce the risks of failure through our framework for healthcare innovation.

These will also form part of our research strategy in our quest to further develop CHEF and apply it to various healthcare products and services and derive core evaluation metrics. There is a clear correlation between Connected Health functionality and health-care net benefits from multiple perspectives. CHEF will be further validated through continued industry engagement and Connected Health technologies to accommodate the rapid growth of healthcare ICT solutions.

7   Conclusion

With significantly greater shifts in demographics and longevity, the cost of healthcare will show a corresponding increase. In an attempt to reduce these growing costs, govern-ments typically attempt to reduce healthcare overheads, including staffing, patient contact time, consultation and scheduling various appointments. This can also create service bottlenecks which jeopardises the quality and safety of healthcare.

There is evidence that a paradigm shift to empower people to take more control of their own health is occurring. Technology innovation enables and aligns with these healthcare shifts, providing greater service efficiencies and effectiveness and supporting the reduction of costs. Connected Health presents an exciting approach towards rede-signed healthcare delivery. However, the success of Connected Health will hinge on evaluation strategies to determine the real value or benefits (healthcare, quality of care, cost, etc.) associated with technological integration into healthcare service systems. This chapter has presented an overview of how existing evaluation frameworks in e-health and information systems can influence Connected Health evaluations and the develop-ment of CHEF.

We propose that the CHEF can be successfully employed through industry engage-ment. We present a case study on the Irish Blood Transfusion Service (IBTS) and describe the challenges that this has presented. CHEF is a first step in offering a holistic view of Connected Health and is a step towards an evaluation of healthcare technological innovations. As part of our future work, we will continue to collaborate with industry and academic members within ARCH - Applied Research for Connected Health Technology Centre. Through our multidisciplinary research team, we will extend this work and validate CHEF with various healthcare stakeholders and IT providers.

Acknowledgements. This work was supported, in part, by ARCH - Applied Research for Connected Health Technology Centre (www.arch.ie), an initiative jointly funded by Enterprise Ireland and the IDA, SFI Lero Grant (www.lero.ie) 13/RC/2094 and Science Foundation Ireland (SFI) Industry Fellowship Grant Number 14/IF/2530.

References

1. Carroll, N.: In search we trust: exploring how search engines are shaping society. Intl. J. Knowl. Soc. Res. (IJKSR) **5**(1), 12–27 (2014)
2. Carroll, N., Richardson, I.: Aligning healthcare innovation and software requirements through design thinking. In: International Workshop on Software Engineering in Healthcare Systems, 14–15 May, Austin, TX, USA (2016)
3. Carroll, N., Travers, M., Richardson, I.: Evaluating multiple perspectives of a connected health ecosystem. In: 9th International Conference on Health Informatics (HEALTHINF), Rome, Italy, February, pp. 21–23 (2016)
4. Caulfield, B.M., Donnelly, S.C.: What is Connected Health and why will it change your practice? QJM **106**, 703–707 (2013). hct114
5. Christensen, C.M., Bohmer, R., Kenagy, J.: Will disruptive innovations cure health care? Harvard Bus. Rev. **78**(5), 102–112 (2000)
6. Cullen, P.: Blood service not told about problems with testing device, Irish Times, 18 Nov 2015. http://www.irishtimes.com/news/health/blood-service-not-told-about-problems-with-testing-device-1.2433719. Accessed 24 Mar 2016
7. Dávalos, M.E., French, M.T., Burdick, A.E., Simmons, S.C.: Economic evaluation of telemedicine: review of the literature and research guidelines for benefit-cost analysis. Telemed. e-Health, **15**, 933–948 (2009)
8. Davis, F.D.: Perceived usefulness, perceived ease of use, and user acceptance of information technology. MIS Q. **13**, 319–340 (1989)
9. Delbanco, T., Walker, J., Bell, S.K., Darer, J.D., Elmore, J.G., Farag, N., Feldman, H.J., Mejilla, R., Ngo, L., Ralston, J.D., Ross, S.E., Trivedi, N., Vodicka, E., Leveille, S.G.: Inviting patients to read their doctors' notes: a quasi-experimental study and a look ahead. Ann. Intern. Med. **157**(7), 461–470 (2012)
10. DeLone, W.H.: The DeLone and McLean model of information systems success: a ten-year update. J. Manage. Inf. Syst. **19**(4), 9–30 (2003)
11. DeLone, W.H., McLean, E.R.: Information systems success: the quest for the dependent variable. Inf. Syst. Res. **3**(1), 60–95 (1992)
12. Dixon, D.R.: The behavioral side of information technology. Int. J. Med. Inf. **56**(1), 117–123 (1999)
13. ECHAlliance: Connected Health – White Paper (2014). http://cht.oulu.fi/uploads/2/3/7/4/23746055/connected_health.pdf. Accessed 03 Sep 2015
14. Fineout-Overholt, E., Melnyk, B.M., Schultz, A.: Transforming health care from the inside out: advancing evidence-based practice in the 21st century. J. Prof. Nurs. **21**(6), 335–344 (2005)

15. Friedman, C.P., Wyatt, J.C.: Evaluation Methods in Medical Informatics. Springer-Verlag, New York (1997)
16. Gartland, F.: IBTS suspends taking blood donations from women, Irish Times, 16 Nov 2015. http://www.irishtimes.com/news/health/ibts-suspends-taking-blood-donations-from-women-1.2432401. Accessed 24 Mar 2016
17. Giardino, C., Bajwa, S.S., Wang, X., Abrahamsson, P.: Key challenges in early-stage software startups. In: Lassenius, C., Dingsøyr, T., Paasivaara, M. (eds.) XP 2015. LNBIP, vol. 212, pp. 52–63. Springer, Heidelberg (2015). doi:10.1007/978-3-319-18612-2_5
18. Grant, A., Plante, I., Leblanc, F.: The TEAM methodology for the evaluation of information systems in biomedicine. Comput. Biol. Med. **32**(3), 195–207 (2002)
19. Gregor, S.: The nature of theory in information systems. MIS Q. **30**(3), 611–642 (2006)
20. Heathfield, H., Pitty, D., Hanka, R.: Evaluating information technology in health care: barriers and challenges. BMJ **316**(7149), 1959 (1998)
21. Hebert, M.A, Korabek, B.: Stakeholder readiness for telehomecare: implications for implementation. Telemed. J. e-Health **10**, 85–92 (2004)
22. IBTS: IBTS Strategic Plan 2013–2016 (2013). https://www.giveblood.ie/About_Us/Publications_Guidelines/Strategic_Plans/IBTS-Strategic-Plan-2013-2016.pdf. Accessed 24 Mar 2016
23. IBTS: IBTS first national transfusion service to roll out non-invasive haemoglobin check for blood donors (2015). https://www.giveblood.ie/About_Us/Newsroom/Press_Releases/2015/IBTS-first-national-transfusion-service-to-roll-out-non-invasive-haemoglobin-check-for-blood-donors.html. Accessed 24 Mar 2016
24. IBTS: Blood Donation Information Leaflet (2016). http://www.giveblood.ie/Become_a_Donor/Information_Leaflets/Blood_Donation_Information_Leaflet.pdf. Accessed 24 Mar 2016
25. Kaplan, B.: Addressing organizational issues into the evaluation of medical systems. J. Am. Med. Inform. Assoc. **4**(2), 94–101 (1997)
26. Kaplan, B.: Evaluating informatics applications—some alternative approaches: theory, social interactionism, and call for methodological pluralism. Int. J. Med. Inf. **64**(1), 39–56 (2001)
27. Kuhn, K.A., Giuse, D.A.: From hospital information systems to health information systems-problems, challenges, perspective. In: Yearbook of Medical Informatics, pp. 63–76 (2001)
28. Leveille, S.G., Walker, J., Ralston, J.D., Ross, S.E., Elmore, J.G., Delbanco, T.: Evaluating the impact of patients' online access to doctors' visit notes: designing and executing the OpenNotes project. BMC Med. Inform. Decis. Mak. **12**(1), 32 (2012)
29. Mathur, A., Kvedar, J.C., Watson, A.J.: Connected health: a new framework for evaluation of communication technology use in care improvement strategies for type 2 diabetes. Curr. Diabetes Rev. **3**(4), 229–234 (2007)
30. Meltsner, M.: A patient's view of OpenNotes. Ann. Intern. Med. **157**(7), 523–524 (2012)
31. Misuraca, G., Codagnone, C., Rossel, P.: From practice to theory and back to practice: reflexivity in measurement and evaluation for evidence-based policy making in the information society. Gov. Inf. Q. **30**, S68–S82 (2013)
32. Murphy, W.: Interview on Morning Ireland [radio]. RTE Radio 1, 17 Nov 2015. 07:26
33. O'Leary, P., Carroll, N., Richardson, I.: The practitioner's perspective on clinical pathway support systems. In: 2014 IEEE International Conference on Healthcare Informatics (ICHI), pp. 194–201. IEEE (2014)
34. O'Leary, P., Carroll, N., Clarke, P., Richardson, I.: Untangling the complexity of connected health evaluations. In: IEEE International Conference on Healthcare Informatics 2015 (ICHI 2015) Dallas, Texas, USA, 21–23 October 2015

35. O'Neill, S.A., Nugent, C.D., Donnelly, M.P., McCullagh, P., McLaughlin, J.: Evaluation of connected health technology. Technol. Health Care **20**(4), 151–167 (2012)
36. Parasuraman, A., Zeithaml, V.A., Berry, L.L.: Servqual. J. Retail. **64**(1), 12–40 (1988)
37. Postman, N.: Technopoly: The Surrender of Culture to Technology. Vintage Press, New York (1992)
38. Postman, N.: Building a Bridge to the 18th Century: How the Past Can Improve Our Future. Alfred A. Knopf Publishers, New York (1999)
39. Richardson, I.: Connected Health: People, Technology and Processes, Lero-TR-2015-03, Lero Technical Report Series, University of Limerick (2015)
40. Rodrigues, R., Huber, M., Lamura, G.: Facts and figures on healthy ageing and long-term care. Itävalta: European Centre for Social and Welfare policy and Research: Vienna (2012)
41. Rojas, S.V., Gagnon, M.P.: A systematic review of the key indicators for assessing telehomecare cost-effectiveness. Telemed. e-Health **14**(9), 896–904 (2008)
42. Rudin, R.S., Jones, S.S., Shekelle, P., Hillestad, R.J., Keeler, E.B.: The value of health information technology: filling the knowledge gap. Am. J. Managed Care, Special Issue: Health Inf. Technol. **20**(17), eSP1–eSP8 (2014)
43. Schwamm, L.H.: Telehealth: seven strategies to successfully implement disruptive technology and transform health care. Health Aff. **33**(2), 200–206 (2014)
44. Shaw, N.T.: 'CHEATS': a generic information communication technology (ICT) evaluation framework. Comput. Biol. Med. **32**(3), 209–220 (2002)
45. Tuffaha, H.W., Gordon, L.G., Scuffham, P.A.: Value of information analysis in healthcare: a review of principles and applications. J. Med. Econ. **17**(6), 377–383 (2014)
46. Van Bemmel, J.H., Musen, M.A.: Handbook of Medical Informatics. Springer, Heidelberg (1997)
47. Ooteghem, J., Ackaert, A., Verbrugge, S., Colle, D., Pickavet, M., Demeester, P.: Economic viability of ecare solutions. In: Szomszor, M., Kostkova, P. (eds.) eHealth 2010. LNICSSITE, vol. 69, pp. 159–166. Springer, Heidelberg (2011). doi:10.1007/978-3-642-23635-8_20
48. Wilson, E.V., Lankton, N.K.: Interdisciplinary research and publication opportunites in information systems and health care. Commun. Assoc. Inf. Syst. **14**(1), 51 (2004)
49. Yusof, M.M., Paul, R.J., Stergioulas, L.K.: Towards a framework for health information systems evaluation. In: Proceedings of the 39th Annual Hawaii International Conference on System Sciences, HICSS'06, vol. 5, p. 95a. IEEE (2006)
50. Zeithaml, V.A., Parasuraman, A., Berry, L.L.: Delivering Quality Service: Balancing Customer Perceptions and Expectations. Simon and Schuster, New York (1990)

Continuous Postoperative Respiratory Monitoring with Calibrated Respiratory Effort Belts: Pilot Study

Tiina M. Seppänen[1,2(✉)], Olli-Pekka Alho[2,3,4], Merja Vakkala[2,5],
Seppo Alahuhta[2,5], and Tapio Seppänen[1,2]

[1] Physiological Signal Analysis Team, University of Oulu, Oulu, Finland
{tiina.seppanen,tapio.seppanen}@oulu.fi
[2] Medical Research Center Oulu,
Oulu University Hospital and University of Oulu, Oulu, Finland
merja.vakkala@ppshp.fi,
{ollli-pekka.alho,seppo.alahuhta}@oulu.fi
[3] Department of Otorhinolaryngology, Oulu University Hospital, Oulu, Finland
[4] PEDEGO Research Unit, University of Oulu, Oulu, Finland
[5] Department of Anesthesiology, Oulu University Hospital, Oulu, Finland

Abstract. Postoperative respiratory complications are common in patients after surgery. Respiratory depression and subsequent adverse outcomes can arise from pain, residual effects of drugs given during anaesthesia and administration of opioids for pain management. There is an urgent need for a continuous, real-time and non-invasive respiratory monitoring of spontaneously breathing postoperative patients. For this purpose, we used rib cage and abdominal respiratory effort belts for the respiratory monitoring pre- and postoperatively, with a new calibration method that enables accurate estimates of the respiratory airflow waveforms even when breathing style changes. Five patients were measured with respiratory effort belts and mask spirometer. Preoperative measurements were done in the operating room, whereas postoperative measurements were done in the recovery room. We compared five calibration models with pre- and postoperative training data. The postoperative calibration approach with two respiratory effort belts produced the most accurate respiratory airflow waveforms and tidal volume, minute volume and respiratory rate estimates. Average results for the best model were: coefficient of determination R^2 was 0.91, tidal volume error 5.8%, minute volume error 8.5% and BPM (Breaths per Minute) error 0.21. The method performed well even in the following challenging respiratory cases: low airflows, thoracoabdominal asynchrony and hypopneic events. It was shown that a single belt measurement can be sufficient in some cases. The proposed method is able to produce estimates of postoperative respiratory airflow waveforms to enable accurate, continuous, real-time and non-invasive respiratory monitoring postoperatively. It provides also potential to optimize postoperative pain management and enables timely interventions.

Keywords: Airflow waveform · Calibration · Regression · Respiratory airflow · Respiratory rate · Respiratory volume

© Springer International Publishing AG 2017
A. Fred and H. Gamboa (Eds.): BIOSTEC 2016, CCIS 690, pp. 340–359, 2017.
DOI: 10.1007/978-3-319-54717-6_19

1   Introduction

Respiratory complications are common in surgical patients after the general anaesthesia. Postoperative respiratory depression and subsequent adverse outcomes can arise from pain, residual effects of drugs given during anaesthesia and administration of opioids for pain management [1]. Respiratory depression often occurs in association with postoperative opioid analgesia [2–4]. The risk of respiratory depression increases with age [1], morbid obesity [5] and pre-existing sleep apnea syndrome [2].

Inadequate respiration can result in respiratory complications, morbidity, mortality, longer recovery room times and excessive costs. Abnormal respiratory rate has been shown to be a common clinical feature in patients before a major clinical event such as cardiac arrest, onset of sepsis, and in patients experiencing pain, shock, asthma attacks and respiratory infection [6, 7]. Adequate respiration monitoring postoperatively is important, so that respiratory depression can be identified as early as possible [8, 9]. This way, we would minimize respiratory complications, facilitate timely interventions, reduce health care costs, and improve patient safety and satisfaction.

During general anaesthesia, mechanical ventilation with intubation or supraglottic device is used, and consequently, monitoring of respiration and gas exchange can be done accurately. During postoperative care, respiratory status can be assessed, for example, with capnometry, pulse oximetry, oxygen saturation (SpO$_2$) measurements, blood gas measurements, subjective clinical assessment and intermittent, manual measurements of respiratory rate [5, 9]. The problems with current methods are that they have poor accuracy, precision, low patient tolerance and they are liable to false alarms [10, 11]. Additionally, they are slow and especially subjective methods are unreliable and give inconsistent results [12]. Current methods do not provide information about respiratory airflow waveform variability and disorders, either. There is thus a need for a continuous, real time and non-invasive respiratory monitoring of spontaneously breathing postoperative patients.

A couple of studies have been recently published on monitoring postoperative respiration continuously and non-invasively. Drummond et al. [13] have studied respiratory rate and breathing patterns of postoperative subjects using encapsulated tri-axial accelerometer taped to a subject's body. They found that abnormal breathing patterns are extremely common. Voscopoulos et al. [14–16] have studied minute ventilation, tidal volume and respiratory rate of postoperative subjects using impedance-based electrodes placed to a subject's body. In addition to these, Masa et al. [17] showed that respiratory effort-related arousals in sleep apnea patients could be identified satisfactorily by assessing the morphology of the thoracoabdominal bands output signal.

Recently, we published a novel calibration method to produce accurate estimates of respiratory airflow signals from respiratory effort belt signals [18]. The method is an extension to the multiple linear regression method with two predictor variables: rib cage and abdominal respiratory effort belt signals. Here, the method is used in order to produce estimates of postoperative respiratory airflow waveforms to enable accurate, continuous, non-invasive respiratory monitoring postoperatively. Pre- and postoperative measurement data of different patients are used to demonstrate the performance of the method. Additionally, we assess the calibration method by using only one predictor variable (rib cage respiratory effort belt or abdominal respiratory effort belt) at the time. The results are assessed between the belts and also compared to the results from two belts.

2   Materials and Methods

2.1   Materials

The study protocol was approved by the Regional Ethics Committee of the Northern Ostrobothnia Hospital District. Five patients who had lumbar back surgery and were expected to need opioid analgesia postoperatively were recruited to the study after informed consent. Exclusion criteria were the planned surgical wound being in the area where respiratory effort belts were placed and BMI (Body Mass Index) over 40. The characteristics of the volunteers are given in Table 1.

Table 1.   Characteristics of volunteers (COPD = Chronic Obstructive Pulmonary Disease).

Patient	Gender	Age [years]	BMI [kg/m^2]	Disease
1	M	68	22	None
2	M	41	30	None
3	F	77	22	None
4	M	64	28	COPD
5	M	67	27	Sleep apnea

Respiratory effort belt signals were recorded with the polygraphic recorder (Embletta Gold, Denver, Colorado, USA). The recorder had inductive respiratory effort belts for rib cage and abdomen. The belt signals stored by the recorder were in arbitrary units and therefore they had to be calibrated to get respiratory airflow estimates in proper units. For calibrating the respiratory effort belt signals, simultaneous respiratory airflow signal was recorded with a spirometer (Medikro Pro M915, Medikro Oy, Kuopio, Finland). Before each measurement session, the spirometer was calibrated using a 3000 ml calibration syringe (Medikro M9474, Medikro Oy, Kuopio, Finland). A mask covering the mouth and nose (Cortex Personal-Use-Mask, Leipzig, Germany) was attached to the mouthpiece of the spirometer. The spirometer was able to record at most 1 min long signals. The sampling rate of respiratory effort belts and spirometer was 50 Hz and 100 Hz, respectively. The spirometer signals were decimated to 50 Hz for further processing.

2.2   Measurement Protocol

The measurements for each patient were done in two sessions: (1) a short measurement session (5 min) preoperatively; and (2) a longer measurement session (3 h) postoperatively.

The first measurements were done in the operating room just before the operation prior to patients had received any sedative premedication. The rib cage respiratory effort belt was placed on the xyphoid process and the abdominal belt just above the umbilicus. The mask of the spirometer was put on the patient's face and its airtightness was secured.

The signals were recorded until two successful recordings of the 1 min were obtained. After that, the places of respiratory effort belts were marked with drawing ink on the skin, so that it was possible to place the belts to the same places postoperatively. The mask and respiratory effort belts were removed.

Patients were operated in general anaesthesia. As soon as possible, measurements were continued postoperatively in the recovery room. The rib cage respiratory effort belt and the abdominal respiratory effort belt were placed to the preoperatively marked places. They recorded the signals during the whole 3 h measurement period. Every 10 min, the mask with the spirometer was put on the subject's face, its airtightness was secured and 1 min measurement with the spirometer was recorded. Participation to the study did not affect the routine management of the patients.

2.3   Calculation of Respiratory Airflow Estimates

In this study, we applied our recently published respiratory effort belt calibration method [18]. The method was therein tested against various breathing style changes and body position changes, and compared with the state-of-the-art methods. It was found out that our method outperformed the other methods showing highest robustness to the breathing style changes and body position changes.

Our method is an extension to the conventional multiple linear regression method so that (1) it uses number N of consecutive input signal samples and linear filtering for estimation of each output signal sample and (2) it is based on polynomial regression to model different transfer functions between the input and output. The method is based on optimally trained FIR (Finite Impulse Response) filter bank constructed as a MISO (Multiple-Input Single-Output) system between the respiratory effort belt signals and the spirometer signal, see Fig. 1.

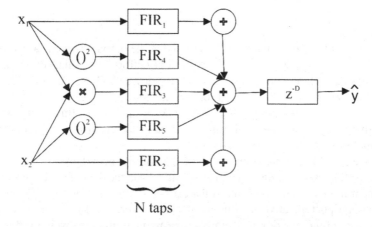

Fig. 1. Calibration method of respiratory effort belt signals used in this study.

The following polynomial transfer functions were tested: linear terms only (M1), linear terms and cross-product term (M2), and linear terms with second order terms

(M3). Additionally, we tested a linear model with only one respiratory effort belt at the time: abdominal respiratory effort belt (M4) and rib cage respiratory effort belt (M5).

Equation 1 shows the realization of the filter bank for model M2. Similar realizations can be derived also for models M1, M3, M4 and M5.

$$y[k] = a_1^T x_1[k] + a_2^T x_2[k] + a_3^T (x_1[k]x_2[k]) + \varepsilon[k], \tag{1}$$

where a_1^T, a_2^T and a_3^T denote the N tap coefficients of filters FIR1, FIR2 and FIR3, respectively:

$$a_i^T = [a_i[1], a_i[2], \dots, a_i[N]], \tag{2}$$

where i = 1, 2, 3. Superscript T denotes matrix transpose. Parameter y (response variable) denotes respiratory airflow from spirometer and ε is zero-mean Gaussian error.

Vectors x_1 and x_2 (predictor variables) include N consecutive signal samples from the rib cage respiratory effort belt signal and abdominal respiratory effort belt signal, respectively:

$$x_j[k] = [x_j[k], x_j[k-1], \dots, x_j[k-N+1]]^T, \tag{3}$$

where j = 1, 2 and k = N, ..., n. Variable n is the number of observations used in the calibration. X is an (n–N + 1) × (3 × N) matrix formed from the vectors x_1 and x_2:

$$\hat{a} = (X^T X)^{-1} X^T y. \tag{4}$$

The length of the vector $\hat{a}$ is 3 × N. Finally, the respiratory airflow signal estimated from the rib cage and abdominal respiratory effort belt signals through the FIR filter bank is

$$\hat{y} = X\hat{a}. \tag{5}$$

In Fig. 1, there is the delay element z^{-D} included at the output. There is always a small delay between the spirometer signal and respiratory effort belt signals due to the physiological reasons and internal delays of measuring devices. Thus, the signals have to be time-synchronized by searching for a proper value for D [18].

Based on our previous study [18], we used 0.15 s and 0.30 s time windows of FIR filters in this study. With the used sampling frequency of 50 Hz, the number of tap coefficients (N) was 8 and 16, respectively.

We made two different test setups: (1) the data of the second preoperative measurement minute were used to train the estimation model and the data of all the postoperative measurement minutes were used to test the estimation model (PRE setup); and (2) the data of the first postoperative measurement minute were used to train the estimation model and data of the rest of postoperative measurement minutes were used to test the estimation model (POST setup).

2.4   Statistics

The similarity of spirometer signals and estimated respiratory airflow signals were assessed by computing R^2 (coefficient of determination) and RMSE (Root Mean Square Error) values. R^2 is the coefficient of determination between the spirometer signal and the estimated airflow. The coefficient of determination is calculated from [19]

$$R^2 = 1 - \frac{SS_{res}}{SS_{tot}}, \tag{6}$$

where SS_{res} is the sum of squares of residuals between the spirometer signal and estimated airflow, and SS_{tot} is the total sum of squares calculated from the spirometer signal. RMSE is a measure of the difference between the spirometer signal and the estimated airflow. Relative RMSE [%] is the proportion of RMSE from RMS (Root Mean Square) of the spirometer signal. Tidal volumes, minute volumes and BPM (Breaths per Minute, respiratory rate) were calculated from the spirometer signals and estimated respiratory airflow signals. Relative errors were calculated.

3   Results and Discussion

Signals were recorded according to the protocol described in Sect. 2.2. A total of 93 simultaneous measurement minutes with spirometer and respiratory effort belts were collected. Five of these had to be discarded due to a malfunction of the spirometer, one of them due to a malfunction of the polygraphic recorder. In addition to that, five postoperative measurement minutes of patient 5 had to be discarded, because of serious difficulties to wake up and stay awake in the recovery room. During the measurements, patients received opioid analgesia as many times as they needed: 3, 2, 1, 6 and 4 times for patients 1–5, respectively.

The following problems related to PRE setup were observed during the measurements. Firstly, there may be a need to tighten or loosen the respiratory effort belts after the operation, because fluids can accumulate in the body or can leave from the body during the operation. This leads to a situation where the estimation model trained with a preoperative data is not valid anymore. Secondly, if there are complications during the operation, the estimation model trained with a preoperative data can be erroneous for the postoperative data. With the POST setup, no problems were observed.

3.1   Accuracy of Airflow Estimates Using Two Respiratory Effort Belts

Results for the models using two respiratory effort belts, namely M1, M2 and M3, are presented first. After that, the results for the models (M4 and M5) using only one respiratory effort belt are presented. At the end of this section, the results of those cases are compared. The selection of the best model and FIR filter length (N value) depended on whether waveform accuracy (R^2, relative RMSE), tidal volume (V_T) error, minute volume (V_{minute}) error or BPM error values were studied.

Table 2 summarizes the best results when PRE setup was used for training and testing estimation models M1, M2 and M3. As seen from Table 2, model M1 produced the best results in all cases. N value 8 produced the best waveform accuracy and the lowest volume errors. N value 16 produced the lowest BPM error, but the respective result with N value 8 was almost the same namely 0.34 ± 0.71.

Table 2. Best results (average value $\pm$ SD) with the best models when PRE setup was used for training and testing estimation models M1, M2 and M3.

Best model	M1	M1	M1	M1	M1
FIR size	N = 8	N = 8	N = 8	N = 8	N = 16
Patient	R^2	Relative RMSE [%]	Abs (V_T error) [%]	Abs (V_{minute} error) [%]	BPM error
1	0.88 ± 0.05	33.4 ± 6.5	12.8 ± 10.7	12.8 ± 10.7	0.03 ± 0.07
2	0.87 ± 0.06	35.2 ± 7.0	26.7 ± 7.7	25.1 ± 8.2	0.01 ± 0.01
3	0.94 ± 0.02	24.4 ± 4.5	10.9 ± 2.9	14.2 ± 5.0	0.09 ± 0.30
4	0.87 ± 0.06	35.8 ± 7.9	15.9 ± 6.0	21.7 ± 8.6	0.09 ± 1.14
5	0.40 ± 0.18	76.7 ± 12.2	71.0 ± 13.7	76.4 ± 17.7	0.64 ± 0.90
Average	0.81 ± 0.20	39.2 ± 18.2	24.9 ± 22.2	27.3 ± 23.7	0.32 ± 0.73

Table 3 summarizes the best results when the POST setup was used for training and testing estimation models M1, M2 and M3. In this case, it is seen that model M2 produced the best results in all cases. N value 8 produced the best waveform accuracy and the lowest BPM error, whereas N value 16 produced the lowest volume errors. As an important difference to the preceding results in Table 2, more accurate waveforms were received since R^2 values were higher, relative RMSE lower and BPM error values lower. Also, the volume error values decreased to the fractions.

Table 3. Best results (average value $\pm$ SD) with the best models when POST setup was used for training and testing estimation models M1, M2 and M3.

Best model	M2	M2	M2	M2	M2
FIR size	N = 8	N = 8	N = 16	N = 16	N = 8
Patient	R^2	Relative RMSE [%]	Abs (V_T error) [%]	Abs (V_{minute} error) [%]	BPM error
1	0.90 ± 0.04	31.4 ± 5.6	11.4 ± 7.9	9.9 ± 6.8	0.01 ± 0.01
2	0.95 ± 0.01	21.3 ± 2.9	5.9 ± 4.5	5.8 ± 5.0	0.01 ± 0.01
3	0.94 ± 0.04	23.9 ± 6.1	5.4 ± 4.7	6.3 ± 4.2	0.10 ± 0.30
4	0.88 ± 0.09	33.7 ± 10.6	8.7 ± 6.0	11.9 ± 10.2	0.83 ± 1.13
5	0.91 ± 0.02	30.1 ± 3.8	10.5 ± 9.3	8.2 ± 6.3	0.01 ± 0.00
Average	0.91 ± 0.06	28.1 ± 8.0	5.8 ± 6.3	8.5 ± 7.1	0.21 ± 0.63

It can be seen from Tables 2 and 3 that the POST setup produces better results than the PRE setup, and that different models yield the best results in them. However, one may wish to use only one model with good overall performance in both setups. The results for that are presented next. Model M3 (using two respiratory effort belt signals and including linear and 2nd order terms) produced clearly worse results than the other models, thus, only the results of using models M1 and M2 are presented here. Table 4 presents the results when the estimation model M1 with N value 8 was used with PRE and POST setups.

Table 4. Results (average value ± SD) for estimation model M1 (N = 8) with PRE setup (upper results) and POST setup (lower results).

Patient	R^2	Relative RMSE [%]	Abs (V_T error) [%]	Abs (V_{minute} error) [%]	BPM error
1	0.88 ± 0.05	33.4 ± 6.5	12.8 ± 10.7	12.5 ± 9.5	0.09 ± 0.27
2	0.87 ± 0.06	35.2 ± 7.0	26.7 ± 7.7	25.1 ± 8.2	0.01 ± 0.01
3	0.94 ± 0.02	24.4 ± 4.5	10.9 ± 2.9	14.2 ± 5.0	0.09 ± 0.29
4	0.87 ± 0.06	35.8 ± 7.9	15.9 ± 6.0	21.7 ± 8.6	1.00 ± 1.06
5	0.40 ± 0.18	76.7 ± 12.2	71.0 ± 13.7	76.4 ± 17.7	0.54 ± 0.75
Average	0.81 ± 0.20	39.2 ± 18.2	24.9 ± 22.2	27.3 ± 23.7	0.34 ± 0.71
1	0.91 ± 0.04	29.2 ± 5.6	11.6 ± 7.8	8.9 ± 7.1	0.01 + 0.01
2	0.94 ± 0.02	23.7 ± 3.5	6.9 ± 7.0	5.7 ± 5.6	0.01 ± 0.01
3	0.94 ± 0.04	24.0 ± 6.2	5.6 ± 4.5	6.0 ± 3.8	0.10 ± 0.30
4	0.87 ± 0.09	34.3 ± 10.5	8.4 ± 4.9	11.9 ± 8.9	0.82 ± 1.13
5	0.90 ± 0.02	30.8 ± 3.9	9.6 ± 5.3	10.9 ± 4.9	0.12 ± 0.35
Average	0.91 ± 0.06	28.4 ± 7.7	8.4 ± 6.1	8.6 ± 6.7	0.23 ± 0.64

Table 5 presents the results when estimation model M2 with N value 8 was used with PRE and POST setups.

Table 5. Results (average value ± SD) of the calibration when estimation model M2 (N = 8) was used with PRE setup (upper results) and POST setup (lower results).

Patient	R^2	Relative RMSE [%]	Abs (V_T error) [%]	Abs (V_{minute} error) [%]	BPM error
1	0.82 ± 0.08	41.9 ± 9.4	17.8 ± 9.8	21.3 ± 8.3	0.10 ± 0.27
2	0.76 ± 0.09	48.0 ± 8.7	42.7 ± 8.8	51.7 ± 9.6	0.04 ± 0.12
3	0.93 ± 0.03	26.6 ± 5.6	13.3 ± 5.1	15.3 ± 6.4	0.09 ± 0.30
4	0.87 ± 0.06	34.8 ± 7.9	15.4 ± 6.7	19.6 ± 8.6	0.91 ± 1.11
5	0.35 ± 0.23	79.2 ± 14.6	55.9 ± 9.4	69.0 ± 18.4	1.72 ± 1.40
Average	0.77 ± 0.21	44.3 ± 18.9	27.4 ± 18.2	33.4 ± 22.6	0.51 ± 0.97
1	0.90 ± 0.04	31.4 ± 5.6	11.4 ± 7.6	10.0 ± 7.0	0.01 ± 0.01
2	0.95 ± 0.01	21.3 ± 2.9	6.1 ± 4.8	6.2 ± 5.3	0.01 ± 0.01
3	0.94 ± 0.04	23.9 ± 6.1	5.3 ± 4.5	6.2 ± 4.0	0.10 ± 0.30
4	0.88 ± 0.09	33.7 ± 10.6	8.8 ± 3.9	11.4 ± 10.1	0.83 ± 1.13
5	0.91 ± 0.02	30.1 ± 3.8	13.1 ± 8.9	9.6 ± 6.4	0.01 ± 0.00
Average	0.91 ± 0.06	28.1 ± 8.0	8.8 ± 6.6	8.7 ± 7.1	0.21 ± 0.63

It is clearly seen in Tables 4 and 5 that the POST setup produced superior results again. Respiratory airflow waveforms are much more accurate, average R^2 increased and relative RMSE decreased considerably with both models M1 and M2. In addition to that, tidal volume errors, minute volume errors and BPM errors were smaller. However, when the average POST setup results of Tables 4 and 5 are compared, it can be seen that models M1 and M2 with N value 8 produced both very good results and that there are small differences between the results. However, the PRE setup produced slightly better results with model M1 than with model M2.

3.2   Accuracy of Airflow Estimates Using Single Respiratory Effort Belt

To study the effects of individual respiratory effort belts, we also made estimations using only one belt at the time. The use of only one respiratory effort belt simplifies the measurements. In addition, it is less likely that a surgical wound would be in the area of the belt than in the case of two belts, thus allowing for a wider variety of surgical operations.

Table 6. Results (average value ± SD) for PRE setup and model M4 with N value 1 (upper results) and N value 8 (lower results).

Patient	R^2	Relative RMSE [%]	Abs (V_T error) [%]	Abs (V_{minute} error) [%]	BPM error
1	0.55 ± 0.08	66.9 ± 5.7	26.6 ± 19.5	21.8 ± 12.4	0.44 ± 0.51
2	0.64 ± 0.05	60.0 ± 4.5	19.8 ± 13.6	18.5 ± 12.5	0.04 ± 0.05
3	0.78 ± 0.09	45.6 ± 9.8	16.2 ± 12.0	17.5 ± 12.3	0.20 ± 0.39
4	0.58 ± 0.10	64.6 ± 7.0	29.9 ± 25.7	19.3 ± 21.7	1.41 ± 1.15
5	0.57 ± 0.06	65.1 ± 4.8	16.1 ± 14.0	20.9 ± 11.4	0.36 ± 0.48
Average	0.62 ± 0.11	60.5 ± 10.1	22.3 ± 18.5	19.6 ± 14.5	0.51 ± 0.81
1	0.87 ± 0.05	35.5 ± 7.2	14.9 ± 11.5	17.9 ± 9.2	0.03 ± 0.07
2	0.95 ± 0.02	23.0 ± 3.6	5.4 ± 5.2	4.8 ± 5.1	0.01 ± 0.01
3	0.88 ± 0.06	33.5 ± 9.6	14.5 ± 12.1	16.5 ± 12.3	0.09 ± 0.30
4	0.83 ± 0.08	40.7 ± 9.2	13.4 ± 9.8	18.0 ± 11.3	1.05 ± 1.02
5	0.84 ± 0.04	40.3 ± 4.9	21.1 ± 9.5	25.0 ± 8.3	0.01 ± 0.01
Average	0.87 ± 0.07	34.4 ± 9.7	13.5 ± 10.8	16.0 ± 11.3	0.26 ± 0.65

Table 6 presents the results when the PRE setup and model M4 (linear model with only abdomen belt) were used with N value 1 and 8. N value 1 refers to the conventional linear regression model, where one sample of each predictor variable is used at a time to predict the response variable. It is clearly seen in Table 6 that N value 8 produced much more accurate waveforms, since average R^2 value increased from 0.62 to 0.87, average relative RMSE decreased from 60.5% to 34.4%, average BPM error almost halved and average volume errors decreased.

Table 7 presents the results when POST setup and model M4 were used with N value 1 and 8. When the average results of Tables 6 and 7 are compared, it can be seen that PRE setup and POST setup with N value 8 produced both very good results and that there are small differences between the results.

Table 7. Results (average value ± SD) for POST setup and model M4 with N value 1 (upper results) and N value 8 (lower results).

Patient	R^2	Relative RMSE [%]	Abs (V_T error) [%]	Abs (V_{minute} error) [%]	BPM error
1	0.55 ± 0.06	66.7 ± 4.2	25.8 ± 14.9	29.9 ± 16.6	0.49 ± 0.54
2	0.69 ± 0.06	55.2 ± 5.7	7.9 ± 6.4	7.9 ± 7.2	0.04 ± 0.06
3	0.74 ± 0.08	50.4 ± 8.2	34.1 ± 11.1	36.7 ± 9.3	0.12 ± 0.29
4	0.60 ± 0.07	63.1 ± 5.6	24.6 ± 20.4	22.8 ± 19.2	1.48 ± 1.12
5	0.55 ± 0.04	66.8 ± 2.9	31.5 ± 11.3	38.1 ± 9.0	0.62 ± 0.61
Average	0.63 ± 0.10	60.3 ± 8.5	24.2 ± 16.3	26.2 ± 17.1	0.57 ± 0.83
1	0.86 ± 0.05	36.3 ± 6.9	16.2 ± 11.6	19.0 ± 9.4	0.03 ± 0.07
2	0.94 ± 0.02	23.8 ± 3.7	6.9 ± 6.6	6.1 ± 6.1	0.01 ± 0.01
3	0.89 ± 0.06	32.8 ± 9.0	16.4 ± 12.8	19.1 ± 12.4	0.10 ± 0.30
4	0.79 ± 0.13	45.3 ± 12.9	11.3 ± 6.6	15.3 ± 13.6	1.23 ± 1.27
5	0.87 ± 0.03	35.8 ± 4.0	10.6 ± 8.0	12.8 ± 7.7	0.13 ± 0.38
Average	0.87 ± 0.09	35.0 ± 10.8	12.4 ± 9.8	14.6 ± 11.2	0.32 ± 0.79

Table 8 presents the results when PRE setup and model M5 (linear model with only rib cage belt) were used with N value 1 and 8. However, in some cases the R^2 values were negative, because the residual was so large that R^2 received negative values. Negative values may occur if the residuals between the spirometer signal and estimated respiratory airflow contain very large values (see section *Statistics*). This may happen when a regression model fails completely to predict data for example from very different breathing styles. A missing R^2 value of a patient means that at least some of the R^2 values of measurement minutes were negative, thus, the average R^2 value of a patient could not be computed. Relative RMSE, volume errors and BPM error are clearly worse in this case than in Tables 6 and 7.

Table 8. Results (average value ± SD) for PRE setup and model M5 with N value 1 (upper results) and N value 8 (lower results).

Patient	R^2	Relative RMSE [%]	Abs (V_T error) [%]	Abs (V_{minute} error) [%]	BPM error
1	–	247.9 ± 86.1	270.0 ± 129.0	234.0 ± 113.1	0.30 ± 0.38
2	–	180.9 ± 17.0	177.6 ± 23.6	175.6 ± 19.9	0.06 ± 0.05
3	0.78 ± 0.12	45.4 ± 10.8	14.0 ± 9.8	16.8 ± 9.7	0.14 ± 0.27
4	–	103.2 ± 29.3	39.3 ± 62.3	25.0 ± 20.6	2.46 ± 1.87
5	–	306.3 ± 52.7	371.5 ± 107.2	336.9 ± 101.0	0.46 ± 0.52
Average	–	171.3 ± 102.7	165.2 ± 152.5	148.9 ± 136.0	0.72 ± 1.31
1	–	259.6 ± 92.5	288.0 ± 142.1	248.3 ± 122.8	0.23 ± 0.31
2	–	131.5 ± 24.2	127.9 ± 24.7	125.8 ± 27.0	0.13 ± 0.30
3	0.91 ± 0.04	29.3 ± 6.6	13.2 ± 7.6	15.8 ± 9.5	0.18 ± 0.40
4	0.51 ± 0.19	68.7 ± 13.2	31.2 ± 12.0	27.3 ± 16.7	2.36 ± 1.97
5	–	245.2 ± 26.7	229.9 ± 31.9	258.2 ± 41.6	1.06 ± 0.94
Average	–	143.5 ± 102.8	135.2 ± 128.9	130.0 ± 119.6	0.80 ± 1.35

Table 9 presents the results when POST setup and model M5 were used with N value 1 and 8. In this setup, N value 1 produced negative R^2 values for patients 1 and 4 in some measurement minutes, therefore, the average R^2 values are missing from them.

However, N value 8 produced much better results, since relative RMSE, volume errors and BPM error decreased considerably. When the average results of Tables 8 and 9 are compared, a clear difference can be seen between the PRE setup and POST setup. The latter with N value 8 produced much more accurate waveforms than the other options. Compared to the PRE setup, relative RMSE and volume errors decreased to the fractions.

Table 9. Results (average value ± SD) for POST setup and model M5 with N value 1 (upper results) and N value 8 (lower results).

Patient	R^2	Relative RMSE [%]	Abs (V_T error) [%]	Abs (V_{minute} error) [%]	BPM error
1	–	73.5 ± 15.0	36.0 ± 30.2	28.6 ± 23.0	0.37 ± 0.48
2	0.61 ± 0.06	62.4 ± 4.6	21.1 ± 7.7	22.0 ± 6.4	0.06 ± 0.05
3	0.74 ± 0.18	48.8 ± 15.2	13.5 ± 8.8	10.0 ± 10.0	0.15 ± 0.28
4	–	83.3 ± 13.3	44.6 ± 25.0	53.8 ± 19.0	2.37 ± 2.23
5	0.55 ± 0.03	66.8 ± 2.1	27.1 ± 14.5	35.3 ± 10.1	0.75 ± 0.61
Average	–	67.6 ± 16.5	29.1 ± 22.5	30.2 ± 21.3	0.74 ± 1.36
1	0.78 ± 0.15	45.0 ± 14.9	17.2 ± 13.0	15.9 ± 15.3	0.37 ± 0.53
2	0.86 ± 0.03	37.8 ± 4.0	17.6 ± 9.6	14.7 ± 9.3	0.02 ± 0.01
3	0.89 ± 0.09	31.0 ± 10.5	11.2 ± 9.2	9.7 ± 8.1	0.11 ± 0.30
4	0.52 ± 0.16	68.8 ± 11.2	37.4 ± 10.6	34.9 ± 18.3	1.76 ± 1.76
5	0.90 ± 0.03	31.4 ± 5.1	9.3 ± 9.4	12.2 ± 7.4	0.23 ± 0.47
Average	0.78 ± 0.18	44.1 ± 17.5	19.5 ± 14.4	18.1 ± 15.6	0.54 ± 1.10

It is clearly seen from average results in Tables 6, 7, 8 and 9 that better results were received with the abdominal respiratory effort belt as the R^2 values were higher and relative RMSE, volume errors and BPM errors smaller. This might be explained by the fact that during quiet breathing most normal subjects are abdominal breathers when in supine position and thoracic breathers when upright [20]. The better estimates are then produced with the dominating abdominal compartment. However, there are differences between the patients. For example, patients 3 and 5 got slightly better results with rib cage respiratory effort belt. A larger data set is clearly needed to confirm the results.

As expected, the results with one respiratory effort belt were worse than with two belts, but not very much. If we compare the results of POST setup with N value 8 from Table 7 (abdominal belt, model M4) and Table 4 (two belts, model M1), it is seen that R^2 increased from 0.87 to 0.91, relative RMSE decreased from 35.0% to 28.1%, tidal volume error decreased from 12.4% to 8.4%, minute volume error decreased from 14.6% to 8.6% and BPM error decreased from 0.32 to 0.23. Conventionally, it is expected that two compartments (rib cage and abdominal) are needed for the estimation of the airflow [21]. With this measurement data, for some patients, the results with one belt were almost as good as with two belts. A possible reason for that could be that the patients were in supine position during the whole measurement sessions and they did not move very much. Caused by this, model training data was quite similar to testing data and the resulting estimated airflow was accurate.

3.3   Case Study 1: Performance Difference Between Setups

Figure 2 demonstrates the performance difference between PRE and POST setups. In this case, model M1 (two respiratory effort belts, N = 16) was used firstly with the PRE setup and secondly with the POST setup. The measurement data was from patient 2. It can be seen from Fig. 2 that there were clear differences with the estimated respiratory airflows (PRE setup with dotted line and POST setup with bold line). Following numerical results demonstrate these differences further. The results for the whole measurement minute with PRE setup were: R^2 was 0.88, relative RMSE 35.0%, tidal volume error 27.4%, minute volume error 26.2% and BPM error 0.04. Equivalently results for the POST setup were: R^2 was 0.94, relative RMSE 24.8%, tidal volume error 1.9%, minute volume error 0.6% and BPM error 0.04. Although, the PRE setup was otherwise remarkably worse than POST setup in this case, BPM was estimated very accurately.

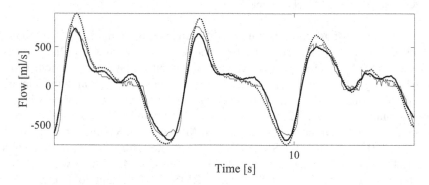

Fig. 2. Short segments of example signals depicting the difference of PRE and POST setups: spirometer signal (thin line), the first estimated respiratory airflow signal (two belts, PRE setup, model M1, N = 16, dotted line) and the second estimated respiratory airflow (two belts, POST setup, model M1, N = 16, bold line).

3.4   Case Study 2: Performance Differences Between Models

Figure 3 demonstrates the performance difference between models M1 and M2. In this case, POST setup (two respiratory effort belts, N = 8) was used with model M1 and secondly with model M2. The measurement data was from patient 1. It can be seen from Fig. 3 that there were very small differences with the estimated respiratory airflows (M1 with dotted line and M2 with bold line). Both models produced excellent estimates of airflow. Following numerical results demonstrate this further. The results for the whole measurement minute with model M1 were: R^2 was 0.95, relative RMSE 21.6%, tidal volume error −1.9%, minute volume error −2.2% and BPM error 0.11. Equivalently results for the model M2 were: R^2 was 0.95, relative RMSE 23.3%, tidal volume error −6.0%, minute volume error 5.3% and BPM error 0.15.

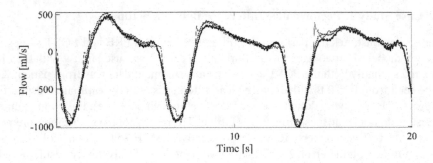

Fig. 3. Short segments of example signals depicting slight differences between model M1 and model M2: spirometer signal (thin line), the first estimated respiratory airflow signal (two belts, POST setup, model M1, N = 8, dotted line) and the second estimated respiratory airflow (two belts, POST setup, model M2, N = 8, bold line).

Figure 4 demonstrates the performance difference between models M1 (both respiratory effort belts), M4 (only abdominal belt) and M5 (only rib cage belt). In these cases, POST setup with N value 8 was used. The measurement data was from patient 1. It can be seen from Fig. 4 that model M1 produced clearly the best results. Following numerical results demonstrate the differences between models. The results for the whole measurement minute with model M1 were: R^2 was 0.93, relative RMSE 26.1%, tidal volume error –5.4%, minute volume error –5.8% and BPM error 0.18. Results for the model M4 were: R^2 was 0.92, relative RMSE 28.6%, tidal volume error –17.2%, minute volume error –15.0% and BPM error 0.00. Results for the model M5, which was clearly the worst in this case, were: R^2 was 0.86, relative RMSE 38.0%, tidal volume error 3.2%, minute volume error 20.9% and BPM error 0.59.

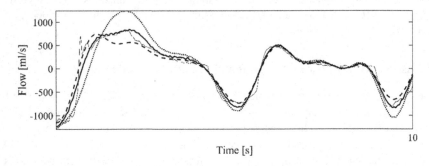

Fig. 4. Short segments of example signals depicting differences between models M1, M4 and M5: spirometer signal (thin line), the first estimated respiratory airflow signal (two belts, POST setup, model M1, N = 8, bold line), the second estimated respiratory airflow (abdominal belt, POST setup, model M4, N = 8, dashed line) and the third estimated respiratory airflow (rib cage belt, POST setup, model M5, N = 8, dotted line).

3.5   Case Study 3: Low Airflow

Patients are often weak and tired after the surgery so their respiration contains very low airflows. It is important to estimate these parts accurately so that they are not interpreted as no-airflow parts and no false alarms are created. Figure 5 depicts short segments of example signals with low airflow. Estimation model M1 (N = 16) was used with the POST setup for the measurement signals of patient 3. Results for the whole measurement minute were: R^2 was 0.93, relative RMSE 25.8%, tidal volume error 0.0%, minute volume error −10.5% and BPM error −0.01. It can be seen from Fig. 5 that the spirometer signal is estimated with good accuracy.

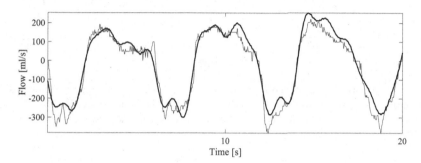

Fig. 5. Short segments of example signals with low airflow: spirometer signal (thin line) and the estimated respiratory airflow signal (POST setup, model M1, N = 16, bold line).

3.6   Case Study 4: Thoracoabdominal Asynchrony

Thoracoabdominal asynchrony refers to the non-coincident motion of rib cage and abdomen and is characterized by a time lag between motion of rib cage and abdomen. It is often observed in many respiratory disorders and/or respiratory muscle dysfunctions. Clinically, it is assessed as a sign of respiratory distress and increased work of breathing. Measurement data of patient 4 included thoracoabdominal asynchrony more or less during the whole measurement session. Figure 6 depicts one example of thoracoabdominal asynchrony. Estimation model M1 (N = 8) was used with the POST setup and the results were the following: R^2 was 0.94, relative RMSE 24.2%, tidal volume error 1.8%, minute volume error −3.1% and BPM error 0.19. These results are consistent with our earlier findings indicating that our method produces very good results with thoracoabdominal asynchrony signals as well [18].

Figure 7 demonstrates the performance difference of models M4 and M5. The same measurement minute from patient 4 as in the previous case was used in these cases as well. Firstly, model M4 (N = 8) and POST setup were used (upper subfigure) and secondly, model M5 (N = 8) and POST setup were used (lower subfigure). It can be seen from subfigures that there were clear differences with the estimated respiratory airflows and model M4 produced better estimates. Numerical results for model M4 were: R^2 was 0.81, relative RMSE 43.0%, tidal volume error 14.7%, minute volume error 21.5% and BPM error 1.25. Equivalently results for model M5 were: R^2 was 0.35, relative RMSE 80.7%, tidal volume error –28.9%, minute volume error –25.7% and BPM error 2.00.

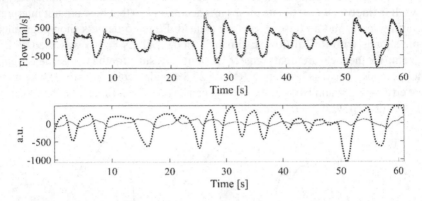

Fig. 6. Example signals of thoracoabdominal asynchrony. Upper subfigure: spirometer signal (solid line) and the estimated respiratory airflow signal (POST setup, model M1, N = 8, dotted line). Lower subfigure: rib cage respiratory effort belt signal (solid line) and abdominal respiratory effort belt signal (dotted line).

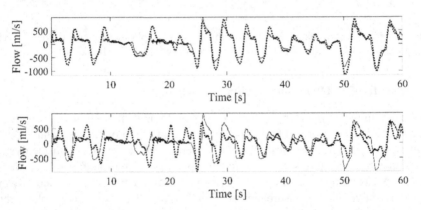

Fig. 7. Segments of example signals depicting the difference of models M4 and M5. Upper subfigure: spirometer signal (solid line) and the estimated respiratory airflow signal (POST setup, model M4, N = 8, dotted line). Lower subfigure: spirometer signal (solid line) and the estimated respiratory airflow signal (POST setup, model M5, N = 8, dotted line).

It can be seen from Figs. 6 and 7 that if respiratory airflow signal is estimated by using both respiratory effort belts, the resulting waveforms are much more accurate: R^2 value 0.94 (relative RMSE 24.2%) compared to 0.81 (43.0%) and 0.35 (80.7%) of models M4 and M5, respectively. In addition to that, volume errors and BMP error are only fractions when model M1 is used.

3.7   Case Study 5: Hypopneic Event

A hypopneic event is commonly defined as greater than 30% reduction in airflow for 10 s or longer. Figure 8 depicts one hypopneic event of patient 4 with COPD. Here, model M1 (N = 8) was used with the POST setup. In this case, R^2 was 0.80, relative RMSE 44.4%, tidal volume error 5.1%, minute volume error 1.4% and BMP error 1.08. It can be seen that the method was able to estimate respiratory airflow very well even in a complicated situation like this.

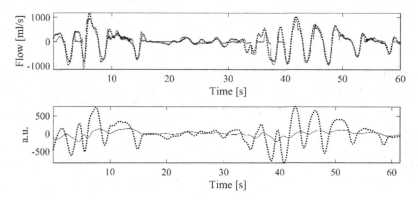

Fig. 8.   Example signals of hypopneic event. Upper subfigure: spirometer signal (solid line) and the estimated respiratory airflow signal (POST setup, model M1, N = 8, dotted line). Lower subfigure: rib cage respiratory effort belt signal (solid line) and abdominal respiratory effort belt signal (dotted line).

Figure 9 demonstrates the performance of models M4 and M5 during the same hypopneic event as in the previous case. Firstly, the model M4 (N = 8) and POST setup was used (upper subfigure) and secondly, the model M5 (N = 8) and POST setup was used (lower subfigure). Numerical results for the model M4 were: R^2 was 0.60, relative RMSE 63.3%, tidal volume error 15.9%, minute volume error 23.1% and BPM error 1.08. Similarly, results for model M5 were: R^2 was 0.44, relative RMSE 74.9%, tidal volume error –32.0%, minute volume error –38.4% and BPM error 2.24. Model M1 produced again the best results. However, model M4 estimated BPM as accurately as model M1 in this case.

3.8   Case Study 6: Apneic Event

An apneic event is usually defined as no detected breaths for periods of longer than 10 s. Figure 10 depicts an example of apneic event from measurement data of patient 5. An apneic event can be seen in the signals from 35 s to 49 s. In this case, model M1 (N = 8) was used with the POST setup. During an apneic event, the rib cage ceases to move, but the abdomen is moving. There is no air exchange, so there is no airflow signal either. It can be seen from Fig. 10 that during the obstruction airflow was zero, but because there

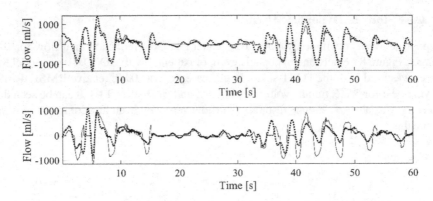

Fig. 9. Segments of example signals during hypopneic event depicting the difference of models M4 and M5. Upper subfigure: spirometer signal (solid line) and the estimated respiratory airflow signal (POST setup, model M4, N = 8, dotted line). Lower subfigure: spirometer signal (solid line) and the estimated respiratory airflow signal (POST setup, model M5, N = 8, dotted line).

was a movement in respiratory effort belts (especially in the abdominal belt), the estimated respiratory airflow signal also showed activity. The same phenomenon was also encountered by Drummond et al. [13] and Voscopoulos et al. [22].

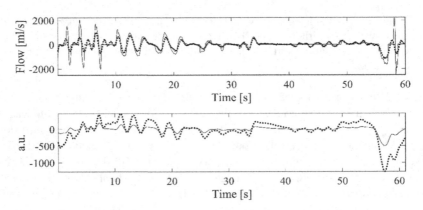

Fig. 10. Short segments of example signals during apneic event. Upper subfigure: spirometer signal (solid line) and the estimated respiratory airflow signal (POST setup, model M1, N = 8, dotted line). Lower subfigure: rib cage respiratory effort belt signal (solid line) and abdominal respiratory effort belt signal (dotted line).

Figure 11 demonstrates the performance of models M4 and M5 during the same apneic event as in the previous case. Firstly, the model M4 (N = 8) and POST setup was used (upper subfigure) and secondly, the model M5 (N = 8) and POST setup was used (lower subfigure). It can be seen from the lower subfigure of Fig. 11 that during the apneic event model M5, which uses only rib cage effort belt produced better estimates.

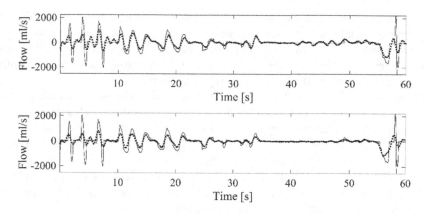

Fig. 11. Segments of example signals during apneic event depicting the difference of models M4 and M5. Upper subfigure: spirometer signal (solid line) and the estimated respiratory airflow signal (POST setup, model M4, N = 8, dotted line). Lower subfigure: spirometer signal (solid line) and the estimated respiratory airflow signal (POST setup, model M5, N = 8, dotted line).

3.9 Limitations of Study

The study had a number of limitations. Firstly, the study included only five patients. The study should be repeated with a larger data set in order to draw more general conclusions. Secondly, respiratory effort belts cannot be used at all if the surgical wound is in the area where the belts are placed. However, the proposed method could be applied to the measurement data acquired with other sensors without this kind of restriction, such as acceleration sensors. Thirdly, as was pointed out in Sect. 3.2, during the apneic event there is no respiratory airflow but still the estimated respiratory airflow exists due to abdominal movements.

4   Conclusions

A novel method was proposed for accurate estimation of continuous respiratory airflow postoperatively. The data from respiratory effort belts were calibrated with a spirometer using an extended multiple linear regression method. Several models for calibration were compared experimentally. The use of two respiratory effort belts was studied in detail and compared with a single belt approach. It was studied whether calibration models should be trained with preoperative data or postoperative data to improve performance. The results showed that training the estimation models with the postoperative data produced much more accurate results than training the estimation model with the preoperative data.

It was demonstrated with data from five patients in a postoperative situation that estimated respiratory airflow signals have accurate waveforms, tidal volume, minute volume and respiratory rate. The method produced good estimates even from challenging respiration signals: low airflows, hypopneic events and thoracoabdominal asynchrony. It was also shown that a single belt measurement, especially the abdominal belt, can be sufficient in some cases. The use of only one respiratory effort belt simplifies the measurements. In addition, it is less likely that a surgical wound would be in the area of the belt than in the case of two belts, thus allowing for a wider variety of surgical operations. However, the study should be repeated with a larger data set in order to confirm the results and in order to draw more general conclusions.

This kind of continuous and real-time respiratory monitoring could provide potential for optimizing pain management and enable timely interventions with a decreased risk of postoperative respiratory complications. For patients at risk for respiratory depression, personalized and more optimal opioid dosing is possible. In addition, for patients with adequate respiratory function, more aggressive and effective dosing could be provided when necessary. All this in its part leads to improved patient safety and satisfaction. Additionally, complications from inadequate pain control are reduced and as a consequence of that health care costs are decreased.

In summary, the proposed method is able to produce estimates of respiratory airflow waveforms to enable accurate, real-time, continuous, and non-invasive respiratory monitoring postoperatively.

Acknowledgements. Finnish Cultural Foundation, North Ostrobothnia Regional Fund and International Doctoral Programme in Biomedical Engineering and Medical Physics (iBioMEP) are gratefully acknowledged for financial support.

References

1. Cepeda, M.S., Farrar, J.T., Baumgarten, M., Boston, R., Carr, D.B., Strom, B.L.: Side effects of opioids during short-term administration: effect of age, gender, and race. Clin. Pharmacol. Ther. **74**(2), 102–112 (2003)
2. Etches, R.: Respiratory depression associated with patient-controlled analgesia: a review of eight cases. Can. J. Anaesth. **41**(2), 125–132 (1994)
3. Gamil, M., Fanning, A.: The first 24 hours after surgery. A study of complications after 2153 consecutive operations. Anaesthesia **46**, 712–715 (1991)
4. Taylor, S., Kirton, O.C., Staff, I., Kozol, R.A.: Postroperative day one: a high risk period for respiratory events. Am. J. Surg. **190**(5), 752–756 (2005)
5. Ramsay, M.A.E., Usman, M., Lagow, E., Mendoza, M., Untalan, E., De Vol, E.: The accuracy, precision and reliability of measuring ventilatory rate and detecting ventilatory pause by rainbow acoustic monitoring and capnometry. Anesth. Analg. **117**(1), 69–75 (2013)
6. Mailey, J., Digiovine, B., Baillod, D., Gnam, G., Jordan, J., Rubinfeld, I.: Reducing hospital standardized mortality rate with early interventions. J. Trauma Nurs. **13**(4), 178–182 (2006)
7. Schein, R.M.H., Hazday, N., Pena, M., Ruben, B.H., Sprung, C.L.: Clinical antecedents to in-hospital cardiopulmonary arrest. Chest **98**(6), 1388–1392 (1990)
8. George, J.A., Lin, E.E., Hanna, M.N., Murphy, J.D., Kumar, K., Ko, P.S., Wu, C.L.: The effect of intravenous opioid patient-controlled analgesia with and without background infusion on respiratory depression: a meta-analysis. J. Opioid Manag. **6**(1), 47–54 (2010)
9. Lynn, L.A., Curry, J.P.: Patterns of unexpected in-hospital deaths: a root cause analysis. Patient Saf. Surg. **5**, 3 (2011)
10. Paine, C.W., Goel, V.V., Ely, E., Stave, C.D., Stemler, S., Zander, M., Bonafide, C.P.: Systematic review of physiologic monitor alarm characteristics and pragmatic interventions to reduce alarm frequency. J. Hosp. Med. **11**(2), 136–144 (2016)

11. Wiklund, L., Hök, B., Ståhl, K., Jordeby-Jönsson, A.: Postanesthesia monitoring revisited: frequency of true and false alarms from different monitoring devices. J. Clin. Anesth. **6**(3), 182–188 (1994)
12. Lovett, P.B., Buchwald, J.M., Sturmann, K., Bijur, P.: The vexatious vital: neither clinical measurements by nurses nor an electronic monitor provides accurate measurements of respiratory rate in triage. Ann. Emerg. Med. **45**(1), 68–76 (2005)
13. Drummond, G.B., Bates, A., Mann, J., Arvind, D.K.: Characterization of breathing patterns during patient-controlled opioid analgesia. Br. J. Anaesth. **111**(6), 971–978 (2013)
14. Voscopoulos, C.J., MacNabb, C.M., Brayanov, J., Qin, L., Freeman, J., Mullen, G.J., Ladd, D., George, E.: The evaluation of a non-invasive respiratory volume monitor in surgical patients undergoing elective surgery with general anesthesia. J. Clin. Monit. Comput. **29**(2), 223–230 (2015)
15. Voscopoulos, C., Ladd, D., Campana, L., George, E.: Non-invasive respiratory volume monitoring to detect apnea in post-operative patients: case series. J. Clin. Med. Res. **6**(3), 209–214 (2014)
16. Voscopoulos, C.J., MacNabb, C.M., Freeman, J., Galvagno, S.M., Ladd, D., George, E.: Continuous noninvasive respiratory volume monitoring for the identification of patients at risk for opioid-induced respiratory depression and obstructive breathing patterns. J. Trauma Acute Care Surg. **77**(3), S208–S215 (2014)
17. Masa, J.F., Corral, J., Martin, M.J., Riesco, J.A., Sojo, A., Hernández, M., Douglas, N.J.: Assessment of thoracoabdominal bands to detect respiratory effort-related arousal. Eur. Respir. J. **22**, 661–667 (2003)
18. Seppänen, T.M., Alho, O.-P., Seppänen, T.: Reducing the airflow waveform distortions from breathing style and body position with improved calibration of respiratory effort belts. Biomed. Eng. Online **12**, 97 (2013)
19. Montgomery, D.C., Peck, E.A., Vining, G.G.: Introduction to Linear Regression Analysis, 3rd edn. Wiley, New York (2001)
20. Verschakelen, J.A., Demedts, M.G.: Normal thoracoabdominal motions. influence of sex, age, posture, and breath size. Am. J. Respir. Crit. Care Med. **151**(2), 399–405 (1995)
21. Konno, K., Mead, J.: Measurement of separate volume changes of rib cage and abdomen during breathing. J. Appl. Physiol. **22**(3), 407–422 (1967)
22. Voscopoulos, C., Brayanov, J., Ladd, D., Lalli, M., Panasyuk, A., Freeman, J.: Evaluation of a novel noninvasive respiration monitor providing continuous measurement of minute ventilation in ambulatory subjects in a variety of clinical scenarios. Anesth. Analg. **117**(1), 91–100 (2013)

Generalizing the Detection of Clinical Guideline Interactions Enhanced with LOD

Veruska Zamborlini[1,3]([✉]), Rinke Hoekstra[1,2], Marcos da Silveira[3],
Cedric Pruski[3], Annette ten Teije[1], and Frank van Harmelen[1]

[1] Department of Computer Science, VU University Amsterdam,
Amsterdam, The Netherlands
{v.carrettazamborlini,rinke.hoekstra,a.c.m.ten.teije,
f.a.h.van.harmelen}@vu.nl
[2] Faculty of Law, University of Amsterdam, Amsterdam, The Netherlands
[3] Luxembourg Institute of Science and Technology - LIST,
Esch-sur-Alzette, Luxembourg
{marcos.dasilveira,cedric.pruski}@list.lu

Abstract. This paper presents a method for formally representing Computer-Interpretable Guidelines. It allows for combining them with knowledge from several sources to better detect potential interactions within multimorbidity cases, coping with possibly conflicting pieces of evidence coming from clinical studies. The originality of our approach is on the capacity to analyse combinations of more than two recommendations, which is useful, for instance, for polypharmacy interactions cases. We defined general models to express evidence as causation beliefs and designed general rules for detecting interactions (e.g., conflicts, alternatives, etc.) enriched with Linked Open Data (e.g. Drugbank, Sider). In particular we show that Linked Open Data sources enable us to detect (suspected) interactions among multiple drugs due to polypharmacy. We evaluate our approach in a scenario where three different clinical guidelines (Osteoarthritis, Diabetes, and Hypertension) are combined. We demonstrate the capability of this approach for detecting several potential conflicts between the recommendations and find alternatives.

Keywords: Clinical guidelines · Semantic Web · Knowledge representation · Ontologies

1 Introduction

Clinical Guidelines (CG) are developed for supporting physicians decision, e.g. specifying what treatment work best in what **situation** [2]. When possible, the **recommendations** provided by CGs are based on **evidences** from clinical

Invited submission as extension of [1].

V. Zamborlini—Funded by CNPq (Brazilian National Council for Scientific and Technological Development) within the Science without Borders programme.

A. Fred and H. Gamboa (Eds.): BIOSTEC 2016, CCIS 690, pp. 360–386, 2017.
DOI: 10.1007/978-3-319-54717-6_20

researches. In this case, there is a direct mapping to the clinical evidence that describes the effects (**transitions**) of certain **care action** (e.g. *do not administer aspirin because of an increased risk of gastrointestinal bleeding*). Since an evidence is not a fact, a multitude of evidence rating systems [3] are adopted by CGs authors. Epistemologically, an evidence reflects a *belief* in the existence of a *causal* relation between e.g. *administering aspirin* and *gastrointestinal bleeding*. Furthermore, CGs are targeted to the treatment of a specific illness. However, it is quite common to have patients with *multiple* diseases (multimorbidity), which requires the combination of different CGs. For example, according to [4], around 40% of 55 years old patients suffer from at least 2 diseases, and 20% of 70 years old patients suffer from at least 4 diseases in Scotland. As with any large volume of regulations, combined guidelines almost inevitably involve intricate **interactions** between the recommendations they describe. Finding interactions (like potential conflicts) requires intensive collaboration in multidisciplinary teams.

When combining guidelines doctors often make use of background knowledge in order to identify potential interactions among recommendations. In particular, the incompatibility among pairs of drugs is largely addressed not only by the medical literature and, as consequence, by external knowledge sources such as Drugbank, DIKB and LIDDI (see Sect. 2). However, the **polypharmacy** issue, i.e. the recommendation of multiple drugs, rise the possibility of incompatibility among several drugs. Although it is a recognized issue [5], it is under-addressed due to the complexity of analyzing interactions among three or more drugs. Few work in the literature provide methods for addressing this issue. Despite the lack of sources providing this knowledge, it can be 'speculated' (with low confidence) from datasets for registering adverse events during treatment, such as AERS from FDA (see Sect. 2).

Computational support can be of great value for supporting physicians to handle all this complexity. Many languages have been proposed for representing "computer-interpretable" guidelines (CIG) and reasoning about it [2]. However, the concepts here discussed are poorly or not addressed by those approaches. The main reason is because much has been devoted to executing guidelines to generate treatment plans rather than other purposes such as combining and updating CGs. In particular, regarding the issue of multimorbidity, existing approaches for combining CGs are limited in their ability to automatically detect the interactions, propose alternatives or combining more than two guidelines [6].

This work follows an incremental methodology. We start by addressing realistic but simplified case studies, and add more complexity according to the lessons learned in each iteration. Therefore, this paper is the continuation of earlier work reported in [6–8,15]. In this series of work, we investigated (i) what knowledge is required to represent and reason about CGs (rather than how to acquire such knowledge), particularly for supporting the multimorbidity issue; (ii) how it can be formalized; (iii) how it can be implemented using Semantic Web technologies, so that (iv) we can exploit the medical knowledge available as Linked Open Data (LOD). As a consequence, the results we obtain are limited to the current expressiveness of the model, e.g. temporal aspects and related interactions will be addressed in future iterations.

This paper reports on improvements to both the models and the evaluation to better address the issue of multimorbidity. The contributions are (C1) a more generic version of the guideline models with respect to recommendations, beliefs and event types. This includes (C2) a formalization of the improved guideline models and guideline interaction rules in FOL; and (C3) a Semantic Web framework for representing and reasoning about recommendations and beliefs using standard vocabularies. This provides (C4) a flexible mechanism for reusing external knowledge bases to extend our ability to detect interactions (showcased using DrugBank, Sider, DIKB, LIDDI, and AERS).

The remainder of this paper is as follows: Sect. 2 presents a case study to illustrate the main concepts, which are further defined in the models and interaction rules (Sect. 3), followed by their implementation by using semantic web technology (Sect. 4). An experimental assessment shows the results obtained for the referred case study in Sect. 5. The related work is discussed in Sect. 6 and the main contributions and future work are discussed in Sect. 7.

2   Case Study

This case study is meant for illustrating the concepts aforementioned and the results of applying the interaction rules (see Sect. 5). Both, concepts and rules, will be detailed in later sections, but we claim that this intuitive example will help the comprehension of the approach. It concerns the combination of parts of three guidelines, namely Osteoarthritis (OA), Diabetes (DB) and Hypertension (HT) and the detection of interactions among them. A graphical representation with the selected recommendations/interactions is presented in Fig. 1. The set of recommendations was extracted/adapted from [9] and represented according to our models. They are:

Diabetes (DB)
1. *Should adm. Dipyridamole to reduce blood coagulation*
2. *Should adm. NSAID to reduce blood coagulation*
3. *Should adm. Insulin to reduce blood sugar level*
Osteoarthritis (OA)
1. *Should NOT adm. Aspirin to avoid increasing risk of gastrointestinal bleeding*
2. *Should adm. Ibuprofen to reduce pain*
Hypertension (HT)
1. *Should adm. Thiazide to reduce the blood pressure*
2. *Should adm. Clopidogrel to reduce blood coagulation*

Among the recommendations some **internal interactions** are identified:

1. DB.1, DB.2 and HT2 are **alternative recommendations** meant for promoting the same effect. Some alternative recommendations are frequently proposed in clinical guidelines to give more freedom to healthcare professionals. Choosing the most appropriated alternative is part of the healthcare

professional tasks. However, this choice can be influenced by other interactions with patient treatments. Thus, we need to highlight these potential interactions.

2. DB.2 and OA.1 are (can be) **contradictory recommendations** since the first might lead to the prescription of *Aspirin* which is non-recommended by the later. Healthcare professionals must be aware about this critical case and evaluate the risks/benefits of these two recommendations. When possible, alternatives need to be proposed/finded to reduce or eliminate the risks.

3. DB.2 and OA.2 are (can be) **repeated recommendations** since the first might lead to the prescription of *Ibuprofen* which is already recommended by the later. Repeated recommendations can be an wasteful expenditure or it can cause overdoses. Thus, healthcare professionals need to be alerted about this type of interaction.

However, in practice, healthcare professionals are expected to be are aware of several implicit information such as side-effects or incompatibilities between recommendations. Indeed, adding all implicit information to paper-based clinical guidelines is not practical However, missing such extra information can lead to underestimating potential interactions. One common practice is to search for it in other sources. Many websites were created to make this search easier for healthcare professionals. With the advances of the Informatics and the semantic technologies, extra information started to be available as linked open data (LOD). In this work, we selected several LOD sources to improve our analysis:

Drugbank[1] contains comprehensive and curated information about drugs, their categories and (pairwise) interactions (DDI)[10];

Sider[2] contains information on medicines and their recorded adverse drug reactions [11];

DIKB[3] contains curated information about drugs and their (pairwise) interactions (DDI), mostly regarding depression treatments [12].

LIDDI[4] regroups information from different datasources using the nanopublication format; contains comprehensive information about drugs and their (pairwise) interactions (DDI)[13].

AERS[5] contains registries about adverse effects involving several drugs reported by patients during treatment. We use it as a source of weak beliefs about the incompatibility among several drugs. In this work we use a LOD version of 2012 published in [14].

When considering the knowledge acquired from LOD sources we could detect different potential interactions, for example:

---

[1] http://drugbank.ca/.
[2] http://sideeffects.embl.de.
[3] http://dbmi-icode-01.dbmi.pitt.edu/dikb-evidence/front-page.html.
[4] https://datahub.io/dataset/linked-drug-drug-interactions-liddi.
[5] http://www.fda.gov/Drugs/GuidanceComplianceRegulatoryInformation/Surveillance/AdverseDrugEffects/.

1. DB.1, DB.2 and HT2 have as **alternative** *Administer Epoprostenol* to achieve the same desired effect according to *DrugBank*. It can complement the alternatives already provided in guideline, if so.
2. DB.2 and OA.2 recommend potentially **incompatible actions**, because the *NSAID* can be selected as *Aspirin*, which is believed to be incompatible with *Ibuprofen* according to *Drugbank*.
3. DB.3 and HT.1 potentially interact because the situation that is meant to be changed by the first recommendation, i.e. *high blood sugar level*, is promoted as **side-effect** if the second is prescribed as *bendroflumethiazide*, according to *Sider*.
4. HT.1 and OA.2 potentially interact because the situation that is meant to be changed by the first recommendation, *high blood pressure*, is promoted as **side-effect** by the second one, *Ibuprofen*, according to *Sider*.
5. DB.2, HT.2 and OA.2 recommend (potentially) **incompatible actions**, since they are registered in an adverse event report according to *AERS*. This case shows the capacity of our approach to detect interactions that (possibly) happens when three or more recommendations exist. Approaches that evaluate pairwise interactions are unable to detect this case.

Without coming into details about the formalism behind the knowledge representation and detection rules, we present in Fig. 1 a summary of the recommendations and interactions described before. The big rectangles in both left and right sides represent **beliefs** regarding the **care actions** (*administering Dipyridamole*). The latter is represented as dotted ellipses inside the beliefs. The **causation beliefs** are about a **transition** between **situations**

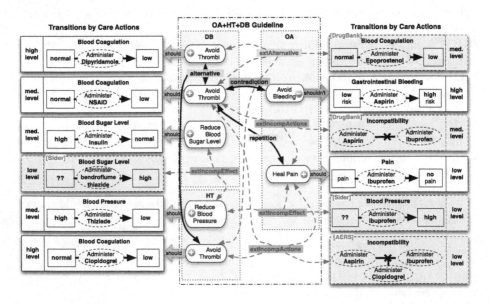

Fig. 1. Case study on combining guidelines for OA+HT+DB.

(*blood coagulation goes from normal to low*) that are believed to be promoted by executing a care action type). The causation belief has a **frequency** e.g. *administer Dipyridamole always reduce the blood coagulation.* For sake of simplicity, we consider in this work only *always* as frequency for all causation beliefs. They also have a **strength** associated, which corresponds to the evidence level (e.g. *high level*), according to the quality attributed to the **sources** (or studies) that provide such knowledge. The beliefs in gray shade represent the knowledge imported from an external source described in the top left (e.g. DrugBank). The strength in this case will depend on the reliability of each data source. The external sources here considered describe two types of beliefs: causation belief or **incompatibility belief**. The latter represents action types that should not be recommended together, e.g. Administer Aspirin is incompatible with Administer Ibuprofen (the reason is not provided in structured way from the sources).

The dotted rectangles in the middle represent the **guidelines**. The more external one is the merge of the three guidelines for OA+HT+DB. They comprise both the **recommendations** (e.g. *avoid thrombi*) and the **interactions** (or *alternatives*) among them. The former is represented as rounded rectangles, and the latter is depicted by labelled thin arrows connecting the interacting recommendations and beliefs. Solid arrows are for internal interactions and dotted arrows for external ones. A positive (or negative) recommendation is indicated by a thick arrow labeled with "should" (or "should not").

3   Conceptual Model and Interaction Rules

In this section we present the adapted version of the TMR (Transition-based Medical Recommendation) models and the rules for detecting internal and external interactions among recommendations [15].

3.1   Conceptual Model

TMR$_{Event}$. Figure 2 presents a UML class diagram for the TMR_{Event} model describing some relevant concepts and relations regarding **event types** in the scope of this work. The concepts introduced in previous versions of the models are depicted in gray-shade (same for the next diagrams). This model is inspired in UFO (Unified Foundational Ontology) [16] that is a formal theory describing some of the general concepts used here, namely Type (Universal) and Category, as well as Object, Event, Action and Situation Types. The model regards mostly types of things[6] since it is meant for modeling, for example, the type of event that is expected as consequence of another one, rather then the particular event that was the consequence of another particular one. In other words, we do not want to say that John's pain was relieved due to the administration of aspirin, but that administering aspirin often relieves the pain of patients[7].

---

[6] For sake of simplicity we can omit the word 'type'.

[7] For a deeper explanation see [7].

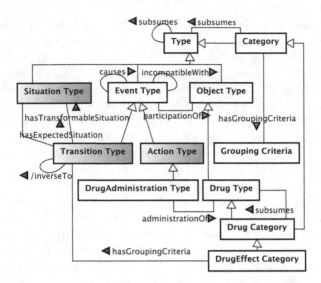

Fig. 2. UML class diagram for the TMR_{Event} model.

While **action types** concern event types to be performed by an intentional agent (omitted in the model), **transition types** concern (deterministic) event types for which pre and post **situation types** can be defined[8]. In other words, it represents the transformation of a **situation type** into another (transformable & expected situations). An event type can be defined as the **participation of** a certain **object type**, e.g. a *DrugAdministration type* is the administration of (participation of) a *Drug type*. An event type can **cause** another event type to happen (occurrence of one causes the occurrence of the other). Moreover, an event type can also be **incompatible with** one (or more) events when they can not or should not occur together. In other words, either happening together is not possible or it would bring results/transitions that are not the expected ones.

A **category** is a type that (transitively) subsumes (or regroups) other types according to a **grouping criteria**, e.g. *ThiazideDrug* is a category of drugs that contains the molecule *thiazide*, e.g. *bendroflumethiazide*. In this case the grouping criteria regards a structural property. However, it can also concern the effect expected to be promoted, e.g. *NSAID* is the category for (non-steroidal) drugs expected to promote the transition *reduce inflammation*, e.g. *Aspirin*.

FOL rules are provided for deriving relevant relations in the context of this work that will be further used for defining the interaction rules (Sect. 3.2). Some relations are defined in terms of other relations, for example, *inverseTo* between transition types is one transition that 'undo' the effect of the other. These relations are preceded by a slash in the models (previous and forthcoming).

---

[8] Detailed discussion about (non-)deterministic or (non-)intentional event types is out of scope of this work.

Inverse Transitions: one transition type *t1* transforms situation *s1* into *s2* while another transition *t2* transforms *s2* into *s1*.

R.1 $\forall t1, t2, s1, s2$ TransitionType(t1) $\land$ TransitionType(t2)
  $\land$ SituationType(s1) $\land$ SituationType(s2) $\land s1 \neq s2$
  $\land$ hasTransformableSituation(t1,s1)
  $\land$ hasExpectedSituation(t1,s2)
  $\land$ hasTransformableSituation(t2,s2)
  $\land$ hasExpectedSituation(t2,s1))
  $\rightarrow$ inverseTo(t1,t2)

TMR_{Belief}. Some relations can be difficult to be precisely defined either for epistemic or ontological issues. For instance, event types might not have a precise definition of their consequences, e.g. *administering aspirin* sometimes relieves the pain, sometimes it does not. We address this issue in the TMR_{Belief} model by representing those relations through beliefs, presented in a UML class diagram in Fig. 3. In this work, **beliefs** allow to express a 'degree of confidence' for assertions about **things**/entities according to a source. It is represented as belief **strength**, accounting for the certainty/quality of the belief, such as the evidence level classification in clinical guidelines. We are particularly interested in beliefs about the relations *causes, subsumes* and *incompatibleWith* between event types, for which we provide 'epistemic/doxastic' versions (represented as dotted lines in the model). In other words, they are relations dependent on the existence of a belief to ground their truthfulness (in practice they have a belief as a third argument). Therefore, they are not the same as the ones in Fig. 2.

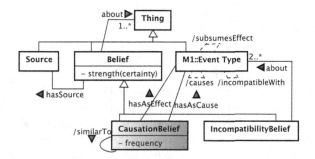

Fig. 3. UML class diagram for the TMR_{Belief} model. 'M1' is used as short reference for TMR_{Event}.

The **causation belief** between event types express the **frequency** in which it was observed. In this context, it also reflects the likelihood/probability of one causing the other according to a source. It allows for complementary beliefs, e.g. *aspirin relieves the pain in 80% of the cases (therefore it does not in 20%)*, but also for inconsistent ones, e.g., *administering aspirin always relieves the pain* and *it never relieves the pain*. This is a desired feature since for some assertions there can be no common agreement from different sources (see [15]). However, we consider that one (merged) guideline that rely on incompatible beliefs is then inconsistent. In this work we focus on beliefs about action (*hasAsCause Action Type*) causing transitions (*hasAsEffect Transition Type*) as justification for the clinical recommendations. Moreover, for the interaction rules we consider only the positive causation beliefs, i.e., an action type always cause a transition, since it is not on the scope of this work both (i) the negative causation, which only appears as sub-justifications of recommendation (discussed in [15]) and (ii) the intermediate frequency values (often, rarely, etc.)[9]. The **subsumption** of event types due to expected effect also relies on the causation beliefs.

Causation - an event type *e1* causes another one *e2* with a certain frequency *f* according to a belief *cb*:

R.3 $\forall e1, e2, cb, f$ (EventType(e1) $\wedge$ EventType(e2)

   $\wedge$ CausationBelief(cb) *land* hasAsCause(cb, e1)

   $\wedge$ hasAsEffect(cb, e2)) $\wedge$ frequency(cb, f)

   $\rightarrow$ causes(e1,e2, f, cb)

Similar Causation Beliefs: two beliefs *cb1, cb2* about different event types *e1, e2* promoting with same frequency *f* another event type *e3*.

R.4 $\forall cb1, cb2, e1, e2, e3, f$ causes(e1, e3, f, cb1)

   $\wedge$ causes(e2, e3, f, cb2) $\wedge cb1 \neq cb2 \wedge e1 \neq e2$

   $\rightarrow$ similarTo(cb1,cb2)

Subsumption Via Causation and Grouping Criteria: if an event type *e1* causes a transition *t1* that is the grouping criteria of another event type *e2* then *e2* subsumes *e1* according to the causation belief.

R.6 $\forall e1, e2, t, cb1$ (EventType(e1) $\wedge$ EventType(e2)

   $\wedge$ TransitionType(t) $\wedge$ causes(e1, t, 'always', cb1)

   $\wedge$ hasGroupingCriteria(e2, t) $\wedge$ e1 $\neq$ e2)

   $\rightarrow$ $\exists cb$ subsumes(e2, e1, cb1)

In its turn, the **incompatibility** between event types is considered in this work to be given as an assertion. Therefore it is also represented as a belief, although it could be explained/derived at a certain level of granularity. One important aspect to consider is **polypharmacy**. As previously discussed, it increases the chances of having the classical interactions between pair drugs, but also among several drugs. Since the model allows for incompatibility beliefs *about* 2 or more Event Types (drug administration), the referred multi-drugs aspect is covered.

TMR_Norm. In general terms, a **Regulation** is composed of a set of **Norms** about the execution of action types based on a causation belief. The norm **strength** can vary from obligation to prohibition. For the specific case of clinical domain, norms are specialized as **Recommendations** and regulations as

---

[9] This approach exclude endless assertions about all the effects an event is not expected to produce since the beliefs are defined in CGs or scientific papers by a community of experts, e.g. cancer is not an effect of a certain drug.

Clinical Guidelines. Since the clinical guidelines are mostly considered a reference for best practices, the strength of recommendations in this work will be considered as 'should' (positive) and 'should-not' (negative), any other variation of strength is out of scope. Finally, among norms there can be **Interactions** of different types. In this work we formalize internal interactions discussed in [15] and we extend and formalize external interactions (introduced in [6]). Figure 4 presents the UML class diagram for the TMR_{Norm} model.

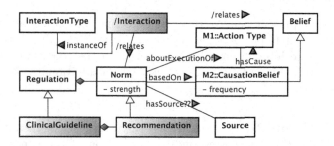

Fig. 4. UML class diagram for the TMR_{Norm} model. 'M1' and 'M2' are used as short references for TMR_{Event} and TMR_{Belief} respectively.

The following FOL rule defines the derivable relation *regulates* while the interactions are defined in the next subsections. Some interaction types have a cumulative behavior, like *Repeated Action* and *Alternative Actions* (introduced in [8]). For example, if three norms recommend the administration of aspirin, there should be one single interaction of type *Repeated Action* among them, rather than three different interactions among pairs of them. External interactions also accumulate, like *External-Alternative Action*. Although the subsumption relation in these rules can be also the epistemic one, derivable through causation beliefs, we adopt the simplified notation *subsumes(a1, a2)* since it does not change the meaning of the rules.

Regulation - a norm n from a regulation r over an action type a has strength st based on a causation belief cb:

R.8 $\forall r, n, a, st, cb$ $\big(($Regulation(r) $\land$ Norm(n)
$\land$ partOf(n, r) $\land$ ActionType(a) $\land$ CausationBelief(cb)
$\land$ aboutExecutionOf(n, a) $\land$ strength(n, st)
$\land$ basedOn(n, cb))
$\rightarrow$ regulates(r, n, a, st, cb)$\big)$

Comparing to our previous work, this section present more generic version of the models with respect to norms, beliefs and event types, (mentioned contribution C1). It allows, for instance, to better handle the hierarchies of action types (or event types) possibly deriving them from hierarchies of drug types, which is commonly found in the existent datasets and terminologies. Particularly the hierarchies concerning effects believed to be promoted (e.g. *Adm. Aspirin specializes*

Adm. Anti-Inflammatory) are handled as beliefs. This is indeed compatible with the discourse of not having certainty about causation relations. FOL formulas are adapted/introduced for the derivable relations. Furthermore, the incompatibility belief is introduced in the TMR_{Belief} model, as well as the strength of beliefs and causation frequency. Finally, the recommendations strength is also introduced in the TMR_{Norm} model.

3.2 Interaction Rules

In this section we present *generic* rules for detecting recommendations interactions (i.e. they are not specific for a guideline/disease, action or drug). We first discuss the internal interaction rules only based on the recommendations, followed by the external interaction rules which exploit additional knowledge from Linked Open Data.

Internal Interaction Rules. Considering the modifications in the model, we propose in this section the corresponding adaptation of the generic **internal interaction** rules presented in [15] (contribution C2). The following types of interactions are defined:

Repeated Action: two positive norms about the same action or about actions in a subsuming relation. The second rule is for the cumulative behavior, i.e. when two different interactions of this type relate the same norm ($n2$) then those interactions are the same.

(I.1.1) $\forall r, n1, n2, a1, a2, cb1, cb2$ (
    regulates(r, n1, a1, 'should', cb1)
    $\wedge$ regulates(r, n2, a2, 'should', cb2)
    $\wedge$ (a1 = a2 $\vee$ subsumes(a1,a2) $\vee$ subsumes(a2,a1)))
    $\rightarrow \exists i$ (**RepeatedAction(i)** $\wedge$ relates(i,n1)
    $\wedge$ relates(i,n2))

(I.1.2) $\forall i1, i2, n1, n2, n3$ (RepeatedAction(i1)
    $\wedge$ RepeatedAction(i2) $\wedge$ relates(i1,n1)
    $\wedge$ relates(i1,n2) $\wedge$ relates(i2,n2) $\wedge$ relates(i2,n3)
    $\wedge$ n1 $\neq$ n3 $\wedge$ n1 $\neq$ n2 $\wedge$ n2 $\neq$ n3)
    $\rightarrow$ **i1 = i2**

Alternative Actions: two positive norms about different actions for achieving the same transition, i.e. they are based on similar causation beliefs for different actions. The second rule is for the cumulative behavior.

(I.2.1) $\forall r, n1, n2, a1, a2, cb1, cb2$ (
    regulates(r, n1, a1, 'should', cb1)
    $\wedge$ regulates(r, n2, a2, 'should', cb2)
    $\wedge$ similarTo(cb1, cb2) $\wedge$ a1 $\neq$ a2)
    $\rightarrow \exists i$ (**AlternativeActions(i)** $\wedge$ relates(i,n1)
    $\wedge$ relates(i,n2))

(I.2.2) $\forall i1, i2, n1, n2, n3$ (AlternativeActions(i1)
$\wedge$ AlternativeActions(i2) $\wedge$ relates(i1,n1)
$\wedge$ relates(i1,n2) $\wedge$ relates(i2,n2) $\wedge$ relates(i2,n3)
$\wedge$ n1 $\neq$ n3 $\wedge$ n1 $\neq$ n2 $\wedge$ n2 $\neq$ n3)
$\rightarrow$ **i1 = i2**

Contradictory Norms: (i) two norms, positive and negative, about the execution of the same action (or actions in a subsuming relationship) or (ii) two norms, positive and negative, about different actions promoting the same transition or (iii) two positive regulations about different actions for achieving inverse transitions.

(I.3) $\forall r, n1, n2, a1, a2, cb1, cb2, t1, t2($
regulates(r, n1, a1, 'should', cb1)
$\wedge$ regulates(r, n2, a2, str, cb2)
$\wedge$ causes(a1,t1,'always',cb1)
$\wedge$ causes(a2,t2,'always',cb2)
$\wedge$ ((str = 'should-not'
$\wedge$ (a1 = a2 $\vee$ subsumes(a1,a2) $\vee$ subsumes(a2,a1))
$\vee$ (str = 'should-not' $\wedge$ a1 $\neq$ a2 $\wedge$ t1 = t2)
$\vee$ (str = 'should' $\wedge$ a1 $\neq$ a2 $\wedge$ inverseTo(t1, t2))))
$\rightarrow \exists i$ (**Contradiction(i)** $\wedge$ relates(i,n1) $\wedge$ relates(i,n2))

Repairable Transition: two norms, positive and negative, about different actions that are believed to cause inverse transitions, i.e. if the undesired effect cannot be avoided, it can be repaired by another action.

(I.4) $\forall r, n1, n2, a1, a2, cb1, cb2, t1, t2($
regulates(r, n1, a1, 'should', cb1)
$\wedge$ regulates(r, n2, a2, 'should-not', cb2)
$\wedge$ causes(a1,t1,'always',cb1)
$\wedge$ causes(a2,t2,'always',cb2)
$\wedge$ a1 $\neq$ a2 $\wedge$ inverseTo(t1, t2))
$\rightarrow \exists i$ (**RepairableAction(i)** $\wedge$ relates(i,n1)
$\wedge$ relates(i,n2))

External Interaction Rules. Beliefs from other sources provide interesting information to enrich the system and improve the accuracy of the interaction detection tasks. We propose hereafter *generic* rules for detecting **external interactions**. In particular the rule for detecting incompatible actions, designed to consider interaction between several recommendations and applied to detect multi-drug interaction (for polypharmacy cases).

External-Alternative Actions: actions believed to promote a desired effect according to external sources are considered alternative to other actions recommended to achieve that effect. The second rule is for the cumulative behavior, i.e. two different interactions of this type relating the same (external) causation belief are considered as the same.

(E.1.1) $\forall r, n1, a1, a2, cb1, cb2($
  regulates(r, n1, a1, 'should', cb1)
  $\wedge$ similarTo(cb1, cb2) $\wedge$ hasAsCause(cb2,a2)
  $\wedge \neg(\exists n2\text{regulates}(r, n2, a2, 'should', cb2))$
  $\wedge \neg$subsumes(a1, a2))
  $\rightarrow \exists i(\textbf{ExternalAlternativeAction(i)}$
  $\wedge$ relates(i,n1) $\wedge$ relates(i,cb2) $\wedge$ relates(i,a2))
(E.1.2) $\forall i1, i2, r, n1, n2, cb(\text{ExternalAlternativeAction(i1)}$
  $\wedge$ ExternalAlternativeAction(i2) $\wedge$ CausationBelief(cb)
  $\wedge$ relates(i1,cb) $\wedge$ relates(i1,n1)
  $\wedge$ relates(i2,cb) $\wedge$ relates(i2,n2)
  $\wedge$ Regulation(r) $\wedge$ partOf(n1,r) $\wedge$ partOf(n2,r))
  $\rightarrow \textbf{i1} = \textbf{i2}$

External-Incompatible Actions: each incompatibility belief relates a number of actions suggesting that they should not be performed 'together'. Therefore, if all actions (or actions in subsumption relationship) are recommended within one (merged) guideline, then exists an interaction relating all the recommendations and the belief.

(E.2) $\forall ib, r\Big(\big(IncompatibilityBelief(ib) \wedge Regulation(r)$
  $\wedge \forall a \, ((Action(a) \wedge about\,(ib,a)) \rightarrow \exists n, a', cb(\text{regulates(r, n, a', 'should', cb)}$
  $\wedge \, (a' = a \vee \text{subsumes(a',a)} \vee \text{subsumes(a,a')})))\big)$
  $\rightarrow \exists i(\textbf{ExternalIncompatibleActions(i)} \wedge relates(i, ib1)$
  $\wedge \forall n, a, a', cb \, ((Action(a) \wedge about\,(ib, a) \wedge \text{regulates(r, n, a', 'should', cb)}$
  $\wedge \, (a' = a \vee \text{subsumes(a',a)} \vee \text{subsumes(a,a')}))$
  $\rightarrow relates(i, r)))\Big)$

External-Incompatible Effects: actions recommended in a guideline are believed to promote certain (side)effects (or situations) according to external sources. If in turn the situation is recommended to be either avoided or changed according to the guideline, then there exist an interaction among the recommendation about the referred action, and the one about the referred effect.

(E.3) $\forall r, n1, n2, a, a1, a2, cb, cb1, cb2, s1($
  regulates(r, n1, a1, st, cb1)
  $\wedge$ causes(a1, t1, 'always', cb1)
  $\wedge$ ((st= 'should') $\wedge$ hasTransformableSituation(t1,s1))
    $\vee$ (st= 'should-not') $\wedge$ hasExpectedSituation(t1,s1)))
  $\wedge$ causes(a, t, 'always', cb) $\wedge$ a $\neq$ a1
  $\wedge$ hasExpectedSituation(t, s1)
  $\wedge$ regulates(r, n2, a2, 'should', cb2) $\wedge$ cb $\neq$ cb2
  $\wedge$ (a2 = a $\vee$ subsumes(a,a2) $\vee$ subsumes(a2,a)))
  $\rightarrow \exists i(\textbf{ExternalIncompatibleEffects(i)} \wedge$ relates(i,n1)
  $\wedge$ relates(i, n2) $\wedge$ relates(i,s1) $\wedge$ relates(i,cb2))

This section concludes the contribution C2 by providing *generic* rules for detecting external interactions. In [6] the rules were specific for a dataset (namely

DrugBank). Now the generic rules apply to beliefs imported from any dataset. At this point its also important to observe that both models and rules are defined in a domain-independent way. An implementation using Semantic Web Technology is provided in the next section.

4   Semantic Web Implementation

This section presents a Semantic Web implementation[10] for the proposed app-roach. The proposed models have a straightforward mapping to OWL2 (ommited in the paper). However, for instantiating the models with the clinical knowledge, we propose the use of a framework as RDF graph structure based on the open formats Nanopublication, Provenance and Open Annotation (see Sect. 6). It is applied for representing the recommendations (norms) and beliefs as assertions connected to their sources, besides other meta-information. In the sequence, the implementation of the FOL rules using SWI-Prolog is exemplified, together with a procedure adopted to import clinical knowledge from LOD (e.g. DrugBank, Sider, etc.) as assertions via generic predicates (beliefs).

4.1   Framework

The framework, illustrated in Fig. 5, follows the *Nanopublication* structure, which presupposes the use of *Prov* vocabulary, and is enriched with (optional) *Open Annotation* vocabulary. The latter is meant for representing assertions that are (somehow) extracted from textual documents.

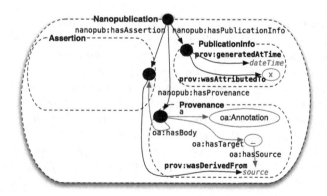

Fig. 5. Nanopublication Schema proposed for representing Beliefs & Norms.

Rounded-dotted boxes represent named graphs containing triples, which in turn are represented as directed-named arrows among resources. The black cir-cles represent the named graphs themselves as subject/object, while the other

---

[10] Accessible at http://rapgmsbgym.github.io.

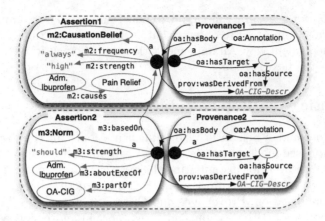

Fig. 6. Nanopublication schema for causation beliefs & norms extracted from guidelines. 'M2' and 'M3' are used as prefixes for TMR_{Belief} and TMR_{Norm} respectively.

resources are represented as ellipses with a description inside (where underline stands for blank nodes) or as an expected data-value (e.g. source or date).

The more external named graph, called Nanopublication, connects the following three named graphs: The *assertion* is a named graph where some knowledge is described using suitable vocabularies, in our case the TMR models. The other named graphs are meant to provide the meta-information about both the assertion and its publication as rdf-data: (i) the *provenance* graph can contain information such as the source (*prov:wasDerivedFrom*) of the assertion (e.g. clinical guideline, study or dataset), and text-annotations (*oa:Annotation*) when the assertion is extracted from a piece of text; and (ii) the *publicationInfo* graph provides meta-information such as when the publication was created and by whom. For sake of readability, henceforth we omit from the figures part of the framework that is not relevant for the discussion here conducted.

Figure 6 illustrates the representation of both a causation belief (at the top) and a recommendation (at the bottom). The *Assertion1* is a *CausationBelief* with *high* strength level, about the action type that *Adm. Ibuprofen always* causes the transition type *Pain relief* according to *OA-CIG-Description*. The *Assertion2* is a *Norm*, part of *OA-CIG* that states *Adm. Ibuprofen should* be executed based on the *evidence* stated in *Assertion1* according to *OA-CIG-Description*. Beliefs taken from external sources are similarly represented (see Fig. 10).

Conclusion. The framework favor data reusability as LOD, since it is compatible with Semantic Web standards proposed for expressing and annotating knowledge extracted from (scientific) publications. It comprises part of contribution C3 (Semantic Web implementation of model and rules).

4.2   Interaction Rules

The proposed FOL interaction rules have the typical format of Prolog rules, what makes its implementation very straightforward. The implemented rules are here illustrated as: function F.1 implements R.8 while functions F.2.1, F.2.2 implement the rules I.1.1 and I.1.2 for interaction *RepeatedAction*. For the purpose of this application, the existential quantifier in the consequent of interaction rules is implemented as a Prolog function called *existsInteraction*. This function uses the *rdf_assertion* built-in-function to insert the respective interaction in the dataset in case it does not exist.

(F.1) regulates(Reg, Norm, ActT, Str, CBelief):-
  instanceOf(Norm, m3:'Norm'),
  rdf(Norm, m3:'partOf', Reg),
  rdf(Norm, m3:'aboutExecutionOf', ActT),
  rdf(Norm, m3:'strength', literal(type(xsd:string,Str))),
  rdf(Norm, m3:'basedOn', Belief, Norm).
(F.2.1) **forall**((regulation(Reg),
  regulates(Reg, N1, ActionT1, 'should', _),
  regulates(Reg, N2, ActionT2, 'should', _),
  different(N1,N2),
  (same(ActionT1, ActionT2)
  ; subsumes(ActionT1, ActionT2)
  ; subsumes(ActionT2, ActionT1))),
  existsInteraction('RepeatedAction', N1, N2)).
(F.2.2) **forall** ((interacts('RepeatedAction', N1, N2, I1),
  interacts('RepeatedAction', N2, N3, I2),
  different(N1,N3), different(I1, I2)),
  rdf_assert(I1, owl:sameAs, I2)).

This section comprises part of contribution C4 (flexible mechanism for reusing LOD to detect interactions) and together with the framework, it concludes contribution C3.

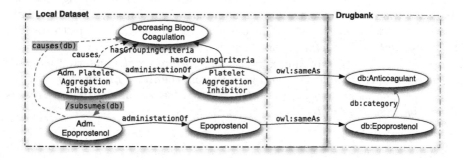

Fig. 7. Importing from Drugbank.

4.3   Using External Knowledge - Linked Open Data

Several medical-datasets are available online, some as LOD, among which we have selected 5 for this work: Drugbank, Sider, DIKB, LIDDI and AERS (see Sect. 2). Each dataset uses its own vocabulary to describe side effects, incompatible and alternative drugs, etc. It means that different datasets describe the same type of knowledge in different ways and it would require, in principle, specific interaction rules to be designed for each dataset. To avoid it, their knowledge is *reinterpreted* according to the TMR model[11].

Importing Rules. The knowledge in the external sources are *reinterpreted* as either causation beliefs or incompatibility beliefs via rules implemented as SWI-Prolog rules. This "importing rules" are indeed specific for each external dataset, as described in this section. However, once the knowledge is imported as a belief, then it will be used by *generic* rules for detecting interactions. On doing so, adding new datasets will not require designing new interaction rules. In order to illustrate the aforementioned *reinterpretation*, we explain three importing rules from three datasets:

Drugbank - Alternative drugs: for all drugs belonging to a *drugCategory* regarding an effect/transition, the causation beliefs are asserted about the actions of administering those drugs promoting the referred effect.

```
forall( rdf(DrugCat, model:'hasGroupingCriteria', Trans),
    rdf(Trans, rdf:type, model:'TransitionType'),
    rdf(DrugCat, owl:sameAs, DrugbankCategory),
    rdf(DrugbankDrugURI, drugcategory:'category', DrugbankCategory),
    rdf(Act, model:'administrationOf', DrugType),
    rdf(DrugType, owl:sameAs, DrugbankDrugURI)
    ),
    (assertCausation(Act, Trans, 'always', 'drugbank', NanopubURI),
    assertProvResourceUsed(NanopubURI, DrugbankDrugURI),
    assertProvResourceUsed(NanopubURI, DrugbankCategory))).
```

Figure 7 depicts an example of how new data is inferred according to Drugbank data (inferred predicates are highlighted in gray-shade). The drugbank category *Anticoagulant* is related in our dataset to the transition *Decreasing Blood Coagulation*. Therefore, since that is the category assigned to the drug *Epoprostenol*, the corresponding action *Adm. Epoprostenol* is inferred as *causing* the referred transition. Due to this causation, the *Adm. Platelet Aggregation Inhibitor*, which as grouping criteria the referred transition, can be inferred as *subsuming Adm. Epoprostenol*.

---

[11] The Drug and Situation Types are mirrored and mapped to the to the external knowledge sources via owl:sameAs.

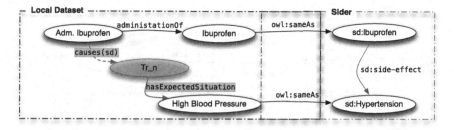

Fig. 8. Importing from Sider.

Sider - Side-effect: for all the DrugSider that are the same as the DrugTypes playing role in a administration Action, and whose sideEffects are mapped the same as Situations in the local dataset, then a causation belief is inferred for that action causing a transition that brings about the referred situation.

> **forall**((
>   rdf(DrugSider, sider:'sideEffect', SideEffect),
>   rdf(Situation, owl:sameAs, SideEffect),
>   rdf(DrugSider, owl:sameAs, DrugbankDrugURI),
>   rdf(DrugType, owl:sameAs, DrugbankDrugURI),
>   instanceOf(DrugType, vocab:'DrugType'),
>   rdf(Action, vocab:'administrationOf', DrugType)
>   ),
>   (assertPartialTransition(Situation, NewTr),
>   assertCausation2(Action, NewTr, 'always', 'sider', NanopubURI),
>   assertProvResourceUsed(NanopubURI, DrugSider),
>   assertProvResourceUsed(NanopubURI, SideEffect)))

Figure 8 depicts an example of how new data is inferred according to Sider data (inferred predicates and entities are highlighted in gray-shade). The drug *Ibuprofen*, which plays a role in *Adm Ibuprofen*, has as side effect *hypertension*, which is the same as the Situation *high blood pressure*. Therefore, the referred action is inferred to *cause* a *transition* that *has as expected situation* the referred situation.

AERS - Incompatible Drugs: This rule is more complex than the previous ones, since the number of involved entities, in this case drugs, is not fixed. To address this issue we need to use an external and an internal universal quantifiers **forall**. We also use an aggregate function to filter only the reports that involve more than one drug annotated with *drugbankDrugURI*, used to link to the drugs in our dataset. Therefore, for all the reports that involves more than one drug and whose all involved drugs are mapped via *drugbankDrugURI* to our dataset, then the incompatibility belief among the corresponding Action Types is inferred.

> **forall**(*%external forall*
>   (rdf(Report, rdf:type, d2s:'Report'),

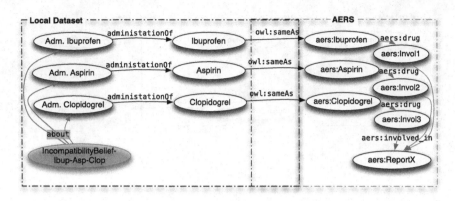

Fig. 9. Importing from AERS.

```
aggregate_all(count, (distinct([Report,DrugbankDrugURI], ,
      (rdf(_1, d2s:'involved_in', Report),
      rdf(_1, d2s:'drug', Drug),
      rdf(Drug, skos:'closeMatch', DrugbankDrugURI))))), Count),
Count > 1,
forall( %internal forall
   (rdf(Involment, d2s:'involved_in', Report),
    rdf(Involment, d2s:'drug', Drug),
   ), %end of the antecendent in the internal forall
   (rdf(Drug, skos:'closeMatch', DrugbankDrugURI),
    rdf(DrugType, rdf:type, vocab:'DrugType'),
    rdf(DrugType, owl:sameAs, DrugbankDrugURI)
   )) %end of the internal forall
), %end of the antecendent in the external forall
( %retrieves the list of all distinct pairs Action+Drug involved in the report
   findall((ActionType,Drug), (distinct([ActionType, DrugAERS],
      (rdf(Involment, d2s:'involved_in', Report),
      rdf(Involment, d2s:'drug', DrugAERS),
      rdf(DrugAERS, skos:'closeMatch', DrugbankDrugURI),
      rdf(DrugType, rdf:type, vocab:'DrugType'),
      rdf(DrugType, owl:sameAs, DrugbankDrugURI),
      rdf(ActionType, vocab:'administrationOf', DrugType)
      ))), EventTypeList),
%uses the list to assert incompatibility belief among several actions (events)
%the correspongin DrugAERS' are used to assert Provenance
assertIncompatibilityMultiEvents(EventTypeList, 'aers')
)).
```

Figure 9 depicts an example of how new data is inferred according to
AERS data (inferred predicates and entities are highlighted in gray-shade). The
ReportX involves more than one drug mapped to our dataset (via drugbankID),
and actually all of them, namely Ibuprofen, Aspirin and Clopidogrel. Therefore,

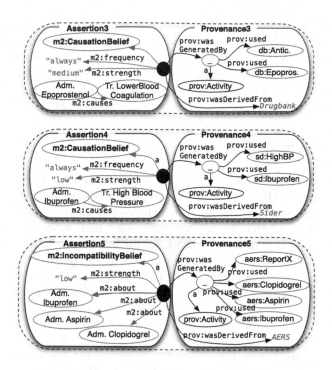

Fig. 10. Nanopublication schema for representing beliefs extracted from Drugbank, Sider and AERS. 'm2' is used as prefix for TMR_{Belief}.

an incompatibility belief is inferred among the administration actions regarding those drugs.

Finally, Fig. 10 illustrates the proposed nanopublication representation for the beliefs extracted from Drugbank, Sider and AERS. The ones imported from DIKB and LIDDI follow a similar structure. The *Assertion3* at the top is a *CausationBelief* with *medium* strength level, stating that the action type *Adm. Epoprostenol always* causes the transition type *Lower Blood Coagulation* according to *Drugbank*. The *Assertion4* in the middle is another *CausationBelief* with *low* strength level, stating that *Adm. Ibuprofen always* causes to *Higher Blood Pressure* according to *Sider*. Finally, *Assertion5* is a *IncompatibilityBelief* with *low* strength level, stating that *Adm. Ibuprofen*, *Adm. Aspirin* and *Adm. Clopidogrel* can be incompatible according to a report in *AERS*. For all of them, the provenance graphs contain, besides the source dataset, the external resources based on which the assertions were generated.

Conclusion. We can enrich the detection of interactions with LOD external knowledge sources in a scalable way, i.e. the number of interaction rules do not increase with the number drugs/actions, guidelines or datasets. For each new dataset only one or two importing rules are required. This section concludes contribution C4 together with the formalisation and implementation of rules for external interactions. Medical guidelines as well as external clinical knowledge can be expressed by means of the conceptual model and can be implemented in a SemWeb-based Framework for automatically detecting interactions. In the next section we provide an experimental assessment by discussing the implementation the aforementioned case study on detecting recommendations interactions enriched by external knowledge sources.

5   Experimental Assessment

This section presents the results obtained by implementing the case study on combining OA+HT+DB guidelines (Sect. 2). The following activities where performed in the experiment: (i) the guideline knowledge was (manually) introduced in a RDF dataset according to the implementation here proposed; (ii) the rules for importing LOD were fired; (iii) the rules for inferring relations and interactions were fired. An 'interactive' documentation describing the experiment and the prolog code is available online[12]. Figure 11 summarize the obtained results. It describes the type of interaction, the interaction and its source (derived from internal or external knowledge). The first 8 lines are the interactions described in Sect. 2. The last two lines illustrates that more external interactions can be detected, actually much more given the large volume of clinical LOD. However, since excess of information can become a disadvantage, we intend to provide filters, such as the causation frequency or the strength of the evidence.

Comparing to the previous implementation [6], the following explicit improvements are observed: (i) reasoning over action type hierarchy allows for detecting non-straightforward interactions (e.g. *DO administer NSAID and DO NOT administer Aspirin* are in contradiction because *Aspirin* specializes *NSAID*); (ii) new datasets (e.g. LIDDI, DIKB) are added without need for writing specific rules for detecting external interactions; and (iii) causation frequency, belief strength (evidence level) and recommendation strength can be represented. Implicit improvements are: (i) a more maintainable and reusable implementation that will favor new features and datasets to be introduced in future work; (ii) the more reliable and/or relevant information can be select.

6   Related Work

Formal languages proposed for representing clinical guidelines as "computer interpretable" ones [2,17] were not designed to handle the combination of multiple CIGs [7]. An alternative solution is the development of alert systems that are independent of the CGs. Such Computerized Physician Order Entry systems (CPOE), are used to alert physicians about drug interactions [18]. Despite the usefulness of these systems, a lot can be gained by tackling interactions between general recommendations on the outset, rather than employing drug-interaction alerts on the hospital floor.

---

[12] http://rapgmsbgym.github.io.

AlternativeActions'	['Avoid thrombi - by should Administer Clopidogrel', 'Avoid thrombi - by should Administer Dipyridamole', 'Avoid thrombi - by should Administer NSAID']	Internal
Contradiction'	['Avoid gastrointestinal bleeding - by should-not Administer Aspirin', 'Avoid thrombi - by should Administer NSAID']	Internal
RepeatedAction'	['Avoid thrombi - by should Administer NSAID', 'Reduce pain - by should Administer Ibuprofen']	Internal
ExternalAlternativeActions'	['Administer Epoprostenol always causes Low Blood Coagulation', 'Avoid thrombi - by should Administer NSAID', 'Avoid thrombi - by should Administer Dipyridamole, 'Avoid thrombi - by should Administer Clopidogrel']	drugbank
ExternalIncompatibleActions'	['Incompatibility among Administer Aspirin, Administer Ibuprofen', 'Avoid thrombi - by should Administer NSAID', 'Reduce pain - by should Administer Ibuprofen']	drugbank
ExternalIncompatibleActions'	['Incompatibility among Administer Ibuprofen, Administer Clopidogrel, Administer Aspirin', 'Reduce pain - by should Administer Ibuprofen', 'Avoid thrombi - by should Administer NSAID', 'Avoid thrombi - by should Administer Clopidogrel']	AERS
ExternalIncompatibleEffects'	['Administer Bendroflumethiazide always causes High Level Blood Sugar', 'Reduce blood sugar level - by should Administer Insulin', 'Reduce blood pressure - by should Administer Thiazide']	sider
ExternalIncompatibleEffects'	['Administer Ibuprofen always causes High Blood Pressure', 'Reduce blood pressure - by should Administer Thiazide', 'Reduce pain - by should Administer Ibuprofen']	sider
ExternalIncompatibleActions'	['Incompatibility among Administer Clopidogrel, Administer Dipyridamole', 'Avoid thrombi - by should Administer Clopidogrel', 'Avoid thrombi - by should Administer Dipyridamole']	liddi
ExternalIncompatibleActions'	['Incompatibility among Administer Dipyridamole, Administer Aspirin', 'Avoid thrombi - by should Administer Dipyridamole', 'Avoid thrombi - by should Administer NSAID',]	AERS

Fig. 11. Case study on combining guidelines for OA+HT+DB.

We have investigated this issue in a series of work. In [7] we analyzed related work that addresses recommendation interactions in different levels. Our research focuses on what we called the CIG level, i.e. it accounts for the need to combine guidelines and handle interactions before applying them to a specific patient. This is the case when common co-occurring diseases are considered during guideline development, but could also be needed for uncommon co-occurring diseases in the practice setting. The related work [9,19,20] has as their main drawback the need for defining specific rules for each interaction, e.g. *give aspirin & don't give aspirin* requires a specific rule and *give ibuprofen & don't give ibuprofen* requires another rule (a more detailed analysis in [6]). As a consequence, they do not provide 'scalable' support for combining guidelines, particularly more than two. Piovesan et al. [21] propose guideline-independent algorithms based on ontologies for detecting interactions, restricted to types "concordance" and "discordance". The use of intentions associated to recommendations for detecting "intention interactions" is close to our approach on verifying transitions related to recommendations. To the best of our knowledge, none of the related work provides means to express negative norms, nor negative causation beliefs, neither they infer interactions among several recommendations, for example due to polypharmacy. Moreover, they do not explore action type hierarchies, nor reuse clinical knowledge available online in order to enrich the detection of interactions.

However, they do address other aspects that we do not address yet, such as intentions, temporal aspects and qualitative transitions.

Our earlier work highlighted the importance of having the recommendations formally represented with a high level of detail. Explicit description of local constraints and impact of recommendations is considered an important source of information for increased reasoning capabilities and improved explanation of conflicts in [22]. The model described in [7] introduced clinical recommendations as governing care actions that cause state transitions; an extended version of this model presented in [8] defines different ways in which recommendation can interact according to the referred actions and transitions. The implementation and evaluation of the model using Semantic Web languages was proposed by us in [6]. We argued that the detection of interactions using *external* knowledge sources (in our case drug interactions modeled in the Linked Data version of DrugBank [10]) can provide more precise information. A Web-based application for browsing the guideline interactions was made available online[13]. Extending this model to introduce the notion of causation beliefs (for evidence) and the subsumption relations among actions was presented in [15]. It was a first formal exercise with the goal of providing a systematic view on possible internal interactions among recommendations.

The emphasis on evidence means that care recommendations are ultimately grounded in domain knowledge (generalizations over facts). The evidence that underlies the recommendations is weighed depending on the quality, depth and breadth of the study: guidelines are part of a larger network of hypotheses, claims and pieces of evidence that span across multiple publications [23]. However, only few CIG languages offer means to link to evidence [2], and they generally are targeted to very concrete and *procedural* guidelines, akin to medical protocols. In [24], the authors describe a lightweight ontology that represents the relations between a guideline, its recommendations, and underlying evidence, as annotations on the guideline and evidence texts using a combination of the Open Annotation[14] and PROV[15] formats. Huang et al. [25] propose an even more lightweight semantic representation of evidence based clinical guidelines, but automatically extract it from guideline texts. It includes UMLS identifiers for medical terms appearing in the text and use proximity, and the types of terms to infer the type and strength of the evidence that underlies recommendations. The Nanopublication model [26] seems to be a natural fit to modeling the evidence that underlies guidelines. It represents a publication as three RDF graphs, that respectively capture an assertion (the finding or evidence), the provenance of the assertion (e.g. an experiment) and publication information about the nanopublication (when was the assertion published and by whom).

The work presented here combines the pragmatic approaches of [24,25] and [26] in a model that takes the epistemological stance that the evidence underlying a recommendation expresses a *belief* that a care action causes a certain state

---

[13] See http://guidelines.hoekstra.ops.few.vu.nl.
[14] See http://www.openannotation.org.
[15] See http://www.w3.org/TR/prov-o.

transition. This strategy allows for using classical logic-based languages for handling inconsistent knowledge, such as conflicting findings published in different clinical studies.

7   Discussion and Conclusion

The work reported on in this paper improves over our previous work by offering a more generic and scalable way to represent clinical guidelines and detecting interactions. This is done by adapting and extending both the conceptual model and the Semantic Web-based implementation. The TMR models and rules are made more generic so that they can be more easily extended to incorporate new features such as hierarchies of transition types and causal chains. Incorporating the epistemological nuance of beliefs in the Semantic Web representation, improves the ability to (i) handle knowledge from different sources and (ii) select the reliable ones; (iii) to allow different, possibly incompatible, beliefs about the same event to co-exist; and (iv) to provide reusable formal rules that are applicable regardless of specific regulations, guidelines or external sources. This has a favorable effect on reusability, maintainability, and scalability beyond the guidelines we currently covered.

We furthermore show the power of using the extensive domain knowledge available on the Semantic Web for enhancing the ability to automatically perform new tasks, such as suggesting alternative drugs. Our use of open standards and vocabularies, such as the nanopublication format, makes that the knowledge accumulated in our own models is shareable and reusable in a similar fashion. We implement inferencing using expressive SWI-Prolog rules that execute over RDF graphs. The adoption of SWI-Prolog was an improvement over the implementation in [6], as it gave us a single environment for expressing our inference rules benefiting understandability and maintainability. This, of course, at the cost of Semantic Web standards compliance for that specific part of our model. In [6] the limitations of OWL2 for detecting the interactions, forced the use of multiple knowledge representation languages. We had to resort to a combination of expressive OWL2 inferencing, Stardog SPARQL rules (a SWRL dialect) and custom SPARQL update queries to perform reasoning.

The experimental assessment shows that interactions can be automatically detected among three guidelines and enriched by knowledge from DrugBank, Sider, LIDDI and AERS, from each of which the relevant knowledge was imported as beliefs. It also shows the ability to detect interactions among several recommendations, either due to their cumulative aspect or due to polypharmacy. Although the case study comprises only drug administration as action types, the approach is designed to address interactions among other types of interventions, such as surgeries and exercise therapy. More complex case studies will be addressed in future work. We faced some issues regarding the integration with these external knowledge sources, particularly on deciding which identity criteria we should use to map to the external datasets. For example, we could choose between PubChem ID, UMLS code, dbpedia and so on, where each choice would bring about different coverage and reliability.

Although this work is applied to clinical guidelines, its potentially of more general application, since both the model and the rules are defined independently of a particular domain. We plan to investigate the applicability of the models and rules to other domains such as disaster management. As ongoing work, we plan to address four limitations: (i) temporal validity for the assertions; (ii) quantification of beliefs and norms (i.e., frequency and strength); (iii) qualification of transitions (e.g. increasing or decreasing a property value); and (iv) considering goals and intentions;

Acknowledgments. We would like to thank colleagues from NEMO-UFES/Brazil for fruitful discussions about transitions, causation beliefs and regulations, and also prof. md. Saulo Bortolon for the nice discussions about medical domain; Jan Wielemaker and Wouter Beek (VU Amsterdam) for helping with SWI-Prolog implementation; Wytze Vliestra (Erasmus Rotterdam) for fruitful discussions about the biomedical domain; and Paul Groth (Elsevier) for fruitful discussions about the potential generality of the model and the use of nanopublications. The first author is funded by CNPq (Brazilian National Council for Scientific and Technological Development) within the program Science without Borders. This work was partially funded by the Dutch National Programme COMMIT.

References

1. Zamborlini, V., Hoekstra, R., Silveira, M., Pruski, C., Teije, A.: Generalizing the detection of internal and external interactions in clinical guidelines. In: Proceedings of the 9th International Conference on Health Informatics (HEALTHINF2016), Rome, Italy (2016)
2. Peleg, M.: Computer-interpretable clinical guidelines: a methodological review. J. Biomed. Informatics **46**, 744–763 (2013)
3. Lohr, K.N.: Rating the strength of scientific evidence: relevance for quality improvement programs. Int. J. Qual. Health Care **16**, 9–18 (2003)
4. Barnett, K., Mercer, S., Norbury, M., Watt, G.: Epidemiology of multimorbidity and implications for health care, research, and medical education: a cross-sectional study. The Lancet (2012)
5. Guthrie, B., Makubate, B., Hernandez-Santiago, V., Dreischulte, T.: The rising tide of polypharmacy and drug-drug interactions: population database analysis 19952010. BMC Med. **13**, 74 (2015)
6. Zamborlini, V., Hoekstra, R., da Silveira, M., Pruski, C., ten Teije, A., van Harmelen, F.: Inferring recommendation interactions in clinical guidelines: casestudies on multimorbidity. Seman. Web J., Open Acess (2015, accepted)
7. Zamborlini, V., Silveira, M., Pruski, C., Teije, A., Harmelen, F.: Towards a conceptual model for enhancing reasoning about clinical guidelines. In: Miksch, S., Riaño, D., Teije, A. (eds.) KR4HC 2014. LNCS (LNAI), vol. 8903, pp. 29–44. Springer, Cham (2014). doi:10.1007/978-3-319-13281-5_3
8. Zamborlini, V., Hoekstra, R., Silveira, M., Pruski, C., Teije, A., Harmelen, F.: A conceptual model for detecting interactions among medical recommendations in clinical guidelines. In: Janowicz, K., Schlobach, S., Lambrix, P., Hyvönen, E. (eds.) EKAW 2014. LNCS (LNAI), vol. 8876, pp. 591–606. Springer, Cham (2014). doi:10.1007/978-3-319-13704-9_44

9. Jafarpour, B.: Ontology merging using semantically-defined merge criteria and owl reasoning services: towards execution-time merging of multiple clinical workflows to handle comorbidity. Ph.D. thesis, Dalhousie University (2013)

10. Law, V., Knox, C., Djoumbou, Y., Jewison, T., Guo, A.C., Liu, Y., MacIejewski, A., Arndt, D., Wilson, M., Neveu, V., Tang, A., Gabriel, G., Ly, C., Adamjee, S., Dame, Z.T., Han, B., Zhou, Y., Wishart, D.S.: DrugBank 4.0: shedding new light on drug metabolism. Nucleic Acids Res. **42**, 1091–1097 (2014). D1091–7, PubMed ID: 24203711

11. Kuhn, M., Letunic, I., Jensen, L.J., Bork, P.: The SIDER database of drugs and side effects. Nucleic Acids Res. **44**, D1075–D1079 (2016)

12. Boyce, R., Collins, C., Horn, J., Kalet, I.: Computing with evidence part I: a drug-mechanism evidence taxonomy oriented toward confidence assignment. J. Biomed. Inform. **42**, 979–989 (2009)

13. Banda, J.M., Kuhn, T., Shah, N.H., Dumontier, M.: Provenance-centered dataset of drug-drug interactions. In: Arenas, M., et al. (eds.) ISWC 2015. LNCS, vol. 9367, pp. 293–300. Springer, Cham (2015). doi:10.1007/978-3-319-25010-6_18

14. Hoekstra, R., Magliacane, S., Rietveld, L., Vries, G., Wibisono, A., Schlobach, S.: Hubble: linked data hub for clinical decision support. In: Simperl, E., Norton, B., Mladenic, D., Della Valle, E., Fundulaki, I., Passant, A., Troncy, R. (eds.) ESWC 2012. LNCS, vol. 7540, pp. 458–462. Springer, Heidelberg (2015). doi:10.1007/978-3-662-46641-4_45

15. Zamborlini, V., Silveira, M., Pruski, C., Teije, A., Harmelen, F.: Analyzing recommendations interactions in clinical guidelines. In: Holmes, J.H., Bellazzi, R., Sacchi, L., Peek, N. (eds.) AIME 2015. LNCS (LNAI), vol. 9105, pp. 317–326. Springer, Cham (2015). doi:10.1007/978-3-319-19551-3_40

16. Guizzardi, G., Wagner, G., Almeida Falbo, R., Guizzardi, R.S.S., Almeida, J.P.A.: Towards ontological foundations for the conceptual modeling of events. In: Ng, W., Storey, V.C., Trujillo, J.C. (eds.) ER 2013. LNCS, vol. 8217, pp. 327–341. Springer, Heidelberg (2013). doi:10.1007/978-3-642-41924-9_27

17. ten Teije, A., Miksch, S., Lucas, P. (eds.): Computer-Based Medical Guidelines and Protocols: A Primer and Current Trends. Technology and Informatics, vol. 139 (2008)

18. Ammenwerth, E., Schnell-Inderst, P., Machan, C., Siebert, U.: The effect of electronic prescribing on medication errors and adverse drug events: a systematic review. J. Am. Med. Inform. Assoc. **15**, 585–600 (2008)

19. López-Vallverdú, J.A., Riaño, D., Collado, A.: Rule-based combination of comorbid treatments for chronic diseases applied to hypertension, diabetes mellitus and heart failure. In: Lenz, R., Miksch, S., Peleg, M., Reichert, M., Riaño, D., Teije, A. (eds.) KR4HC/ProHealth -2012. LNCS (LNAI), vol. 7738, pp. 30–41. Springer, Heidelberg (2013). doi:10.1007/978-3-642-36438-9_2

20. Wilk, S., Michalowski, M., Tan, X., Michalowski, W.: Using first-order logic to represent clinical practice guidelines and to mitigate adverse interactions. In: Miksch, S., Riaño, D., Teije, A. (eds.) KR4HC 2014. LNCS (LNAI), vol. 8903, pp. 45–61. Springer, Cham (2014). doi:10.1007/978-3-319-13281-5_4

21. Piovesan, L., Molino, G., Terenziani, P.: An ontological knowledge and multiple abstraction level decision support system in healthcare. Decis. Anal. **1**, 8 (2014)

22. Bonacin, R., Pruski, C., Da Silveira, M.: Architecture and services for formalising and evaluating care actions from computer-interpretable guidelines. IJMEI Int. J. Med. Eng. Inform. **5**, 253–268 (2013)

23. de Waard, A., Shum, S.B., Carusi, A., Park, J., Samwald, M., Sándor, Á.: Hypotheses, evidence and relationships: the hyper approach for representing scientific knowledge claims. In: Proceedings of the 8th ISWC, Workshop on Semantic Web Applications in Scientific Discourse. Springer, Berlin (2009)

24. Hoekstra, R., de Waard, A., Vdovjak, R.: Annotating evidence based clinical guidelines - a lightweight ontology. In: Paschke, A., Burger, A., Romano, P., Marshall, M.S., Splendiani, A. (eds.) Proceedings of the 5th International Workshop on Semantic Web Applications and Tools for Life Sciences, Paris, France, 28–30 November 2012. CEUR Workshop Proceedings, vol. 952 (2012). CEUR-WS.org

25. Huang, Z., Teije, A., Harmelen, F., Aït-Mokhtar, S.: Semantic representation of evidence-based clinical guidelines. In: Miksch, S., Riaño, D., Teije, A. (eds.) KR4HC 2014. LNCS (LNAI), vol. 8903, pp. 78–94. Springer, Cham (2014). doi:10.1007/978-3-319-13281-5_6

26. Mons, B., van Haagen, H., Chichester, C., Hoen, P.B., den Dunnen, J., van Ommen, G., van Mulligen, E., Singh, B., Hooft, R., Roos, M., Hammond, J., Kiesel, B., Giardine, B., Velterop, J., Groth, P., Schultes, E.: The value of data. Nat. Genet. **43**, 281–283 (2011)

Online Stress Management for Self- and Group-Reflections on Stress Patterns

Åsa Smedberg[1](✉), Hélène Sandmark[1], and Andrea Manth[2]

[1] Department of Computer and Systems Sciences, Stockholm University, Kista, Sweden
{asasmed,helene}@dsv.su.se
[2] The Swedish Post and Telecom Authority, Stockholm, Sweden
Andrea.Manth@pts.se

Abstract. In today's society, many people suffer from unhealthy levels of stress. A result of being exposed to a lot of stressors for a longer period of time, and showing strong stress reactions to these, is decreased wellbeing and eventually sick leaves. To intervene at an early stage is therefore important. However, learning about one's stress reactions and struggling for empowerment can be a challenge. Different types of applications for self-help are available on the Internet, but the ones for stress management are still in their early phase. In this article, we emphasize the social aspects of stress management. We present an artifact designed for analyzing and reflecting upon stress patterns, through both self- and group-reflections. This artifact is a result of further research work towards an integrated holistic stress management platform. The holistic platform includes a mobile application for storing data about events that cause stress reactions and a web-based system in which different actors and functions can complement each other, through self-help exercises, evidence-based information and learning through interaction with peers and experts. The research is based on traditional system development methods and interdisciplinary research in the area of e-health.

Keywords: Online self-help · Online stress management · Preventive care · Self-reflection · Stress patterns · Social support

1 Introduction

Sick leave due to mental illness is a growing problem in Sweden and other countries in the Western world. High levels of stress have become an increasingly common condition in people's everyday lives. Work life has become more complex than before and puts demands on our cognitive abilities to keep up with technological developments, increased competition and constant change. Factors in certain areas of the work environment such as work organization, management, hierarchy, and interpersonal relations can trigger stress reactions which are associated with psychosocial strain.

Great demands from the outside world can lead to unhealthy stress levels in individuals. Grandey and Cropanzano describe how the requirements of the job and at home have grown for the whole family [1]. There is also a historical expectation that women take care of the family. Multiple demands from family and work can increase negative stress and be

© Springer International Publishing AG 2017
A. Fred and H. Gamboa (Eds.): BIOSTEC 2016, CCIS 690, pp. 387–404, 2017.
DOI: 10.1007/978-3-319-54717-6_21

a challenge especially to women's health and wellbeing, and a determinant for long-term sickness absence and less wellbeing. Several studies have shown that those who have multiple roles and demands are more exposed to negative stress, resulting in physical and psychosocial dysfunctions [2–4].

Earlier studies found that domestic workload, mainly connected with own children, has increased for men and women, but to a greater extent for women during the past twenty years [5, 6]. The load in family life and in the workplace is considered to have a correlation. Grandey and Cropanzano point out that the strain grows significantly when the discrepancy between the expectations of the family and the expectations from the workplace increases, and if any of the areas have to stand back [1].

However, it is a rather complex issue, since people react differently and to different stressors. People who work and live under the same conditions can therefore react in different ways. The individual's subjective experience of a condition and a situation affect whether and what symptoms are developed [7]. Stress is an individual combination of external reality, individual perception of the situation and the estimated capacity to solve the problem.

Some people require a change in lifestyle to deal with problems of stress reactions. Continuous and social support has shown helpful when managing stress and change of lifestyle [8]. Studies have shown that the social aspect of learning is also important to consider, and to reflect together with others. Online support groups and reflection tools have shown to influence people with the symptoms of stress positively [9].

Tools that address e-health come in many different shapes and technologies. During recent years, mobile phones and mobile apps have changed the scene quite a lot. In Sweden, for example, it was shown in 2013 that 80% of the population in the age group 12 to 35 years were using the mobile phone on a daily basis [10]. Also apps for health and training have increased rapidly in number [10]. Regarding apps related to physical training and exercises, the development has led to interacting functions for collection, presentation and evaluation of data related to the training. These widely used apps are often connected to web-based systems for visualization of data and support for analysis of the data. Available technologies differ in strengths; mobile phones can be used independently of space, apps tend to have defined tasks, while computers are normally better suited for processing information and doing analysis. Regarding systems for stress management, there is no consensus regarding how such a system should be designed in order to combine on the one hand the advantages of different technologies, and on the other hand complementary areas of e-health, such as self-help, social and individual reflections, communication and information. This article will discuss the challenges related to stress and stress management and how IT can support; it is based on [11]. Special focus will be paid to IT-support for self- and group reflections on patterns of stressful events, and the article will present a prototype system that is the result of some practical research work in this field. Design Science was used as a framework to iteratively build the prototype. The system is a development of the web-based stress management platform described in previous publications (see [12], e.g.). In the article, some early test results from the usage of this web-based stress management platform will also be presented.

The next section (Sect. 2) discusses stress management with a certain focus on stress patterns, reflections and online self-help. The following sections (Sects. 3, 4 and 5) describe the design considerations, some preliminary test results concerning the whole web-based platform and the design of the proposed system for reflections on stress patterns, followed by a user scenario to illustrate the use of the whole platform, including the proposed system for reflections (Sect. 6). Finally, conclusions and future work are presented (Sect. 7).

2 Background

2.1 Stress Management

Stress is about biological and psychological responses to situations that are perceived as threatening or challenging. A stress reaction is an individual combination of external reality, the human perception of the situation and the capacity to solve the problem.

To manage work-related stress, effective interventions connected to the workplace are necessary. Interventions regarding stress in the workplace could focus on the individual or on the organizational level [13]. Thus there are two different approaches to manage work-related stress. At the individual level, relaxation and cognitive-behavioural techniques have been applied to improve an individual's mental resources and responses. At an organizational level, job adjustment and workplace communication activation have been applied in order to improve the occupational context.

However, in earlier intervention studies regarding the organization level the effect was limited. Individual level interventions such as cognitive-behavioural approach comprising coping techniques has so far been found to be more effective, although the long term effects are not yet fully known [13, 14].

2.2 Stress Patterns for Intervention

Patterns and Double-loop Learning. In order for people to learn to change behavior they need to have empowering views of reality, insights about the current situation [15]. According to Senge, there are three different levels of reality; the first one consists of single events, the second of patterns of behavior and the third the systemic structure. For the single events it is enough to develop reactive responses to take care of the events as they appear. The patterns of behavior are about repetitiveness and demands for reflections and analysis of the past events. The systemic structure is about understanding why the patterns occur, to prevent negative patterns (or enhance positive patterns).

The different levels of reality can be compared with single- and double-loop learning [16]. According to Argyris, both single- and double-loop learning have to be present to achieve true learning. The single-loop is about observation and implementation, i.e., to take care of situations as they appear to us, while double-loop learning is concerned with assessment and designing, i.e., questioning why certain situations reoccur and to develop the mental model of situations in order to learn to prevent negative situations from being repeated [16]. In systems thinking, this is also referred to as developing responses to certain disturbances and thereby amplifying variety vs. conceptualize to attenuate variety [17].

In stress management, the three levels of reality distinguished by Senge can be exemplified through a person who experiences difficulties getting a good night's sleep. If this happens once, the person can deal with this as a single event, and gets to bed earlier the night after. If sleeplessness occurs several times and affects the person's quality of life, it has become a negative pattern. The last level, the systemic level, is to try to understand why the bad sleeping pattern has occurred and to question the current values and priorities, i.e., one's mental model.

Stress Patterns in Stress Management. To identify sources of stress, it is crucial to look closely at habits, attitudes and estimates of stressful events. It can help individuals to recognize the underlying problem and seek measures to cope with it. Diaries or log books can give insight on peoples' perceptions of experienced stressors and stress reactions and help to identify the regular stressors in work life and how to deal with them. To keep a daily log makes it possible to see patterns and common themes in order to enhance the management of stress problems [18, 19].

2.3   Empowerment and Self-help Online

Care processes are becoming more patient-centered than before (see e.g., [20, 21]). To pay more attention to the individual patient's needs, and to have him or her included in the decision-making process, are a couple of things mentioned in relation to the concept of patient-centered care; however, Barry and Edgman-Levitan takes a leap forward by claiming that patient-centered care is not achieved until "an informed woman can decide whether to have a screening mammogram and an informed man can consider whether to have a screening prostate-specific–antigen test" [21, p. 781].

From this follows a discussion about the role of the patient as being actively involved in the processes and managing on their own according to capacity. New IT-tools is an enabler of increased self-management. This section will discuss the concepts of empowerment and self-management through IT-tools.

Empowerment. Empowerment can be explained as the experience of personal growth through developing skills and abilities along with a more positive self-image [22]. It refers, for example, to the ability to make personal decisions, to exercise critical thinking and to access relevant resources [23]. For a person who suffers from lifestyle issues, it concerns managing everyday life experiences and exercise lifestyle changes, for example. Empowerment is related to the concept of power, such as in 'power to' act that refers to the ability to make decisions, 'power over' a situation and to affect other people, and 'power within' that has to do with gaining a sense of control [24].

IT-tools for Self-management. The importance of encouraging patients to be more self-managing, and thereby more actively involved in his or her care processes, is part of empowering the patients. This has been recognized in relation to the achievement of care goals. For example, it is shown that patient satisfaction, development of healthy behaviors and wellbeing of the patients are positively affected by the self-management of the patient. Self-management has been described as how "the individual engages in activities that protect and promote health, monitors and manages symptoms and signs of illness, manages the impacts of illness on functioning, emotions and interpersonal relationships and adheres to treatment regimens" [25, p. 1].

To what extent and how self-management can be manifested depends on the health status and situation of the patient. Different IT-tools are available to support different forms of self-management. In the group of IT tools for self-management are everything from sensor-based monitoring systems for patients with great care demands to online self-help communities that give patients the freedom to explore and learn on their own through peer interaction. Sensor-based monitoring systems give that patients take a more subordinated role due to their health status and severe care needs, but at the same time the systems allow patients to continue to live at home. There are also IT tools for people to self-monitor blood pressure and blood sugar levels, weight, etc., and to educate themselves through online programs, access health information, and to communicate on a more continuous basis with health professionals as well as cooperate with them online [26].

As mentioned, the self-management tool that allows for most autonomous patients is the one for self-help groups [26, 27]. This kind of tool enables social support in groups and lets patients communicate, learn and act independently of the healthcare professionals. In self-help groups, the participants can help each other develop new skills and attitudes.

Empowering Online Self-help Groups. Barak et al. emphasize that the communication in online communities influences people's wellbeing in a positive way, and that the writing process is a good technique to structure thoughts and feelings [23]. When sharing experiences one has also begun to reflect on these. Furthermore, the online forums can increase people's personal empowerment and may affect their ability to make decisions: "Reliance on self and peers rather than on authoritative professionals contributes to gaining a sense of personal competence" [23, p. 1869]. The observed positive effects depend, among other things, on confirmation from the group. The individual wants to be understood and have the experienced situation acknowledged by the others. Also the fact that participation is voluntary and that you can help others with your own experiences is beneficiary when personal empowerment is concerned. To share burdensome emotions with others and to feel socially involved in the group have shown to result in a sense of relief [23].

A positive self-image and social identity affect people's mental health [28]. Wenger describes how social identity can be shaped and social learning take place in communities [29]. He argues that individuals are influenced by others through the group views and competences presented in the community. In order to have a healthy community, learning should also take place at the boundaries, according to Wenger. This means that we need to be open for new ideas, 'experience in the world', and have them combined with the competence of the community. In the area of online communities, this is also referred to as strong-tie vs. weak-tie relationships [30].

Positive effects such as greater empowerment have been seen in patients who engage in online support groups [23, 31]. Through the group, experiences, advice and recommendations can be shared.

One possible disadvantage of these types of peer-to-peer groups is that they can spread misinformation which then can lead to unhealthy behaviors. However, previous studies have shown that online self-help groups rather complement information distributed by the healthcare (see [32], e.g.). It can also take some time before people feel comfortable in the group [23]. Barak notes that online groups do not replace the necessary therapy.

2.4 Self- and Group-reflections

Self-reflection is one tool among other to manage stress reactions and deal with situations that cause stress. Reflections on situations can help people understand what is beneficiary and what causes disadvantages to their health.

Reflection is about discussing situations that are perceived to be complex or uncertain [33]. By reflecting, one can illuminate a situation from different angles and correct the mental image that was constructed initially. Self-reflection can then reform the perspective of a situation or action [34].

Mezirow et al. describe how critical reflection triggers learning in individuals [34]. Among other things, they describe how critical self-evaluation helps people to re-evaluate their experiences, knowledge, beliefs, feelings and actions. They believe that critical reflection aims at the question "Why?", and that the answer will be the causes and consequences of one's actions in the future. They also point out that the critical self-reflection is a highly significant method for learning among adults.

Reflection can be seen as a subjective awareness and assessment of events and their significance for the individual and the environment [33]. One can, however, change the perspective of experiences. It means that one becomes aware of how and why our assumptions limit the world as we experience it. To change the perspective, we need to reframe assumptions and apply these new values in future decisions and actions [34].

Westberg and Jason ask the question why reflection and self-evaluation are important [35]. Results from studies of medical care students show that reflections help students build new skills. Reflection helps to identify gaps in knowledge and correct the error in reasoning. By reflecting and be self-critical, learning time can also be shortened. As an example, Westberg and Jason refer to trainers who let athletes look at their own performances and evaluate them critically [35]. The method has shown to increase the rate of learning in athletes.

3 Design Considerations

The practical work with the design of a system prototype for self-reflection was done iteratively outgoing from interdisciplinary research in the area of stress management online. The holistic idea of combining ICT tools for self-management, evidence-based information and learning through feedback and communication in groups and with experts have been investigated in previous studies and manifested in a web-based platform [32, 36]. The prototype for self-reflection was designed through four iterations, each with the following steps: (1) a discussion in the online stress management team to clarify the problem to be solved and the requirements, followed by (2) development and demonstration, and then (3) evaluation of the prototype version by the online stress management team. Below, the basic considerations for the design of the prototype are presented.

3.1 Preventive Care

The design should consider the aspects of preventive care. Since the target group is people with stress symptoms, the system is to support them in their daily working or student life, and to help them reach increased wellbeing. Stress exists in all persons in a varying scale. It is manifested differently in people, and it also accelerates through different phases, according to Selye [37]. It is therefore beneficiary for the user if one can be aware of unhealthy stress levels and associated problems as early as possible.

3.2 Analysis of Data and Trends for Self-reflection

The user must have the ability to process and analyze data in order to change behavior in the future. As discussed earlier, reflection has shown to be an effective learning method [34]. It is therefore important to be able to see trends and historical data of the stressful situations that the user experiences.

To serve this purpose, the prototype should support the collection and presentation of different pieces of data related to the stressful situations of the user. Stress levels, textual descriptions, images, time and a geographical context are examples of types of data that are related to the stressful situations of the users.

To customize the visualization of trends is also important, and to be able to focus on a certain period of time. It is important to note trends to be able to reformulate one's perception so that new values can be applied in future decisions and actions [34].

In the prototype, it should be possible to highlight a situation as private or work-related. Stress-related problems often increase when stressful situations occur in both private and work-related contexts [1]. If this data is made visible, trends and patterns can be analyzed and changed.

3.3 Sharing for Social Support

A system should take into account the good influences that online social groups have demonstrated. Support from other members can create a social identity [38], increase self-confidence and contribute to a positive self-image. Communication in social groups and online forums has shown positive effects, such as increase in self-respect and empowerment among the users [23]. The prototype should therefore also support the need to gain social support and not only focus on individual reflections.

3.4 Holistic Design - Complementary Knowledge and Actors

Different actors, both experts and peers, have shown to be able to share complementary understanding of health-related issues in online systems [32]. While peers contribute with their experiences and practical advice, for example, medical experts can offer in-depth knowledge and understanding of different health-related issues. For the prototype, this means that both experts and peers are to be regarded helpful for social support and group-reflections.

The prototype is also to be integrated with the existing web-based platform that facilitates the combination of different knowledge and experiences. In this platform, there are functions for communication with both stress experts and peers implemented already. There are also exercises that can be performed, as well as real-life cases and research results to read about. The user can navigate freely in the platform and choose among the different types of functions and support to make combinations as needed. The existing functions that this whole web-based platform offers are the following: Ask-the-expert, Forums for peer communication, Online consultation for group counselling session (chat), Practical exercises (text as well as sound and video recordings) and Stories told and research results (see e.g., [12]). The different functions are organized in four different stress management areas: Sleep, Work and studies, Balance in life and Physical wellbeing.

3.5 Simplicity

The design must be simple, especially since the target group is people with stress symptoms, it is important not to increase the level of stress caused by system complexity. It is important to remember that people with symptoms of stress may have physical and mental constraints. The prototype uses accepted design principles according to Nielsen [39] and the recommendations of the Swedish Government's Workgroup Use Forum [40]. Particular emphasis has been placed on the principles of "Simplicity" and "Understanding the context" in these recommendations.

In rule 53 of the design guidelines recommended by Nielsen, it is said that the system should offer the user direct access to high-priority information [39]. This is also relevant for the design of the prototype.

3.6 Flexibility

Considering the differences among people with stress issues, their lives, experiences and stress reactions, the design of online stress management systems need to be flexible. Since stress is also affected by the level of control, stress management online has to let the user experience a sense of control and be empowered. This includes being able to freely navigate in the system, to get the information wanted and choose whom to talk to. The user should be able to decide if individual reflections should be in focus one day, and social interactions and group reflections another day.

4 Preliminary Test Results of the Stress Management Platform

As mentioned in previous section, the whole web-based platform for stress management is based on holistic thinking that offers the users the possibility to combine different functions and to access different actors and knowledge sources (see Sect. 3.4). This web-based platform is the starting point for the prototype work that is the focus of this article. The aim of the prototype is to support self-reflections of stress-related events, but also, as will be demonstrated later in this article, to connect to the whole web-based platform in order to include group-reflections and other types of stress management support. The idea of the platform is to allow the users to access the different functions and sources of knowledge in a way they find useful.

Previously in our research work, some preliminary user studies have been done. One of the most significant studies that the web-based stress management system as a whole has undergone is a preliminary feasibility study. The web-based stress management intervention was evaluated through observations of user activity (log data) and interviews with the users to get information about their experiences and opinions. The evaluation was done with 14 participants who experienced stressful lives and who were also students in computer and systems sciences at the time. To simulate a natural setting, the participants were encouraged to navigate the system and use the functions in a spontaneous way, according to their own preferences. Through the study setting, the users' spontaneous combinations of the functions in the web-based platform could be analyzed. The participants used the platform for two weeks; their usage was logged and later coded for analysis, together with the interview transcripts, by using a software tool for data coding. The web log contained information about the function being accessed, time of access and duration time.

The log results showed that the users were mostly interested in communicating with peers and accessing information and exercises stored in the system, and they chose to communicate less with experts. Only four out of the 14 participants made use of the expert service in the system, while the system's archived information/exercises and communication with peers were used by 9 and 10 participants respectively; two participants made no significant use of any of the system functions. In the interviews, one explanation for not using the expert service was lack of time, and that they intended to get in contact with the expert but that other things came in the way. Furthermore, the results indicated that the participants who used both archived information and communication with peers had sought out different types of knowledge unconsciously. One of the participants reflected in the interview that the reason for viewing the exercise videos and accessing the peer forum was because he/she wanted to get different types of practical advice. Another participant explained that the reason for accessing different sources in the system was to get reliable information from research documents and then more practical information from forum posts. It turned out that in most cases, the participants used one type of function as a complement to another, for example when complementing archived general information with communicative personalized information, or vice versa, but that the participants did it unconsciously.

5   The Prototype for Self-reflection

This section outlines the prototype system for self-reflection that is the result of the iterative design process and based on the design considerations presented in Sect. 3.

5.1 Prototype System Boundaries

The prototype system for self-reflection aims to combine the strength of the mobile phone that it is always at hand in everyday situations, with the more complex web-based platform where various areas of knowledge, expertise, exercises and peer forums are available. The prototype proposed in this article is a web-based system for self-reflections on stress and patterns of events, and for initiating reflections in groups. It is fed by data from a mobile phone app in which events that cause stress reactions are registered. Data about these events are then used in the prototype system to visualize patterns of stress reactions – frequency, scope and time - over shorter and longer time periods. By putting data in relation to each other, the user can reflect on unwanted events from a broader perspective. The aim is also to have the prototype integrated with the larger web-based stress management system where social interactions, reflections and advice from stress experts and peers can take place. The prototype, together with these other functions, will create a system that offers different kinds of support and in which the user can navigate according to his or her needs.

5.2 Functionality

Stress situations that the user has registered through the mobile phone app can be displayed in different ways in the prototype. The prototype should support analyses and reflections, with customized display of data. As can be seen in Fig. 1, historical data of events can be presented according to both frequency and level. The user chooses a time period for the presentation. In the example in Fig. 1, the time period is one winter month and the results show that during this time frame 22 stress situations were registered by the user, of which some were classified as causing high stress levels (80–100) on the scale of 1 to 100. By clicking on a registration presented in the chart, the date of the registration will be visible. Figure 1 shows also stress levels of the registered situations from the last 15 days so that the user can monitor the latest development of his or her stress levels.

The prototype offers also a visualization of all registered situations on a map (see Fig. 2). It shows how many times there have been situations registered in a particular geographical place. It is also possible to view the data as a geographical trend. The places in question are marked out on the map and different colours are used to illustrate the frequency of reported stress situations.

When looking at the registered stress situations, one by one or in a certain time frame, the user can add his or her reflections. In Fig. 3, the time frame chosen is five days in December of 2014. In the example, ten situations with different stress levels occurred in this time frame. When looking at the levels and descriptions of the situations, the user makes the reflection that the period is characterized by recurring sleeping problems. In the example, the reflection is tagged as belonging to the stress area called "Sleep". This makes it possible to share the stress chart and self-reflection, and use it as a basis for conversations and group-reflections with other actors, both experts and peers, in the larger web-system. The user's own reflections can eventually be posted in the forum on

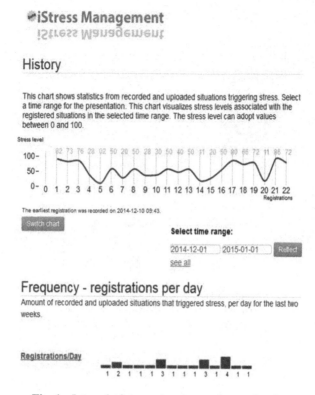

Fig. 1. Interval of stress situations and stress levels.

sleep (for peers) and also referred to in conversations with the experts in the larger web-system.

Figure 4 shows an overview of the prototype with its different functions. To the left, a chart of past registered stress situations are seen. When clicking on one of the situations, detailed information about the situation appears. A description about the situation is seen below the chart, and it is possible to edit the text, if the user for example was not able to write so much at the time when the situation was registered. If the user has taken a picture in connection to the stress situation, this image is shown to the right, together with the geographical position. The reflections made by the user are seen to the right. Different themes and concerns can be seen in the list of reflections. Behind each reflection there is a certain stress situation or interval of stress situations. New reflections can also be added from here. Also, in order to work actively with managing the stress situations and to learn how to deal with them, the user can define goals (placed at the bottom), and there are also links to the larger web-based stress management platform where the user can engage in conversations with peers and experts, do exercises, read about research results, and so on.

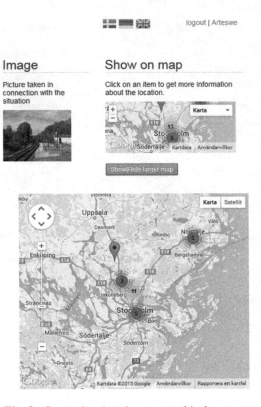

Fig. 2. Stress situations in a geographical context.

5.3 Technical Solution in Brief

The mobile phone app, developed earlier, is integrated with the prototype for self-reflection in the sense that it sends data to the prototype. Through scripting language (AngularJS) the logic could be handled, and HTML was used for the graphical presentations. The code is executed in a browser when it is actually being used, and there are no requirements on the underlying platforms or software. The prototype can thus be used in all modern browsers and easily integrated with other systems. The coding work follows the recommendations from W3C. The idea was to have the technical solution as generic as possible to be able to easily integrate the surrounding systems. The prototype consists of the markup language HTML 5, style sheets (CSS) for the graphical presentation and a dynamic JavaScript based framework (AngularJS).

As data carrier between the mobile phone app and the prototype, JSON files were used. These are good for demonstrating the features of the prototype but will not be used in the final system. JSON files are platform independent and can easily be packed together with code and transferred between different environments. This makes it easy to further develop the prototype anywhere without losing functionality. When having done necessary user tests of the prototype, it will be transferred, using the same database system as the overall web-based stress management system. The overall web-based

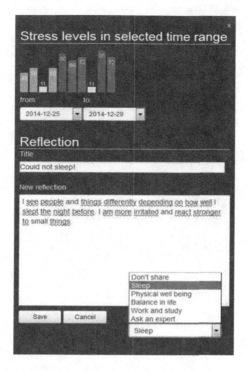

Fig. 3. Sharing of reflection and stress diagram.

system is built on the PHP framework Zend framework 1.11.12 that includes a division of components according to the MVC (Model-View-Controller) architecture.

6 User Scenario

Different technologies have different advantages, and through the research and development of the stress management platform, we show how the user can benefit from the different technologies and features linked together to create a more holistic system solution. This section presents a user scenario in which the system prototype for self-reflection and its relations with the mobile app and also the larger web-based system for stress management are illustrated. The user in this scenario is called "Linda".

Linda, who is an employee at an insurance company, experiences stress reactions that she believes are related to her situation with the new job and her family duties. She downloads the mobile phone app for self-reflection on stress situations to her phone and surf around a bit on the related web-based system to get familiar with the systems. Some days later, Linda experiences symptoms of stress when facing a tough meeting with her boss. She grabs her phone and starts to describe the situation in a few words and saves it. The same evening at home, the kids are being very active and the stress level increases. Again, Linda records the situation in the phone.

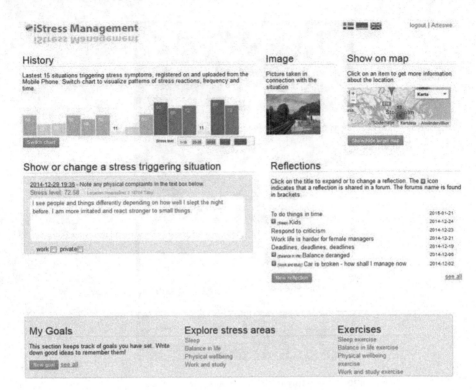

Fig. 4. Prototype overview

A few hours later, Linda loads the data from her phone app to the web-based system. Uploaded data then appear as two registered situations. From here, Linda marks a situation as private and the other one as job-related. She also writes some brief notes about the physical symptoms that arose in connection with the job-related incident. Then, she compares the situations with each other and tries to reflect on the causes and effects.

In the beginning, there is not much data for analysis. But as the registered situations increase in number, the comparisons and visualizations of trends become more meaningful. Eventually, Linda checks the map to see how often situations have occurred in different geographical contexts and decides to write a short reflection on a number of geographical context.

Linda recognizes also that the stress-related situations usually seem to occur at home and at work on the same day. She thinks it would be interesting to hear what others can tell about similar correlations. Linda decides to post a message in the forum "Balance in Life". But first, in the popup window where Linda wrote her reflection, she marks that she wants to share the reflection in the forum "Balance in Life". This means that the reflection together with the underlying stress diagram become available from the web-based system. When Linda formulates her question in the forum, the shared reflection is attached to the message.

Eventually, Linda gets a number of tips from users who have experienced similar situations. The answers help her relax since she understands that it is ok to feel the way she does and that she is not alone. She also gets some practical advice that she starts to practice. But some advice is difficult to understand, and she recalls that there is also a possibility to ask questions to a stress expert. Linda then formulates a question to the expert and shares this time a longer sequence of data from the past stress situations. The stress expert reconnects with some tips and explanations and also informs about an upcoming online chat session in which Linda is welcome to participate, if she wants to reflect together with others under the supervision of the expert.

7   Conclusions and Future Work

In this article, we have presented the design of a web-based prototype system that supports individual and social reflections for people with stress reactions. The interdisciplinary research that forms the basis of the prototype design was elaborated.

Negative effects of long and intense exposure to stressors have been extensively studied over the years. How to deal with stress reactions and to avoid ending up in patterns of stressful events is still, however, regarded a difficult matter. The picture is sometimes complex and involves many interacting factors. Detecting patterns and not just single events is one important step in the learning process. Another one is to start to reflect upon the patterns and try to understand why they occur, and, then, start planning for how to avoid the patterns from continuing. The prototype presented in this article was anchored in the interdisciplinary field of stress management online, with specific focus on detection of patterns and different ways to reflect.

Visualization of events and patterns of events plays a central role in the design of the prototype. It is manifested in the form of graphs presenting events in relation to each other, events over time, and the geographic locations of events. Previous studies have shown that visualization and evaluation of events that cause undesirable stress reactions provides a good support for reflections.

In the proposed solution, the user has also the option to specify whether an event is related either to a personal or a work-related situation. It has been shown in earlier research that people are more likely to fail to cope with stress exposures when both areas are affected. This might result in a decrease in their wellbeing. Being more conscious of the different potential sources of conflicts in work and family life respectively can help individuals to find effective strategies.

Both individual reflections and social support have shown important for changing a negative trend of stress exposure as well as reactions to stressors. Therefore, the proposed system presented in this article includes the possibility to share one's reflections, and also data sequences of stressful events, with others, both peers and medical experts. The prototype has been prepared to enable the sharing of reflections in social forums and with experts, because communication in online groups has been shown to have a positive impact on human well-being. In the proposed system, the user gets a clear indication when a reflection has been shared in the forum provided by the larger web-based system.

As discussed, the prototype we have presented is part of a larger online stress management platform. It is integrated with both the mobile phone app where the stress events are first documented, and the web-based system that allows the users to communicate with others, and to access information and exercises. All together, these integrated subsystems create a dynamic and combined online stress management platform, which could give individuals better opportunities to reduce mental problems due to stress exposure.

The system is intended for preventive and self-managing care, and it aims to increase quality of life for individuals. Although the proposed system is not a patient care system, the registered events and reflections on these may be useful in personal meetings with health professionals, a coach or a therapist. When put in practice, the system will contain valuable documentation of events, patterns of events, consequences and thoughts on those that are useful in such contexts.

The next step is to ensure technical robustness of the whole system. It is especially important in regard to security and integrity issues. Since the system stores data about individual experiences of stress levels, together with personal stories and reflections on private and work-related events, authentication and security issues need to be rigorously dealt with. In addition, further user evaluations are needed. So far, user studies have been restricted to students and during shorter time periods. Future empirical evaluations will include larger groups of employees and students who experience negative stress, together with stress experts.

References

1. Grandey, A.A., Cropanzano, R.: The conservation of resources model applied to work–family conflict and strain. J. Vocat. Behav. **54**, 350–370 (1999)
2. Scharlach, A.E.: Role strain among working parents: implications for workplace and community. Commun. Work Family **4**, 215–230 (2001)
3. Nordenmark, M.: Balancing work and family demand. Do increasing demands increase strain? A longitudinal study. Scand. J. Publ. Health **32**, 450–455 (2004)
4. Hallsten, L., Bellagh, K., Gustafsson, K.: Utbränning i Sverige – en populationsstudie (Burnout in Sweden – a national survey). In: Arbete och Hälsa (Work and Health), 6. National Institute for Working Life, Stockholm (2002)
5. Lundberg, U., Krantz, G., Berntsson, L.: Total workload, stress and muscle pain from a gender perspective. J. Soc. Med. **3**, 245–254 (2003)
6. Voss, M., Floderus, B., Diderichsen, F.: Physical, psychosocial, and organisational factors relative to sickness absence: a study based on Sweden post. Occup. Environ. Med. **58**, 178–184 (2001)
7. Henderson, M., Harvey, S.B., Overland, S., Mykletun, A., Hotopf, M.: Work and common psychiatric disorder. J. Roy. Soc. Med. **104**, 198–207 (2011)
8. Karasek, R.A., Theorell, T.: Healthy Work: Stress, Productivity, and the Reconstruction of Working Life. Basic Books, New York (1990)
9. Sandmark, H., Smedberg, Å.: Stress at work: developing a stress management program in a web-based setting. In: Proceedings of The Fifth International Conference on eHealth, Telemedicine, and Social Medicine, pp. 125–128. IARIA (2013)
10. Findahl, O: The Swedes and the Internet 2014 (Original reference in Swedish: Svenskarna och Internet 2014), SE. The Internet Foundation in Sweden (2014)

11. Smedberg, Å., Sandmark, H., Manth A.: Online stress management: design for reflections and social support. In: Proceedings of the 9th International Joint Conference on Biomedical Engineering Systems and Technologies, pp. 117–124. Scitepress (2016)
12. Smedberg, Å., Sandmark, H.: Dynamic stress management: self-help through holistic system design. In: E-Health Communities and Online Self-help Groups: Applications and Usage, pp. 136–154. IGI Global (2012)
13. Czabała, C., Charzyńska, K., Mroziak, B.: Psychosocial interventions in workplace mental health promotion: an overview. Health Promot. Int. **26**(1), 70–84 (2011)
14. Giga, S.I., Cooper, C.L., Faragher, B.: The development of a framework for a comprehensive approach to stress management interventions at work. Int. J. Str. Man **10**(4), 280–296 (2003)
15. Senge, P.: The leader's new work: building learning organizations. In: Starkey, K., Tempest, S., McKinlay, A. (eds.) How Organizations Learn: Managing the Search for Knowledge, 2nd edn., pp. 462–486. Thomson Learning, London (2004)
16. Argyris, C.: Teaching Smart People How to Learn. Harvard Business Review Press, Boston (2008)
17. Espejo, R., Schuhmann, W., Schwaninger, M., Bilello, U.: Organizational Transformation and Learning – A Cybernetic Approach to Management. Wiley, Chichester (1996)
18. Almeida, D.M., Wethington, E., Kessler, R.C.: The Daily inventory of stressful experiences (DISE): an interview-based approach for measuring daily stressors. Assessment **9**, 41–55 (2002)
19. Cooper, L.C., Payne, R.: Causes, Coping and Consequences of Stress at Work. Wiley, New York (2008)
20. Green, S.M., Tuzzio, L., Cherkin, D.: A framework for making patient-centered care front and center. Perm J. **16**(3), 49–53 (2012). Summer
21. Barry, M.J., Edgman-Levitan, S.: Shared decision making — the pinnacle of patient-centered care. N. Engl. J. Med. **2012**(366), 780–781 (2012)
22. Staples, L.H.: Powerful ideas about empowerment. Adm. Soc. Work **4**, 29–42 (1990)
23. Barak, A., Boniel-Nissim, M., Suler, J.: Fostering empowerment in online support groups. Comput. Hum. Behav. **24**(5), 1867–1883 (2008)
24. Spencer, G.: Empowerment, Health Promotion and Young People: A Critical Approach. Routledge, New York (2014)
25. Gruman, J., Von Korff, M.: Indexed Bibliography on Self-management for People with Chronic Disease. Center for Advancement in Health, Washington, DC (1996)
26. Barrett, M.J.: Patient self-management tools: an overview. California Healthcare Foundation (2005). http://www.chcf.org/publications/2005/06/patient-selfmanagementtools-an-overview. Accessed 20 Sept 2014
27. McGowan, P.: Self-management: a background paper. Centre on Aging, University of Victoria (2005)
28. Aneshensel, C.S., Phelan, J.C., Bierman, A.: Handbook on the Sociology of Mental Health, 2nd edn. Springer, Dordrecht (2013)
29. Wenger, E.: Communities of practice and social learning systems. In: Starkey, K., Tempest, S., McKinlay, A. (eds.) How Organizations Learn: Managing the Search for Knowledge, 2nd edn., pp. 238–258. Thomson Learning, London (2004)
30. Haythornthwaite, C.: Social networks and online community. In: Joinson, J., et al. (eds.) The Oxford Handbook of Internet Psychology, pp. 121–137. Oxford University Press, Oxford (2007)
31. van Uden-Kraan, C.F., Drossaert, C.H., Taal, E., Shaw, B.R., Seydel, E.R., van de Laar, M.A.: Empowering processes and outcomes of participation in online support groups for patients with breast cancer, arthritis, or fibromyalgia. Qual. Health Res. **18**, 405–417 (2008)

32. Smedberg, Å.: To design holistic health service systems on the Internet. In: Proceedings of World Academy of Science, Engineering and Technology, pp. 311–317, November 2007
33. Creek, J., Lougher, L.: Occupational Therapy and Mental Health, 4th edn. Churchill Livingstone Elsevier, Edinburgh (2008)
34. Mezirow, J. and Associates (eds.): Fostering Critical Reflection in Adulthood. Jossey-Bass, San Francisco (1990)
35. Westberg, J., Jason, H.: Fostering Reflection and Providing Feedback – Helping Others Learn from Experience, 1st edn. Springer, New York (2001)
36. Smedberg, Å., Sandmark, H.: A holistic stress intervention online system - designing for self-help through multiple help. Int. J. Adv. Life Sci. 3(3–4), 47–55 (2011). IARIA
37. Selye, H.: History and present status of the stress concept. In: Monat, A., Lazarus, R.S. (eds.) Stress and Coping, 2nd edn. Columbia University, New York (1985)
38. Wenger, E.: Communities of practice and social learning systems: the career of a concept. In: Blackmore, C. (ed.) Social Learning Systems and Communities of Practice, pp. 179–198. Springer, London (2010)
39. Nielsen, J.: Design guidelines for homepage usability (2001). http://www.nngroup.com/articles/113-design-guidelines-homepage-usability/. Accessed 2 Feb 2015
40. Swedish Government's Workgroup Use Forum: Design Principles for public digital services. (Original reference in Swedish: Regeringens arbetsgrupp Användningsforum, 2014. Designprinciper för offentliga digitala tjänster) (2014). http://www.anvandningsforum.se/designprinciper/. Accessed 1 Dec 2014

Requirements, Design and Pilot Study of a Physical Activity Activation System Using Virtual Communities

Lamia Elloumi[1]([⊠]), Margot Meijerink[1], Bert-Jan van Beijnum[1],
and Hermie Hermens[1,2]

[1] University of Twente, Enschede, The Netherlands
{l.elloumi,m.s.meijerink,b.j.f.vanbeijnum}@utwente.nl
[2] Roessingh Research and Development, Enschede, The Netherlands
h.hermens@rrd.nl

Abstract. Increasing researches support the importance of physical activity in maintaining health and preventing from diseases. Nowadays, researchers from different disciplines and health organizations are putting a lot of effort on interventions and strategies to increase the motivation of people in being more physically active. Following the same incentive, we introduce a novel system "ICT-based Community Coaching" as a strategy to motivate people to be physically active. It is a strategy based on human-to-human feedback where we transform the physical activity into a social activity. As the system is a new approach we focused on the elicitation of the requirements. In order to elicit the requirements of this system, we elaborated a scenario in order to make explicit the ideas of the Community Coaching system. Then we elaborated a questionnaire to be able to define important functionalities that can be useful for potential users.

Keywords: Virtual community · Physical activity · Community Coaching · Requirements elicitation · Design · Usability study · Usefulness study

1 Introduction

Physical activity is important to maintain health and prevent chronic diseases. Physical inactivity is the fourth leading risk factor for global mortality. Increasing levels of physical inactivity are seen worldwide. Globally, 1 in 3 adults is not active enough [22].

To stimulate people to become more active, many interventions have been developed. The first ideas emerged in the early seventies which encourage people to fill in certain questionnaires. These questionnaires aimed at getting more insights in the daily activity pattern so that people become more aware of how they behave. However, since the introduction of various measurement devices, it could be done faster and more accurate which was seen as a big step forward.

© Springer International Publishing AG 2017
A. Fred and H. Gamboa (Eds.): BIOSTEC 2016, CCIS 690, pp. 405–425, 2017.
DOI: 10.1007/978-3-319-54717-6_22

Common device to monitor daily activity are step counters. Only wearing such a device can already increase the daily number of steps made [2].

Nowadays it is often used in combination with a smartphone since it allows people to continuously access their activity data and to receive appropriate feedback any time needed. Additionally, recent ICT-based interventions use persuasive technologies in order to help people to be regularly physically active. UbiFit [3] and ActiveLifestyle [16] are examples of systems using persuasive technology in order to change physical activity behaviours.

Although many of these interventions have shown to be successful [2], a drop of use is noticed after a relatively short period [17]. One of the reasons could be because of the "one size fits all" approach, meaning that not much attention is paid to personal preferences and environment factors [1]. In case feedback is given at any arbitrary time and not personalized, people will perceive the feedback given as annoying and not as really supportive [1]. This feedback is based on system-to-human interactions. Additionally, these system-to-human systems are limited in terms of provision of social support, they are focusing on the appraisal support.

Social support from family and friends has been consistently and positively related to regular physical activity. Various studies showed a positive relationship between social support and physical health outcomes [18]. These interventions are based on face-to-face meetings and recently implemented in e-coaching systems [8]. Social networks and virtual communities are also used in physical activity support to provide mainly the emotional and informational support (some examples are WebMD [19], PatientsLikeMe [15] and MedHelp [13]).

From existing solutions to support in physical activity and enhancing compliance we are missing a more intelligent system that is more cleaver in maintaining and mediating between humans in order to provide a human-to-human feedback.

To enhance compliance on long-term using virtual community and help in being physically active, this research aim is to improve the provision of feedback by the introduction of the ICT-mediated Community Coaching functionality. The functionality activates the immediate social environment and use it instead of computer-tailored messages to encourage the users to do more physical activity. The hypothesis is that having to do physical activities together as social activity would have a higher impact on the motivation to be physically active and to enhance the compliance of use of the system. The system tracks the physical activity of the user and whenever he/she is not complying to the daily recommended level of physical activity, it would notify who are close by and available in the social environment and ask them to invite him/her for a social activity involving physical activity. These notifications are the main communication stream in the system.

To further work out the idea, we elaborated a scenario in order to make explicit the ideas of the Community Coaching system. The scenario describes the various activities performed by the users of the system with the Community Coaching. Based on the scenario we elaborated a questionnaire to get input from people to elicit important functionalities that can be used and useful for potential users. Based on the resulting requirements, a design could be made which makes clear the functional architecture of the system and its interface.

Previously, we developed the virtual community TogetherActive [5] and we will use it as platform for the Community Coaching system. Based on prioritization of functionalities, we implemented the part of the system and integrated it in the TogetherActive system. The Community Coaching can be generalized to any physical activity platform where other kind of social support can be provided.

To evaluate the system, a study with real users was performed. The study took 11 days in total. At the end of the period all participants were asked to fill in a questionnaire, with the focus on the system usability and usefulness. Additionally, the system logs were used and analyzed to investigate the real system use.

This paper is organized as follow. Section 2 presents the Community Coaching system requirement elicitation where we described the scenario, the questionnaire and its results, and the list of required functionnalities for the community coaching concept. Section 3 describes Community Coaching design, implementation and integration in the TogetherActive system. The evaluation protocol and the results are presented in Sect. 4. Finally, Sect. 5 is a conclusion and discussion about the present paper, where we present lessons learned and recommendations for future.

2   Community Coaching Requirements

To explore the ideas of community coaching to promote physical activity, the requirements need to be made clear. Hence we need to elicit the functionalities needed and desired by the users. In order to achieve this, we used the concept of a scenario is basically two ways. Firstly, the concept of a scenario was used to express and communicate the basic ideas of community coaching, hence expressing how a day using this concept would look like. Secondly, we used the scenario to identify and question the various aspects of this concept. In return, these questions (together with the scenario) were used in an on-line survey to elicit the opinions about the functional aspects of this concept for two potential users roles. These outcomes and their analysis are then used to formulate the functional requirements for a system implementing the concept. In the next subsections these steps in the requirements elicitation process are described.

2.1   Community Coaching Scenario and Questionnaire

In order to get the requirements of the new system, the scenario-based approach was used. We started by writing a scenario describing the various activities that can be performed by the users of the system. Since we have no basic reference of what the system and potential users may need, we used the scenario in order to make explicit the ideas of the Community Coaching system. Based on scenario we build a questionnaire to get input from people to elicit important functionalities that can be used and useful for potential users. The scenario used to elicit the requirements of community based coaching is shown in the boxed text below.

Oscar is 26 years old. He is PhD at the University Twente and in his spare time he likes to read thriller novels. Due to his sedentary lifestyle, he doesn't get the minimum amount of thirty minutes moderate activity a day. However, especially for him it is very important because overweight is a serious problem in his family and he is not averse of eating croquettes during the lunch break.

On advice of the Health and Safety Consultant of the University, he is now using TogetherActive. It offers the social support needed to get motivated to be physically active. This system is a physical activity support system based on a physical activity monitoring sensor and a virtual community. He can wear the sensor on his hip or put it on his pocket and check the portal via his Smartphone or laptop. Using the system, Oscar can get informed about last news regarding the benefits of physical activity, about tips and suggestions. He can also see experiences of others and share his own. He is also able to see his daily physical activity goal (set by the Health and Safety Consultant to 10000 steps a day), his progress and accomplishments over time. He belongs to a virtual group (suggested by the Health and Safety Consultant) where they share common goals and they have to accomplish them together and compete against other groups. Those group members are able to track each other activity level.

To get motivated and be able to accomplish daily personal and group goals, there is a social feedback component built in. This means that feedback is given by people instead of the ordinary computer-tailored feedback. Those people can be group members wearing a sensor or people invited by the main user (Oscar) without a sensor. Although these people are not able to track their activity level, they can help support Oscar with being active. To ensure that everyone involved will participate actively to the support process a competition element is built in. For each of the 'helpers' Oscar has to assign a certain role, which can be a friend, colleague or family member for example, depending on the area/group he should be active in. Oscar invited to the system Emma, Peter and Amy as his colleagues and Lucy, Sam and David as his family members. Additionally, to each of the different areas, Oscar has to add a certain number of activities appropriate for that specific location. He added among others a five-minute walk, Frisbee and Petanque for office activities and running and dog walking for free time activities. Once into use, the list can be completed by the other people concerned. Because Peter is a really big fan of FC Twente, he adds playing football to the list.

At a normal workday Oscar gets out of bed at 7:00. He fixes the physical activity monitoring sensor to his jeans, packs his lunch and drives to work. Once arrived, he changes his status to 'at work'. In this way the system knows who to contact when necessary. He says good morning to his colleagues and continues with his work he left the previous day. At 10:00 he still didn't come off his chair and the system didn't notice any physical activity. Therefore the system sends a text notification via the mobile application to all group members and helpers who have been recognized as being a colleague of Oscar. In this message they are asked to accept the notification and choose one out of a number of different activities to perform with Oscar.

Because the weather is very sunny, Oscar's colleague Emma decides to play a Frisbee game outside. After she selected the activity 'Frisbee', the invitation is automatically forwarded to Oscar. Once Oscar has accepted the activity a chat screen is automatically opened to discuss time and place of the meeting. To avoid that Oscar receives a lot of invitations, the notifications become invisible after someone has chosen an activity. Oscar is typing that he has to finish one thing and that they will meet in ten minutes near the entrance of the building. To provide others the possibility to join the activity, Oscar has to fill in when and where the activity will take place in order to display it on a list online. Next to the future activities, also current and completed activities are shown, to keep others informed about their status in the competition.

Half an our after the planned activity, the system checks whether the activity has actually taken place by comparing the data from the sensor with the estimated effort, which is based on the intensity and duration of the activity. If the system approves the activity, all participants get the number of points associated with the activity. The one who initiated the activity receives bonus points. Also points can be earned each time a person is adding an activity to the list. In this way, every month an award is given to the best supportive helper.

At 17:30 Oscar is already at home. But his wife Lucy is still at the office. Before she leaves she likes to check Oscar's activity level. Therefore she logs in to the application with a special code, received from Oscar himself. With this code she is able to check Oscar's activity level. The application shows that Oscar still didn't reach the recommended number of steps. Therefore Lucy sends Oscar a message with the request to buy some groceries in the supermarket nearby, so he will bike or walk for some time.

The questionnaire consisted of 26 questions (21 multiple choice and 5 open questions), addressing various aspects of the community coaching concept and different potential user roles. Question areas concerned: (1) Demographics; (2) Creation of support group and data sharing; (3) Context of use and activity types; (4) Use and management of activities; and (5) competition elements for giving support.

2.2 Questionnaire Results

In total 60 people replied to the questionnaire of which 28 men and 32 women and the majority (82%) was younger than thirty-years-old. The total duration for both reading the scenario and filling in the questionnaire was estimated to be 10–15 min. The results of the multiple choice questions are shown in Tables 1, 2, 3, 4 and 5.

Table 1 shows the demographics, as can be seen most of the respondents are students. Gender is balanced, and reflects the population of students at our institute.

The first important result from the questionnaire is about the two potential roles of the community coaching concept described in the scenario: Main User and Helper. For the role Main User, the majority (88%) of the respondents liked the idea of getting help to improve physical activity but on the precondition that they can decide themselves who to invite. 96% of them would invite friends, 79% family, 49% colleagues and 21% would invite neighbors. For the role Helper, 60% of the respondents were willing to help the Main User being more active. Two important reasons for saying yes is because the care about that person and/or because it is beneficial for their own health too. 37% of the respondents said maybe, depending on the relation between them, the Main User's attitude towards improving his health, costs and their own availability (agenda).

The second important set of results from the questionnaire are the potential functional requirements for a system implementing the community coaching concept. An important aspect of the coaching concept is to create the support group and helper group, and the sharing of activity data. The results are shown in Table 2. These results show that users want to have a high degree of control over the composition of their support group, they want to have control over their helper group. In addition, their seems to be some reluctance is sharing physical activity data. The latter suggests that the system should provide control mechanisms so as to let the user determine which information is shared with whom. Furthermore, users see friends and family members as their most important groups to receive support from and colleagues are at a good third place with 49% (note that multiple answers were possible in this question).

As shown in Table 3, users have a higher preference to use the system at home than in a working or studying environment. On the other hand, as shown

Table 1. Demographics.

Item	Result
Gender	Male: 47%
	Female: 53%
Age	18–30 years: 82%
	30–40 years: 10%
	40–50 years: 3%
	50–99: 5%
Occupation	Student: 62%
	Employee: 27%
	Other: 11%

Table 2. Support group creation and data sharing.

Item	Result
Is creation of support group by system acceptable?	Yes: 27%
	No: 73%
Do you want to invite your helpers yourselves?	Yes: 88%
	No: 12%
Who would you like to invite as helper?	Friend: 96%
	Neighbour: 21%
	Family member: 79%
	Colleague: 49%
	Other: 6%
Do you want to share your physical activity data?	Yes: 62%
	No: 38%
Would you be willing to be a helper to someone else?	Yes: 60%
	May be: 37%
	No: 3%
Would you like to have access to the activity data of the helped user?	Yes: 53%
	May be: 47%
	No: 3%

in Table 4, users would like to use the system always (60%). Key issue seems to be how well the system takes context information into account, especially agenda items. Regarding the type of activities users have in mind, different activities are identified in different setting (study and work versus home). Given the results shown in Table 4, it is clear that invitations should not be generated too many, and at the right time. There is a high willingness to extend an activity to a group activity (87%).

Regarding a competition element between helpers (see Table 5), the opinions are mixed: 53% do see an added value in this, whereas 47% of the users do not. There are different opinions regarding the nature of an award for helpers. In general, users prefer a longer period of time for a competition among helpers.

2.3 Required Functionality for Community Coaching

Based on the results of the questionnaire, the following roles and concepts within the Community Coaching system were introduced:

– Main User role: user of the system that needs help to be physically active.
– Helper role: user of the system that supports the Main User in being physically active.

Table 3. Context of use and activity types.

Item	Result
Do you want to use the system at the university/office?	Yes: 48%
	No: 52%
What types of activities do you like at the office?	5 min walk: 97%
	Frisbee: 24%
	Petangue: 3%
	Football : 10%
	Walk to coffee machine: 90%
	Rope jumping: 7%
	Other: 31%
Do you want to use the system at home?	Yes: 83%
	No: 17%
What type of activities do you like to perform at home?	Running: 66%
	Walking with the dog: 66%
	Climbing stairs: 54%
	Gardening: 56%
	Housekeeping: 68%
	Other: 36%

Table 4. Using and managing physical activities.

Item	Result
When would you like to use the system?	Weekend: 12%
	Working days: 22%
	Always: 60%
	Other: 6%
How often would you like to receive an invitation for an activity?	Every 2 h: 0%
	During breaks: 18%
	Depends on agenda: 68%
	Anytime: 5%
	Other: 8%
Is it acceptable for you that other can see your planned activities?	Yes: 53%
	No: 47%
There is a possibility for others to enroll in a planned activity, what is the maximum number of participants?	Only with inviter: 13%
	3–5 persons: 28%
	6–10 persons: 12%
	No restriction: 47%

Table 5. Competition elements for giving support.

Item	Result
Does a competition element for helpers have added value for you?	Yes: 53%
	No: 47%
What types of competition elements would you recommend?	Points for completed activity: 90%
	Points for initiated activity: 53%
	Points for adding new types of activity: 39%
	Other: 97%
At the end of a competition period, what kind of reward would you as helper motivate most?	Virtual award: 35%
	Physical award (e.g. challenge cup, chocolate bar): 50%
	Other: 15%
What would be a suitable competition period?	One day: 0%
	One week: 27%
	One month: 29%
	When a specified goal is reached: 35%
	Other: 8.8%

- Support group: A support group is composed by Helpers of the Main User. The support group can be a family group, a friends' group, or a colleagues' group or even peers.
- Notification: Message sent to either the Main User or Helper.
- Activity: It represents the main communication content between the Main User and Helpers and it is represented in lively actions planned and performed to increase physical activity level. Each activity has an activity type.

Additionally, based on the results of the questionnaire, we defined the following main requirements:

- **Notify the Main User' Inactivity**
  To get the Main User's number of steps, the system should synchronize the physical activity data to the portal first. If the physical activity data is below the recommended level, and the Main User's agenda allows, the system send notifications to the Helpers nearby.
- **Manage Activities**
  To arrange an activity between a Main User and a Helper, the system should provide a functionality to start up activities easily and quickly. Furthermore it should be possible for other Helpers to participate to a planned activity

(enrolling). The systems sends notifications whenever actions are performed (activity created, activity accepted, and enrollment to activity).

- **Manage Competition**
  In order to motivate the support groups, a competition between support group and between Helpers should be created. A reward would be given depending on the outcome of the competition.
- **Manage Support Groups**
  Main Users have the ability to manage their support groups. They should have the opportunity to approve or deny access to their personal physical activity data (personal or group access).
- **Manage Activity Types**
  The system should allow the users (Main Users or Helpers) to create their own activity types. Activity types created by Helpers should be approved by the Main User.
- **Manage Reward Types**
  The system should allow the users (Main Users or Helpers) to create their own reward types. Reward types created by Helpers should be approved by the Main User.

3   Community Coaching Design and Implementation

3.1   Functional Architecture

The architecture of the TogetherActive portal is based on the concepts of a Service Oriented Architecture.

Figure 1 shows the functional architecture of the Community Coaching. Portlets are divided into Main User portlet, Helper portlets, or shared portlets. Portlets may use services. Services are either Main User services or Helper Services.

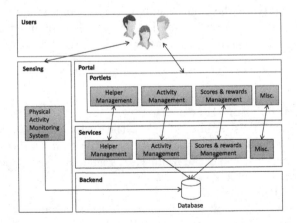

Fig. 1. Community Coaching functional architecture.

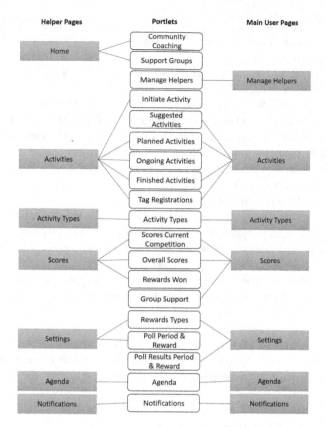

Fig. 2. Community Coaching portal pages and portlets.

The Community Coaching portal is composed of a set of pages hosting one or more portlets. These are categorized into Main User pages and Helper pages. Each portal page contains one or more portlets. The organization of the portlets within the pages is shown in Fig. 2.

Based on the requirements of the Community Coaching, we designed the following set of portlets:

- Helper Management Portlets:
  - Community Coaching portlet gives a short introduction about the Community Coaching system
  - Support Groups portlet gives an overview of the Main Users helping and to introduce other Helpers in the group
  - Manage Helpers portlet allows the Main User to manage the members in his support group
- Activity Management Portlets:
  - Initiate Activity portlet allows the Helpers to invite a Main User for an activity, without receiving a notification first

- Planned Activities portlet gives an overview of the planned activities for that day and to allow Helpers to enroll to activities that were not initiated by themselves
- Ongoing Activities portlet gives an overview of the activities currently going on
- Finished Activities portlet gives an overview of all the completed activities.
- Tag Registrations portlet gives an overview of all the tag registrations for the current day. When an activity is finished, the Main User tags (example with NFC tag) the joined Helper as a confirmation
- Activity Types portlet checks the list of activity types and to allow the users the update the list
- Competition Management Portlets:
  - Scores Current Competition portlet informs the Helpers about their position in the competition and the scores of their competitors
  - Overall Scores portlet gives an overview of the number of rewards won versus the total number of activities performed
  - Rewards Won portlet reminds the Helper to his success
  - Group Support portlet shows the Main User what group of relations give the most support
  - Reward Types portlet checks the variety of reward types in a certain support group and to allow the users to create new ones
  - Poll Reward and Period portlet gives Helpers the opportunity to give their preference with respect to the reward type and the period of the competition
  - Poll Results Reward and Period portlet keeps the Main User informed about his Helper's votes
- Miscellaneous Portlets:
  - Agenda portlet allows people to put their meetings and busy time to avoid that they get overwhelmed with notifications. The system takes into account Main User and Helper's agendas
  - Notification portlet gives an overview about received notifications

3.2 Community Coaching Implementation

The TogetherActive system [5] is a virtual community system that provides social support to people on their daily physical activities. It supports them in order to get physically active and to maintain an appropriate level of physical activity.

The Community Coaching system was implemented using Liferay [11]. This decision was made based on the fact that the Community Coaching system will be integrated in TogetherActive system (which is already implemented in Liferay). Due to the time constraints before starting the evaluation of the system, we prioritized some portlets to implement and simplified or replaced some other portlets. Some portlets such as the Poll Reward & Period portlet and the Poll Results Reward & Period portlet were omitted in the implementation.

The following additional portlets were designed and implemented:

- Set Sub-Goals portlet to replace the agenda portlet. It allows the Main User to set sub-goals. A sub-goal is characterized by a number of steps to reach by a certain time of the day (example reaching 2000 Steps by 10:00 am). The system checks the Main User's level of physical activity reached during the time of the sub-goals. If the Main User didn't achieve the amount of physical activity set in sub-goals, the system sends notifications to the support groups (Helpers). Three sub-goals are set by default, but the Main User can update the 3 sub-goals (time and number of steps).
- Validate Participation portlet replaces the Tag registration portlet. Instead of using and NFC-like system by the end of the performed activities to validate the participation, this portlet allows the Main User to approve or reject the real participation by simple check-box functionality.

The notification portlet was simplified to be a service of notifications sent via emails. In order to receive notifications Main Users and Helpers have to use the email address that they check often, or the one that gets synchronized to their smartphones. Three types of notifications were created:

- Activity invitation notification: it is sent whenever the Main User is not meeting his physical activity sub-goals (with a maximum of 3 notifications per day)
- Activity suggestion notification: it is sent whenever a Helper suggests an activity to the Main User, as an activity invitation notification received or as an initiative from the Helper
- Activity acceptance notification: it is sent to the Helper whenever the Main User accepts his/her proposed activity

To integrate the Community Coaching system in the current TogetherActive system, the Main User pages described in the Sect. 3.1 are integrated as child-pages (Fig. 3).

Fig. 3. Integration of Community Coaching System in TogetherActiveSystem.

4   Community Coaching Evaluation

4.1   Evaluation Protocol

In order to evaluate the prototype, usability and usefulness studies were planned. A study with Main Users and Helpers for 11 days was designed in order to use the system. Within the study we recruited the Main Users, and each participant was asked to invite at least two Helpers from their social network. The participants and their potential Helpers were asked to use their personal emails in order to be able to receive the notifications. The study was approved by the University Ethical Committee. Participants (Main Users) were recruited from the university via Facebook, emails and flyers. Recruited participants received full information about the system.

One day before the start of the study, all participants were invited for a short introduction meeting. They were asked to sign the informed consent and a borrowing agreement for the sensor. During this meeting, information was provided about what people could do on the portal and how to wear/connect the Fitbit sensor [6]. Each participant was asked to invite at least two Helpers from their network. The participants got their own portal access credentials and were informed how to give their Helpers access to the portal. The remaining part of the day was intended to get familiar with the system and to invite their Helpers.

At the end of the study period all Main User participants were asked to fill in questionnaire. The questionnaire consisted of three parts. The first part of the questionnaire was aimed to get some general information about the participants and their backgrounds. Background related topics were the use of social networking, use of apps for health purposes, and physical activity stages of change. The second part of the questionnaire was aimed at the system usability. To measure usability, the Computer System Usability Questionnaire (CSUQ) was used [10] which is based on a 7-point likert-scale, starting from strongly agree (value 1) to strongly disagree (value 7). The third part of the questionnaire was about the usefulness of the system, with the focus on the Community Coaching aspect on the portal. For measuring the usefulness of the system, no appropriate, standardized questionnaire was found from literature. Therefore, a new questionnaire (for Main User and for Helpers) was conducted with some input from the Technology Acceptance Model (that is focusing on the usefulness of a system device for office workers [4]). Similarly to the usability questionnaire, the usefulness questionnaire is based on a 7-point likert-scale as the usability questionnaire. As a final outcome of the study, we looked into the real use of the system.

4.2   Participants

We recruited 10 participants (7 males and 3 females) aged between 18 and 30 and were studying or working at the University of Twente. Participants were recruited from the University of Twente. Inclusion criterion to participate in the study was that participants should have some time for using the physical activity monitoring system and using the portal.

4.3   Results

For the analysis of the questionnaires, two of the ten Main User subjects were excluded. The reason was insufficient system use: one did actually never log in and the other wore the sensor for just one day. The remaining 8 subjects were three women and five men. Six participants were using social networks (like Facebook) for more than 4 years now, and 50% of the subjects spend around 5 h a week on social networking. Two of them already used social network for health or well-being purposes, and six of them used apps on their phones for health or well-being purposes (informational and/or exercising and schedule compliance). Based on the question on stage of change, we found that the six subjects were in the maintenance stage, which means that they have been sufficiently active for the last six months. One of the subjects was in the precontemplation stage and one in the contemplation. People in these stages are both insufficiently active but the difference is that people in the contemplation stage do think about to become more active.

The number of Helpers that filled in the questionnaire was four: two belonged to Support Group 2, one to Support Group 7 and one to Support Group 9. Because the low number of interactions they had with the system and because Helper 7 did not receive notifications.

Usability Study Results. Following the guidelines from Lewis [10], the results from system usability (Table 6) are summarized into the 4 factors reported as mean values: overall system usability (OVERALL), system usefulness (SYSUSE), information quality (INFOQUAL) and interface quality (INTERQUAL). The table also includes the results from the previous study on the TogetherActive portal without the Community Coaching component [5].

The overall score for the Main User was 4.3 and most outcomes are around 4 which indicates that it is slightly negative. Some Main Users mentioned that it was hard to find the different features and that the interface could be done more intuitive. Two other Main Users noted that they were too busy to consider the system more closely. One person mentioned that the user manual was not sufficient enough. Although the Helper portal contains less pages and portlets, the overall score for the usability is even higher, namely 4.6. Because just two of the four Helper respondents used the possibly to add some extra comments,

Table 6. Usability results.

Score	Main User	Helper	Previous study
OVERALL	4.3	4.6	3.8
SYSUSE	4.7	4.6	3.9
INFOQUAL	4	4.7	3.8
INTERQUAL	4.2	4.3	3.5

it is difficult to figure out the items that need to be improved. Furthermore, the given comments were very generic, so not much information can be obtained.

Usefulness Study Results. For the usefulness study, 8 Main Users and 4 Helpers replied to the related usefulness questionnaire.

The overall score of usefulness for the Main User is 3.4 which is less than average (4.0). Main Users agree on (based on the questions 2, 5–6, 8–9 and 17):

– having the possibility to allow their Helpers to see their data
– allowing Helpers to invite them for an activity when they don't meet their physical activity goal
– allowing other Helpers to join a planned activity
– knowing who is supporting them the best and which support group is supporting them the best
– the Community Coaching is useful.

Main Users are neutral regarding (based on questions 4, 10–11 and 13–15 with the highest score between 4.0 and 4.9):

– creating their own activity type as a motivation to perform that activity
– Community Coaching helps them to reach the daily goal 10000 steps and reduce their sedentary behaviour
– feeling of having someone always looking after them
– the system is meeting their needs

Additionally, the overall score of usefulness for the Helper is 4.2 which is neutral. Helpers agree on (based on questions 1, 3, 4, 6–7, 11–12 and 14):

– helping the Main User to be more physically active
– creating new activity types and that it increases their motivation to do that activity
– winning real reward that are decided by Main User as being best supporter
– knowing scores from other Helpers in the same support group

The Helpers don't agree about the fact Community Coaching could be useful (last question with a score of 5.0). Giving the limited number of Helpers that replied to the questionnaire, we cannot make a conclusion about the results.

In conclusion, the outcome from the usefulness study, especially from the Main Users, is showing that Community Coaching system is a promising feedback modality to be included in the virtual community.

System Use Results. In order to get more insights about the real use of the system and validate the usability and usefulness results, we checked from the portal logs the involvement of Helpers in the portal, the actions made (such as creating activity types, activities and rewards types) and the visits to the portal.

As part of the protocol, the Helper acquisition procedure was fully managed by the Main Users, we could only observe from the system logs how this it

Table 7. The number of invited, registered and real Helpers for the different Main Users.

Main User	Invited Helpers	Registered Helpers	Real Helper
MU2	3	2	0
MU3	3	2	2
MU4	5	1	1
MU5	3	0	0
MU6	3	2	1
MU7	3	1	0
MU8	3	0	0
MU9	3	2	1
MU10	10	3	3
Total	41	13	8

realized. We categorized Helpers into: invited, registered and real. If the Helpers logged in the system for a first time they are categorized as Registered Helper, and if they did more actions in the system, they considered as real Helpers.

Although we can see in Table 7 we can see that 41 Helpers were invited, only 13 Helpers registered, and 8 were real Helpers. From another side, we cannot make sure that it is not the recruited Main Users who used the system as Helpers, since they were the one in charge of the recruitment of Helpers.

Having a closer look on those real Helpers, we noticed that 3 Helpers used fake email addresses so no notifications were sent to them. Thus the setting of Helpers didn't meet expectations and not all Main Users experienced all proposed functionalities and expected added value from the Helpers.

For the 11 days of the study, Helpers were supposed to receive max 33 notifications (type Activity invitation notification) in case that their associated Helper doesn't comply to the 3 daily physical activity sub-goals. Figure 4 represents the total number of activity invitation notifications sent to the Helpers during the period of the study. It shows that in average 22 notifications for activity invitations were sent to the Helpers.

Four Helpers from the real Helpers suggested an activity to their Main Users and two did accept the invitation for the activity. Additionally, one Main User and two Helpers created activity types (4 activity types were created). For the rewards types, only 3 Main Users created reward types; 1 virtual and 2 reals. Regarding the portal visits by the Main User and Helpers, the majority of Main Users accessed the portal more than 1 day and the majority of Helpers accessed the portal only one time in only one day.

5   Discussion and Conclusion

Within this work, we presented a novel way to enhance the compliance and overcome drop experienced in [17]. The proposed ICT-mediated Community Coaching

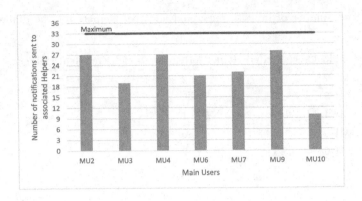

Fig. 4. Total number of activity invitation notifications.

functionality turn the physical activity into social activity, and stimulate the social collaboration to enhance the motivation for physical activity. In this paper we presented the requirements elicitation, design, implementation and evaluation of the Community Coaching system. The Community Coaching system was integrated within the TogetherActive system. To evaluate the system, a study was performed including ten Main Users. Each participant had to recruit his own Helpers. The study was over 11 days. At the end of the period all participants were asked to fill in a questionnaire, with the focus on the system usability and usefulness. Also a system use analysis was performed, to get insights in the real use of the system.

The scenario and the associated questionnaire as a requirement elicitation method gave us a good and clear idea about what are the main roles and the main requirements of such a new concept. We could design the main requirements, and implemented them or part of them (adapted versions due to practical limitations). In order to generalize to different target groups we should be careful and do an extra investigation because the respondents for the questionnaire were in average young, students or working in the university and healthy.

Regarding the usability study, the Community Coaching system can be improved. Extending the TogetherActive system with the Community Coaching component could have increased the complexity of the full system and affected the usability outcomes of the Main User and Helpers especially with the INFO-QUAL and INTERQUAL factors. Results show that there is room for improvements. The focus should be on the integration of the existing portal, to ensure they are more in line with each other. Because the TogetherActive system is designed/implemented in Liferay, the portlets can be easily reused on other pages of the portal and tuned according to the need. Other points to focusing on are the intuitiveness of the system and the look and feel.

Another usability study protocol would be better invested such as the task-oriented usability testing [21] or the walk-through approach. A list of key tasks and sub-tasks within the system should be undertaken in order to achieve the goal of evaluating significant aspects and key functionalities of the system. Example of a task is to invite for an activity. During the study, notifications (for activity invitation) are the main communication stream that goes between the support group and the Main User, but as a reply for these notifications the number of activities that were suggested or accomplished was really low based. If Helpers don't try at least one time during a study to invite to an activity, the vision to the system and its usefulness would be biased. This would be overcome with the different usability protocol.

Regarding the usefulness study, the outcomes, especially from the Main Users, are showing that Community Coaching system is a promising functionality to be included in the virtual community and can be generalized to any physical activity platform where social support can be provided. The outcomes of the Main Users are higher than that of the Helpers. Although the content of the questionnaire was not exactly the same, we consider that Helpers value the usefulness lower than Main Users. Because the outcomes of the first questionnaire pointed out that still 40% of the Helpers was not willing to help the Main User, this is not a very remarkable result. It could be that the Helpers did participate because of the social pressure (because they were asked by a friend for example) rather than being interested in it. Other causes for the low score could be because the Helpers were insufficiently informed about the purpose of the system or because they didn't read the Helper manual on the portal. Furthermore, the current protocol didn't give the chance to Helper to get a physical activity monitoring system. It would enhance their awareness and gives them more motivation if they can also monitor themselves to invite Main Users for activities. Additionally, although the usefulness questionnaire was adapted from the Technology Acceptance Model, it was not validated. This should be done in future experiment with the use of Cronbach Alpha for example, and with the use of the current results in order to validate it.

The hypothesis of this research is that such approach is more motivating and therefore enhances the compliance on long-term as it transforms physical activity into a social activity. Although the outcomes of the system usability and system use were neutral, most subjects liked the idea of Community Coaching (from the usefulness results). Therefore, it is recommended to further investigate on this topic. Extra recommendations should be integrated in a newer version of the system. First, the protocol of inviting Helpers should be more supervised. This supervision will make sure that all invited Helpers are real Helpers. Second, for a short-length study, Helpers should get a similar introduction meeting to the Main Users or all Helpers should come together with their Main Users for the introduction meeting. Finally, the email setting should be supervised, to avoid similar problems with this study, where some Helpers didn't change the default email address or used a fake email address.

Another suggestion would be to change the target group, in which the benefit for such a system is higher. The change of target group should be handled from the requirement elicitation process till the evaluation protocol. One possible target group is people with chronic condition given the increased awareness and evidences about the importance of physical activity for prevention and treatment [9,12,14]. Another target group could be the elderly people, known by feeling lonely [7] and their physical activity level gets influenced by their loneliness [20].

References

1. op den Akker, H., Moualed, L.S., Jones, V.M., Hermens, H.J.: A self-learning personalized feedback agent for motivating physical activity. In: Proceedings of the 4th International Symposium on Applied Sciences in Biomedical and Communication Technologies, ISABEL 2011, p. 147. ACM, New York, October 2011
2. Bravata, D.M., Smith-Spangler, C., Sundaram, V., Gienger, A.L., Lin, N., Lewis, R., Stave, C.D., Olkin, I., Sirard, J.R.: Using Pedometers to Increase Physical Activity and Improve Health (2007)
3. Consolvo, S., Mcdonald, D.W., Toscos, T., Chen, M.Y., Froehlich, J., Harrison, B., Klasnja, P., Lamarca, A., Legrand, L., Libby, R., Smith, I., Landay, J.A.: Activity sensing in the wild: a field trial of UbiFit garden. In: CHI 2008 (2008)
4. Davis, F.D.: Perceived usefulness, perceived ease of use, and user acceptance of information technology. MIS Q. **13**, 319–340 (1989)
5. Elloumi, L., Beijnum, B.J.V., Hermens, H.: Physical activity support community TogetherActive - architecture, implementation and evaluation. In: Proceedings of the International Conference on Health Informatics, pp. 200–211. SCITEPRESS - Science and and Technology Publications (2015). http://www.scitepress.org/DigitalLibrary/Link.aspx?doi=10.5220/0005289102000211
6. Fitbit (2007). https://www.fitbit.com/. Accessed Oct 2015
7. Huitt, W.: Educational psychology interactive: feedback (2004). http://www.edpsycinteractive.org/topics/behavior/feedback.html
8. Kamphorst, B.A., Klein, M.C.A., Wissen, A.V.: Autonomous e-Coaching in the wild: empirical validation of a model-based reasoning system. In: AAMAS 2014 (2014)
9. Kreuter, M.W., Strecher, V.J.: Do tailored behavior change messages enhance the effectiveness of health risk appraisal? Results from a randomized trial. Health Educ. Res. **11**, 97–105 (1996)
10. Lewis, J.R.: IBM computer usability satisfaction questionnaires: psychometric evaluation and instructions for use. Int. J. Hum. Comput. Interact. **7**, 57–78 (1995)
11. Liferay (2000). https://www.liferay.com/. Accessed Sept 2014
12. Lin, J.J., Mamykina, L., Lindtner, S., Delajoux, G., Strub, H.B.: Fish'n'Steps: encouraging physical activity with an interactive computer game. In: Dourish, P., Friday, A. (eds.) UbiComp 2006. LNCS, vol. 4206, pp. 261–278. Springer, Heidelberg (2006). doi:10.1007/11853565_16
13. MedHelp (1994). http://www.medhelp.org/. Accessed Sep 2014
14. Middelweerd, A., Mollee, J.S., van der Wal, C., Brug, J., Te Velde, S.J.: Apps to promote physical activity among adults: a review and content analysis. Int. J. Behav. Nutr. Phys. Act. (2014)
15. PatientsLikeMe (2004). http://www.patientslikeme.com/. Accessed Sept 2014
16. Silveira, P., Daniel, F., Casati, F., van het Reve, E., de Bruin, E.D.: ActiveLifestyle: an application to help elders stay physically and socially active. In: FoSIBLE Workshop at COOP 2012 (2012)
17. Tabak, M.: New treatment approaches to improve daily activity behaviour. Ph.D. thesis, University of Twente (2014)
18. Uchino, B.N.: Social support and health: a review of physiological processes potentially underlying links to disease outcomes. J. Behav. Medi. **29**, 377–387 (2006)
19. WebMD (2005). http://www.webmd.com/. Accessed Sept 2014
20. van Weering, M.G.H., Vollenbroek-Hutten, M.M.R., Tönis, T.M., Hermens, H.J.: Daily physical activities in chronic lower back pain patients assessed with accelerometry. Eur. J. Pain (Lond., Engl.) **13**, 649–654 (2009)

21. Wharton, C., Rieman, J., Lewis, C., Polson, P.: The cognitive walkthrough method: a practitioner's guide, pp. 105–140 (1994)
22. WHO: Physical activity (2014). http://www.who.int/topics/physical_activity/en/. Accessed Sept 2015

Author Index

Adams, Michael 205
Afsarmanesh, Hamideh 122
Alahuhta, Seppo 340
Alho, Olli-Pekka 340
Ali, Hesham 167
Alvarez, Sergio A. 276
Arini, Pedro David 191

Bai, Jie 22, 295
Barone, Sandro 67
Bart, Hayo 122
Besse, F. 145
Betancur, Manuel J. 255
Boutaayamou, Mohamed 236
Brüls, Olivier 236
Bustamante, Samuel 255

Carroll, Noel 319
Cheah, Lam A. 22, 295
Christ, Peter 205
Correa, Julio C. 255
Croisier, Jean-Louis 236
Cuesto, Germán 41

da Silveira, Marcos 360
Demonceau, Marie 236
Denoël, Vincent 236
Descombes, X. 145
Dilling, Thomas J. 56

Ell, Stephen R. 22, 295
Elloumi, Lamia 405

Feygelman, Vladimir 56
Forthomme, Bénédicte 236

Garraux, Gaëtan 236
Gilbert, James M. 22, 295
Gonzalez, Jose A. 22, 295
Green, Phil D. 22, 295

Hao, Zhili 3
Hayashida, Morihiro 108

Heras, Jónathan 41
Hermens, Hermie 405
Hesse, Marc 205
Hlinka, Jaroslav 87
Hoekstra, Rinke 360
Hörmann, Timm 205

Inomata, Akihiro 224

Kalina, Jan 87
Koyano, Hitoshi 108

Latifi, Kujtim 56

Maeda, Kazuho 224
Manth, Andrea 387
Maquet, Didier 236
Mata, Gadea 41
Medioni, C. 145
Meijerink, Margot 405
Menßen, Christian 205
Moonis, Majaz 276
Moore, Roger K. 22, 295
Morales, Miguel 41
Mori, Tatsuya 224
Moros, Eduardo G. 56

Nakata, Yasuyuki 224

Oya, Takuro 224

Paoli, Alessandro 67
Pérez, Vera Z. 255
Pruski, Cedric 360

Razetti, A. 145
Razionale, Armando Viviano 67
Richardson, Ita 319
Romero, Ana 41
Rubio, Julio 41
Rückert, Ulrich 205
Ruiz, Carolina 276

Sakata, Masato   224
Sandmark, Hélène   387
Savignano, Roberto   67
Schwartz, Cédric   236
Seppänen, Tapio   340
Seppänen, Tiina M.   340
Shafahi, Mohammad   122
Smedberg, Åsa   387

ten Teije, Annette   360
Torkar, Drago   191
Travers, Marie   319

Uchida, Daisuke   224

Vakkala, Merja   340
van Beijnum, Bert-Jan   405
van Harmelen, Frank   360
Verly, Jacques G.   236

Wang, Chiying   276
Wang, Dan   3
West, Sean   167

Yaginuma, Yoshinori   224
Yepes, Juan C.   255

Zamborlini, Veruska   360
Zhang, Geoffrey G.   56

Printed in the United States
By Bookmasters